How Scientific Practices Matter

How Scientific Practices Matter

Reclaiming Philosophical Naturalism

Joseph Rouse

THE UNIVERSITY OF CHICAGO PRESS
Chicago and London

Joseph Rouse is professor of philosophy at Wesleyan University. He is the author of *Knowledge and Power: Toward a Political Philosophy of Science* and *Engaging Science: How to Understand Its Practices Philosophically.*

THE UNIVERSITY OF CHICAGO PRESS, CHICAGO 60637
THE UNIVERSITY OF CHICAGO PRESS, LTD., LONDON

Printed in the United States of America

11 10 09 08 07 06 05 04 03 02 5 4 3 2 1

ISBN (cloth): 0-226-73008-5

Library of Congress Cataloging-in-Publication Data

Rouse, Joseph, 1952–
How scientific practices matter : reclaiming philosophical naturalism / Joseph Rouse.
p. cm.
Includes bibliographical references and index.
ISBN 0-226-73008-5 (cloth : alk. paper)
1. Science—Philosophy. 2. Naturalism. I. Title.
Q175.R5668 2002
501—dc21
2002007831

♾ The paper used in this publication meets the minimum requirements of the American National Standard for Information Sciences—Permanence of Paper for Printed Library Materials, ANSI Z39.48-1992.

CONTENTS

ACKNOWLEDGMENTS

ATTRIBUTION OF AUTHORSHIP obscures the highly collaborative character of philosophical work. This book could not have been written without extensive exchange with many colleagues, students, and friends over the past decade. They should not be held responsible for what is written here, but they have made indispensable contributions that I happily acknowledge.

Special recognition and thanks must be given to Karen Barad, Mark Okrent, and Sanford Shieh. Important components of my arguments could only have been developed through ongoing conversation with them and their willingness to share their work in progress. Many others have also contributed to the book in ways that far exceed what can be acknowledged in specific notes. Among them are Steve Angle, Mario Biagioli, Bill Blattner, Bob Brandom, Nancy Cartwright, David Cerbone, Andrew Curran, Bert Dreyfus, Terri Elliott, Brian Fay, Arthur Fine, Lydia Goehr, Gary Gutting, John Haugeland, Steve Horst, Dan Jacobson, Thomas Kuhn, Rebecca Kukla, Mark Lance, Lisa Lloyd, Huib Looren de Jong, Jill Morawski, Lynn Hankinson Nelson, Andy Pickering, Hans Radder, Paul Roth, Hans-Jörg Rheinberger, Bob Scharff, Ted Schatzki, Brian Cantwell Smith, Betty Smocovitis, David Stern, Sonia Sultan, Charles Taylor, Jennifer Tucker, Thomas Uebel, Joanne Waugh, Sam Wheeler, Mark Wrathall, Alison Wylie, and the two anonymous readers for the University of Chicago Press.

The book began to take its current form during a 1997 National Endowment for the Humanities Summer Institute, "How Background Practices Produce Intelligibility," in Santa Cruz, California. I thank the National Endowment for the Humanities, the Institute directors, Hubert Dreyfus, David Hoy, and Jocelyn Hoy, and all of the Institute participants.

The book has also been significantly improved by thoughtful responses from the audiences to whom I have presented various portions of it, including the history and philosophy of science departments or programs at

Northwestern University, the University of Pittsburgh, and the University of Notre Dame; the philosophy departments at Georgetown University, the University of New Hampshire, Wesleyan University, and the University of Southern Connecticut; meetings of the Boston Colloquium for the Philosophy of Science, the International Society for Phenomenological Studies, and the Central Division of the American Philosophical Association; the Center for Cultural and Literary Studies at Harvard University, the Center for the Humanities and the Public Affairs Center at Wesleyan University, and the Center for Ideas and Society at the University of California at Riverside; and various conferences or workshops, including "Sciences as Social Practices" at Carleton University, "Toward a More Developed Philosophy of Experimentation" at the Vrije Universiteit, Amsterdam, "Feminism and Naturalism" at Washington University and the University of Missouri at St. Louis, "Interpretation" at the State University of New York at Stony Brook, "Models of Critique in Science, Politics, and the Arts" at the Van Leer Institute and the Edelstein Center for History and Philosophy of Science, and "Practices and Social Order" at Zentrum für interdisziplinare Forschung, Universität Bielefeld.

Much of what I say here was worked out initially in my courses in the philosophy of science, science studies, and twentieth-century philosophy at Wesleyan University. I am grateful to all of my students for their thoughtful participation in these courses, but special thanks go to Jessica Allen, Abby Bass, Dave Burland, Jessica Cohen, Allison Dimond, Paul Edlefson, Dan Gartner, Chris Hanssman, Lauren Heller, Talia Inlender, John Jacobson, Alex Klein, Felicity Kohn, Evan Leonard, Yannig Luthra, Tess Moore, Sarah Osterhoudt, Colin Patrick, Adam Roessler, Jeremy Saks, Sigrid Schmalzer, Susanne Sreedhar, Joanna Starrels, Vanessa Stubbs, Whitney Trevelyan, Laura Warren, Mary Weaver, Elizabeth Wells, Carla Westen, Heather Wieler, and Lucas Winter.

Preliminary versions of the following chapters were published elsewhere, and I am grateful to the original publishers for the opportunity to incorporate material from these articles in the book: parts of chapter 2 appeared as "Interpretation in Natural and Human Science," in *The Interpretive Turn,* ed. David Hiley, James Bohman, and Richard Shusterman (Ithaca: Cornell University Press, 1991), 42–56; chapter 3 appeared as "Indeterminacy, Empirical Evidence, and Methodological Pluralism," *Synthese* 86 (1991): 443–65; chapter 4 appeared as "Feminism and the Social Construction of Scientific Knowledge," in *Feminism, Science, and the*

Philosophy of Science, ed. Lynn Hankinson Nelson and Jack Nelson (Dordrecht: Kluwer, 1996), 195–215, and is reprinted with kind permission of Kluwer Academic Publishers; and chapter 5 includes material that appeared earlier as "Understanding Scientific Practices: Cultural Studies of Science as a Philosophical Program," in *Science Studies Reader,* ed. M. Biagioli (New York: Routledge, 1999), 442–56, and "Two Concepts of Practices," in *The Practice Turn in Contemporary Theory,* ed. T. Schatzki, K. Knorr-Cetina, and E. von Savigny (London: Routledge, 2001), 189–98.

I have had the good fortune and the pleasure to work with an extraordinary editor, Susan Abrams, in bringing this book to press. I am also grateful to Carlisle Rex-Waller for her careful and thoughtful copyediting.

Sally Grucan has been unwavering in her love, encouragement, and support throughout my years of work on this book. Brian Rouse and Martin Rouse have made my life joyful during these years, and have also helped me keep the book in proper perspective.

The book is dedicated with love and gratitude to my mother, Helen McClimans Rouse, and to the memory of my father, Joseph Thomas Rouse. By word and example, they have helped me understand what it means to be responsible, and encouraged me never to lose sight of what really matters.

INTRODUCTION

THIS BOOK CONCERNS the meaning and significance of naturalism in philosophy. It also claims to vindicate a broadly naturalistic program for the philosophy of science, and more generally for philosophical reflection upon language, thought, and action. Yet many philosophers who think of themselves as naturalists will understandably find this claim astonishing, for the book rejects some of their most prominent theses and some aspects of their defining orientation toward science and philosophy. My task is therefore not merely to articulate and defend a particular philosophical program, but also to justify doing so still under the banner of "naturalism."

Naturalism

Naturalism in the broadest sense is the claim that philosophical reflection should be continuous with (or perhaps a part of) empirical science. Naturalists no longer aspire to provide philosophical grounding for the authority and intelligibility of the sciences; scientific understanding instead provides the basis for whatever remains of philosophy. That naturalism is now the predominant metaphilosophy is perhaps most clearly evident in the homages to naturalism offered by apparently adamant antinaturalists. For instance, Christine Korsgaard introduced her lectures on the sources of moral normativity with this characterization of the contemporary philosophical situation: "For us, reality is something *hard,* something which resists reason and value, something which is recalcitrant to form" (1996, 4). John Haugeland provides another striking example. At first glance, he might seem to dispute Korsgaard's concessions to naturalism, by arguing that the supposedly recalcitrant objects of the natural sciences are themselves constituted by us through an "existential commitment" comparable to faith or love. Yet Haugeland, too, in the end insisted that these "norms of objective correctness [must] be understood in a spirit of naturalism, appropriately construed, [which is] the thesis that people are, though distinctive, still naturally evolved creatures somehow implemented in whatever physics tells us about" (1998, 317, 358 n. 15).

Korsgaard's and Haugeland's remarks invoke a generic sense of naturalism, to which almost everyone in contemporary philosophy subscribes.[1] This generic understanding of naturalism needs to be disambiguated, however. Philosophy can be brought within the ambit of the sciences in different ways, and different conceptions of science thereby become authoritative for philosophy. Naturalism is sometimes conceived as a thesis *about* philosophy, emphasizing its continuity with or dependence upon the best empirical science. Yet naturalism has also been proposed as a thesis *within* philosophy, concerning the dependence of semantic, epistemic, or ethicopolitical norms upon the regularities, necessities, or causal powers expressed in scientific laws and theories. The former ("metaphilosophical naturalism") is a claim about philosophy as an activity or a discipline and its relation to the disciplines of one or more of the sciences. The latter ("metaphysical naturalism") is a philosophical claim about the relation between the domain of natural facts or laws disclosed within the sciences and some other domains (psychological, semantic, epistemic, social, or moral) within which allegedly true claims or binding obligations can be expressed. In this latter sense, a naturalist is one for whom the existence, content, or binding authority of these normative domains is dependent upon their relation to the natural world disclosed by the sciences.[2]

This distinction between naturalism as a metaphilosophical thesis and naturalism as a philosophical thesis readily aligns with different construals of what is authoritative for naturalistic philosophy. In the former case, it is *science* to which philosophy must defer; in the latter, it is *nature*, the world to which the sciences themselves are also taken to be accountable. We should not think of these two approaches as mutually exclusive, however. A coherent naturalism must ultimately be able to reconcile these approaches, since scientific understanding is achieved by natural beings, and

1. In chapter 2, I discuss how this generic sense of naturalism, to which even contemporary antinaturalists subscribe, emerged from the criticisms of the more fundamentally antinaturalist programs of logical analysis, phenomenology, and neo-Kantianism, which dominated much of philosophy in the first part of the twentieth century.

2. Naturalists in this second sense often deeply disagree about how to construe the appropriate relation between natural laws and normative claims, whether as reduction, supervenience, elimination, or mere predictive enhancement.

nature is intelligible to us primarily through scientific practice. Such reconciliation is not often attempted, let alone successfully displayed, for reasons I shall discuss in the chapters to follow. Yet philosophical naturalists are not without resources for making sense of the scientific disclosure of nature as itself "natural." Each of the two strains of naturalism makes an important contribution to understanding what a successful reconciliation would require.

Metaphilosophical naturalisms countenance no arbitrary philosophical impositions upon the sciences. No philosophical account of what science *must* be like or *ought* to be like has the standing to block well-motivated developments within the sciences themselves. As Quine succinctly said, "philosophy of science is philosophy enough." Accord with this Quinean naturalist commitment would require that the sciences be accepted as the defeasible basis for disclosing the natural world. Still, holding oneself accountable to this demand presupposes neither a compelling account of the boundaries of "the sciences themselves" nor a clear distinction between arbitrary and well-motivated constraints upon scientific work. Accounts of "science itself" and of what constitutes an "arbitrary imposition" upon it must be developed together, guided by the concern to allow the sciences to develop freely, without arbitrary constraint. Such accounts can also be guided by past examples in which scientific work succeeded in part by freeing itself from what in retrospect were mistaken methodological or ontological presuppositions.

Metaphysical naturalisms also answer to a concern whose precise content cannot be fully specified in advance but must be articulated in practice. Metaphysical naturalists can countenance no appeals to anything mysterious or supernatural in their own explications and arguments. This constraint is especially telling concerning "us," the beings who account for and are accountable to norms. Naturalists insist that human beings and any other beings to whom (putative) norms might intelligibly apply are natural beings, embodied, causally intra-active, and historically and biologically evolved. Call this the Nietzschean commitment of contemporary naturalism, in honor of perhaps its most relentless adherent.[3] Here also, however, endorsement of this commitment need not presume an already established boundary between the "natural" and the "supernatural." Such boundaries are articulated and shifted through critical discussion of natu-

3. Neurath would also be a plausible contender for this accolade.

ralists' ongoing attempts to distinguish acceptable from unacceptable concepts, claims, and arguments, informed by continuing engagement with scientific work. Ongoing recourse to the rich history of argument concerning what is an unacceptable appeal to something mysterious or inexplicable and what suffices to make a concept, claim, or argument scientifically and naturalistically acceptable averts any need to stipulate once and for all the bounds of the natural.

Think of these two concerns, expressed in the Quinean metaphilosophical commitment to avoid arbitrary impositions upon the development of science and the Nietzschean philosophical commitment not to accept or rely upon what is mysterious or supernatural, as articulating what is at stake in a commitment to philosophical naturalism. Naturalists can disagree about how these concerns should be articulated and applied, but they should agree that it matters to get them both right, and to hold one another accountable to their appropriate application. I suggest that the tradition of naturalism is held together by an underlying commitment to be responsible to these two concerns, and it is in this sense that I take this book to have been conceived in the spirit of naturalism.

Some may worry that a sense of naturalism defined by these Quinean and Nietzschean commitments is too inclusive and tolerant for a decisive resolution of fundamental disagreements about what naturalism amounts to. Yet in both cases, the history of their application suggests how easily their strictures can be unwittingly violated, and thus how stringent are the constraints they impose. From Nietzsche's relentless exposure of theological vestiges within supposedly freethinking inquiries, through Neurath's criticisms of residual metaphysical assumptions, Quine's and Davidson's objections to the dogmas of empiricism, Derrida's parallel complaints about the metaphysics of presence, Dennett's applications of "Darwin's dangerous idea," or Rorty's treatment of objectivity and reality as vestiges of a desire for transcendent authority, naturalists and other antisupernaturalist thinkers have found that doing without supernatural entities or relations is harder than it looks. On the other front, the now familiar conflicts between scientific practice and many philosophical empiricisms are forceful reminders that today's seemingly sensible codification of scientific norms can be tomorrow's unwarranted imposition upon science.

Moreover, these fundamental naturalist commitments sharply delimit the project of reconciling the metaphilosophical and metaphysical strains of naturalism. Metaphysical naturalists have an obligation to show how and why their preferred conception of nature is indeed what scientific inquiry discloses, and to do so without conceiving of science in ways that impose arbitrary constraints upon inquiry. Metaphilosophical naturalists are likewise obligated to indicate how their conception of scientific understanding is itself intelligible as part of the world the sciences disclose, and to do so without invoking anything magical or supernatural. The joint satisfaction of these two obligations places stringent constraints upon a satisfactory articulation of philosophical naturalism.

A central theme of this book is that taking these constraints seriously compels substantial revisions in the most familiar conceptions of philosophical naturalism. Such revisions are needed, I argue, because the issues between naturalists and antinaturalists have been conceived in ways that block a satisfactory understanding of the relation between scientific inquiry and the world that such inquiry makes manifest. The most familiar conceptions of scientific understanding, and of the natural world as disclosed by the sciences, make it difficult to show how sciences like ours could reveal such a world from within it. Moreover, these difficulties affect not only committed naturalists, but also their critics who nevertheless subscribe to a more generic and attenuated sense of naturalism. The appropriate response to these difficulties is not to doubt the capacities of the sciences, but to reconceive scientific inquiry, the world it discloses, and the relation of both to philosophical reflection about scientific understanding. Outlining such a reconception of science, the natural world, and their philosophical interpretation in the spirit of naturalism is the aim of this book.

Normativity and Necessity

To understand contemporary conceptions of science and nature that shape discussions of naturalism, we first need to consider briefly their philosophical origins. In the early decades of the twentieth century, Frege's and Husserl's criticisms of psychologism in logic contributed to a widespread rejection of philosophical naturalism. Along with such diverse figures as Russell, Heidegger, Wittgenstein, Cassirer, Moore, and Carnap, Frege and Husserl insisted that the validity, objectivity, or rationality of judgment could not be explicated on the basis of empirical matters of fact. Logic, as Frege bluntly claimed, is a normative science, and could not be otherwise. Moreover, the meaning and justification of empirical scientific claims were also thought to require a nonempirical grounding for their authority. It is widely recognized that early-twentieth-century antinaturalists often distinguished the logical forms or essential structures that were to secure the normativity of meaning and justification from the empirical content that the natural sciences then supplied empirically. Less well understood is how these distinctions of form and content mapped onto distinctions in modality. What supposedly prevented the results of empirical science from securing their own normative grounding was not their substantive content, but the contingency of that content. Human psychology or historical context could not underwrite the validity of scientific judgments, because our cognitive processes and our historical situation might have been different. By contrast, formal/tautological logical relations, essential structures of consciousness, or transcendental conditions of thought or representation were put forward as necessary conditions for meaning or thought. Normativity was to be grounded in a priori, formal/structural necessities: We ought to think and speak in these ways because we could not think or speak at all otherwise.[4]

Philosophical work in the latter half of the twentieth century was decisively shaped by criticism of these attempts to distinguish logical or essential form from experiential content. Formalist accounts of meaning were opposed by Quine's deconstruction of analyticity, Wittgenstein's criticisms of the autonomy of rules, and Heidegger's relocation of meaning from essential structures of consciousness to historical configurations of practical activity. Experientialist conceptions of semantic content were likewise challenged in Sellars's attack on the myth of the given, Wittgenstein's denial of the intelligibility of a private language, and Heidegger's insistence that public norms of affect, understanding, and discursive articulation are presupposed in any interpretation of something "as" something. These challenges undercut the methodologically foundational role of logical analysis or phenomenological description, and with them, the aspiration to what Quine called "first philosophy." In particular, meaning (whether located in thought, language, or action) no longer comprised a distinctively philosophical domain of inquiry, accessible through a characteristically

4. Appeals to necessity might seem to secure the normative *authority* of such forms or structures, at the cost of their normative *force:* their necessity might render it impossible to go wrong, such that they would not speak to what is at issue in actual thought, speech, or action. Normative force was typically retained, however, by criticizing various pretensions to meaning or justification: prominent examples included the positivist critiques of meaningless metaphysics, phenomenological attacks upon confusions induced by the natural attitude, and the use of logical analysis to expose fallacious reasoning or confusions of surface grammar with underlying logical form.

philosophical method, that could disclose necessary structures demarcating the limits of legitimately empirical science.

Naturalism once again became a serious option within philosophy as a result of these criticisms.[5] A common route to naturalism has been to decouple the earlier association between a priori knowledge and necessity. Normativity is still to be grounded in necessity, but the relevant necessities are not logical or transcendental necessities knowable a priori. The world disclosed by the sciences is not merely a jumble of contingent facts, but is structured by brute, a posteriori necessities. Such necessities supposedly provide a more local, naturalistic grounding for the normativity of scientific understanding. Note well, however, that such strategies leave untouched the presumption that the normativity of meaning and justification must be grounded in some form of necessity.[6] I highlight this point because my arguments in the chapters to follow will challenge this presumption. Before considering this challenge, however, I want to look briefly at the ways in which the normativity of meaning and justification has been grounded in a posteriori necessities that require scientific rather than philosophical discovery.

It is useful to begin with attempts to locate the normativity of meaning and justification in some form of social-historical necessity. These attempts involve appeals to the role of shared presuppositions, paradigms, languages, cultures, or standards in securing the intelligibility and authority of what individual participants in such collectives say and do. Instead of the unqualified necessity of formal structures of language or thought, one looks to the contextualized necessity of particular material presupposi-

5. The full extent of the shift toward naturalism in midcentury becomes more evident when one recognizes that historicism and sociologism are variations upon naturalism. Many philosophers who reject any close alliance between philosophy and physics or biology would instead situate philosophical understanding historically and sociologically. They thereby agree with naturalists (more narrowly defined) in rejecting the philosophical conceptions dominant earlier in the twentieth century, which sought to ground philosophical understanding in logical or transcendental necessities that were knowable a priori.

6. Philosophical readers will be well aware that at least one prominent philosophical naturalist, W. V. O. Quine, takes a quite different approach to naturalizing epistemology and semantics. Quine eschews the use of modal locutions, because he thinks the best scientific understanding of human knowledge acquisition cannot justify epistemic access to modalities. Since my own project is developed in part through a discussion of Quine, I will take up his approach later when I introduce my own. I nevertheless call attention now to one respect in which Quine's posing of the problem may be inadequate. Quine asks whether and how we are justified in adding modal locutions to an antecedently intelligible extensional idiom, but I would suggest that more careful attention to language may raise doubts about the antecedent intelligibility of an idiom bereft of modal concepts.

tions. The modal character of such claims has rarely been made explicit, but recognition of their implicit modality is useful. Why, after all, should one accede to the strictures of a community or a culture as authoritative? Or, from a third-person perspective, how does the social-historical setting of individual performances make their occurrence intelligible, rather than merely causally explicable? The implicit claim is that one should accede to community standards (and should be understood by others as having done so), because one *must*. Such a necessity is usually construed in one of two ways. Either one must accede to the standards of some community or other in order to be subject to norms at all, and hence to be intelligible as making a claim rather than just a noise (on this conception, one must accede to the standards of *this* community only because it is the one in which one happens to find oneself). Alternatively, one must accede to the standards of this community because its members (oneself included, of course) are the ones upon whom one could possibly make claims, and hence the ones to whom one's claiming must be accountable.

However plausible they might have seemed initially, such appeals to social-historical necessity to ground the intelligibility and justification of utterances or actions have encountered fundamental difficulties. Quine's criticism of the analytic/synthetic distinction can be taken as a model for undermining any attempt to distinguish more locally the necessary presuppositions of a language, community, or culture from contingently disputable claims based upon those presuppositions. There is no principled basis for distinguishing a community's necessary presuppositions from those claims that its members contingently happen to accept, or for choosing among different ways of drawing the boundaries between the necessary and the contingent.[7] Similar problems arise in demarcating the boundaries of the communities whose standards and presuppositions would be binding upon a particular speaker. Indeed, the de facto plurality of linguistic, cultural, scientific, and other communities that might govern the performances of a particular individual undercuts any attempt to ascribe even local necessity to the standards and presuppositions of any one of them.[8]

7. Often the difficulty of distinguishing essential presuppositions from contingent convergences in belief has emerged most clearly in the attempt to understand social or conceptual change, by distinguishing revolutionary from incremental change in communities or cultures, changes in language from changes in belief, or changes in standards and norms from changes in how those standards are actually met.

8. This plurality received sustained philosophical attention in the debates over "essentialism" within feminist theory in the 1980s and 1990s, as an emerging gender politics came

The actual availability and familiarity of alternative forms of life is a rather compelling objection to their supposed impossibility or unintelligibility. Finally, appeals to social-historical necessity have notoriously objectionable normative implications. To the extent that particular communities' performances, presuppositions, or standards are necessary conditions for their constituents' intelligibility (and hence their status as persons or agents), these conditions seem to be effectively beyond criticism. Yet the characteristic performances and claims of most actual communities (our own very much included!) seem not merely open to criticism, but seriously in need of it.

An alternative way to ground meaning and justification has been to seek the source of semantic, epistemic, and practical authority in the a posteriori necessity of natural laws or causal relations. The physical sciences, evolutionary biology, or neuroscience and cognitive psychology might reveal essential features of language, thought, or action by human beings in this world, even though the relevant scientific truths are not necessarily true (i.e., not true in all possible worlds), and not knowable a priori. Such appeals to natural necessity have gained plausibility in the past few decades, because two of the most prominent philosophical objections to the idea of natural necessity seem to have been overcome by recent philosophical work.[9] First, criticisms of deductive-nomological theories of explanation and empiricist theories of meaning have strongly undermined empiricist suspicions about the intelligibility of "metaphysical" necessity. These suspicions turned on the doubt that one is ever empirically justified in *adding* modal significance to antecedently intelligible descriptions of what actually occurs. The difficulty here lies first in the presumption that unmodalized descriptions of the world are antecedently intelligible. Even the most minimally adequate scientific or everyday descriptions of many familiar phenomena seem to require modalized characterizations of dispositions and capacities, such that it no longer seems plausible that one could conduct much empirical science or even ordinary conversation without employing modal locutions of some kind. Second, long-standing suspicions

into conflict with the competing claims of race, ethnicity, and class. Fuss 1989, Butler 1990, Spelman 1988, and Witt 1995 provide a good introduction to these debates.

9. Robert Brandom has prominently cited these two factors in the restoration of philosophical respectability to modal locutions in his lecture "Modality and Normativity," given at both the Boston Colloquium for the Philosophy of Science, April 1999, and the American Philosophical Association Central Division, April 2000.

about the inferential plasticity of modal locutions have been allayed by powerful formal work in modal logic and its semantics that has rendered their use philosophically respectable. Strictly speaking, such work only applies directly to alethic modality, and not to the weaker causal modalities that might seem relevant to claims of natural necessity.[10] By assigning causal capacities or lawful regularities to objects and natural kinds, however, one might plausibly hope to construe natural necessities in terms of Kripkean semantics for modal identity claims (e.g., water is essentially H_2O, including whatever laws and causal capacities follow from its having that structure).[11]

How might the empirical discovery of natural necessities then serve to explicate the intelligibility and authority of the sciences, and thereby retrospectively vindicate the very claims that made such explication possible? Two strategies have been prominently proposed. Richard Boyd has argued that "mature" scientific disciplines can be shown to have crossed a crucial threshold marked by their ability to deploy highly theory-dependent methods in ways that consistently yield both predictive success and the ability to improve that predictive success through the reflexive theoretical critique of those methods. The predictive reliability of theories in these disciplines, he argues, should be explained in terms of the causal relations between the natural kind terms employed in the theories and the natural kinds whose causal capacities determine what is possible and impossible. Boyd's approach would not account for the intelligibility of any discursive practices except those of the theoretically "mature" sciences, and it also dims the prospects for explaining past or hastening future disciplinary transitions to maturity: "The theoretical innovations that establish the first successful paradigm within a particular scientific discipline must be thought of as the beginnings of successful methodology within the field, not as consequences of it" (1996, 227). Boyd's approach has in any case been subject to withering criticism, both as requiring a manifestly mistaken historiogra-

10. Alethic modalities are different ways in which a proposition might be true (necessarily or contingently, possibly or impossibly). Natural necessity seems to be weaker than necessary truth (since the natural laws governing the world might have been different), yet stronger than merely contingent truth.

11. Alternatively, David Lewis (1973) has proposed to construe causal relations in terms of a possible worlds semantics for counterfactual conditionals. Apart from the difficulty that his account does not seem readily applicable to the kinds of causal laws that characterize scientific invocations of natural necessity, his approach has also been subject to extensive criticism (e.g., Kim 1973; Horwich 1987; Bennett 1987) that would require substantial new work to accommodate.

phy of science (Laudan 1981) and as turning on a self-defeating, circular line of argument (Fine 1986; Putnam 1983; Rouse 1987). His account does have the virtue, however, of trying to ground the meaning and authority of scientific theories in natural necessities governing the subject matter to which they are accountable.

A different approach to explicating the normativity of thought, language, and action in terms of natural necessities looks instead to the natural laws governing human cognition. By appealing to the best current theories about information-processing in human causal interactions with the environment, to cognitive structures of human language and thought or their causal-functional evolutionary adaptation, or to neurological mechanisms, naturalists in the philosophy of mind and language (e.g., Dretske, Fodor, Millikan, Churchland and Churchland) or the philosophy of science (e.g., Kitcher, Giere, Laudan) hope to respond to characteristic philosophical concerns with empirical methods and results. These are extensive, philosophically and empirically rich programs of research. Although they have been widely discussed in the philosophical literature and remain worthy of further reflection and development, my aim here is not to assess the details and arguments presented in these attempts to develop an adequate philosophical naturalism. The extent to which the scientific fields they consider should inform a naturalistic conception of human cognition and scientific understanding remains open. My concern is instead with the philosophical conception within which these empirical considerations are brought to bear.

My strategy is twofold. I first argue that the attempt to account for the normativity of thought, meaning, and action in terms of natural necessity does not take sufficiently seriously what should have been learned from the most fundamental criticisms of earlier antinaturalist programs that appealed to logical or transcendental necessities of language and thought. Criticisms that have been misunderstood as challenging only the a priori accessibility of necessity and the distinction between essential structures (of consciousness, language, or representation) and their contingent contents instead show the untenability of appeals to necessity generally. Moreover, the conceptions of nature and normativity resulting from this misunderstanding have introduced a debilitating dualism that has afflicted the philosophical discussion of naturalism both by naturalists and by their critics. The "natural world" to be understood scientifically and the normativity of the thoughts, utterances, and actions in which such understanding could be realized have been preconceived in ways that prevent their reconciliation. For naturalists, I argue, this dualism has prevented the joint satisfaction of the two central concerns identified earlier: either they must place arbitrary constraints upon scientific inquiry and its disclosure of nature or they must postulate entities or processes that should be unacceptable to naturalists. In particular, the dualism makes unintelligible the disclosure of nature through scientific practices. Nature so conceived could not have become manifest through scientific practices that are themselves a part of nature so conceived.

The second part of my strategy is constructive. My account begins with scientific *practices* of interaction with the world, rather than scientific *representations* of it. My earlier critical arguments already direct attention toward scientific practices as the focal point of the dualism of nature and normativity. My constructive arguments show how to reconceive practices in a way whose very possibility has been systematically overlooked in the framing of the dualism of nature and normativity. Scientific practices have typically been misconceived philosophically as social-historical regularities in what scientists say and do. Invoking scientific practices as the basis for understanding normativity might then seem to be just a variant of the forlorn project of construing the normativity of scientific understanding in terms of a social-historical rather than natural a posteriori necessity. Such an interpretation is triply mistaken, however. On my account, practices are not meaningful, socially and historically situated actions as opposed to inexorable natural processes; they are identifiable by their normative accountability rather than by any performative or dispositional regularity; and they allow us to understand the modality of causal processes on the basis of the normativity of scientific practices rather than the reverse.

In all three respects, an account of scientific practices with these characteristics may seem difficult to conceive. It is commonly assumed, for example, that social relations and natural objects form relatively autonomous components of the world, such that scientific knowledge must either be socially constructed or objectively grounded.[12] Philosophical discussions

12. The supposed opposition between social construction and objective determination assimilates within the category of "objectivity" both naturalistic appeals to causal reliability and antinaturalistic appeals to rational or objective standards. These alternatives can of course be realigned to oppose realist accounts of knowledge, on the basis of how the world constrains our beliefs, to antirealist accounts of how human knowers can or should form their beliefs (in the latter case, social constructivists, empiricists, and rational methodologists are lumped together in their opposition to realism). I have discussed elsewhere (Rouse

of normativity have usually found it inconceivable except as grounded in necessity or objective regularity; even philosophers who have tried to construe normativity directly have usually reduced it to the necessity of a rule or to quasi-objective regularities ("objective values" or community norms). Finally, philosophers have been inappropriately suspicious of causality, and have tried to explain away causal relations in favor of exhibited regularities or governing laws. The most difficult challenge of this part of the book is show how alternatives to these assumptions are even possible. I need to show how we can understand causality and normativity without reducing them to regularities or rules, and how we can understand *the* world without tacitly splitting it into a meaningful "social world" and an anormative "natural world." Once a conceptual space has been cleared within which such claims are concretely conceivable, it is comparatively straightforward to argue for the constitutive normativity of scientific practices and the world they disclose. In the following section, I provide an overview of how I shall go about making sense of these claims and showing how they respond appropriately to the problems highlighted by my criticisms in the opening chapters.

Preview of the Arguments

My argument begins historically in chapter 1, through a reconstruction of common themes in the criticism of Husserl's and Carnap's antinaturalism by Heidegger and Neurath. Heidegger and Neurath might seem to make very strange philosophical pairing. Yet the radicalism of their criticisms of antinaturalist conceptions of necessity makes their arguments especially interesting for my purposes. Their arguments pose what I call the problem of manifest necessity. The problem is this: any attempt to ground normativity in necessity must be able to show how the alleged necessities are both authoritative and binding upon materially and historically situated agents. The issue is not epistemological, as both Heidegger and Neurath rightly insisted, but ontological. The worry is not that we might fail to know what is or is not necessary, but that its supposed necessity would make no effective contact with the performances whose normativity it was to explicate.

1996) the ways in which all of these oppositions are formed on the basis of mistaken assumptions. Haugeland (1998, chap. 9) beautifully undermines the parallel presumption that mind and objects are discrete components of the world that only interact at well-defined interfaces, and I have extended his argument to challenge familiar distinctions between the social and the natural worlds. (Rouse 2002b)

Appeals to necessity would thereby account for normative authority at the expense of normative force. A familiar example of such a failure can be seen in the problem confronting the formal theories of confirmation and explanation proposed by logical empiricists. These theories supposedly acquired their authority from the formal validity of the inferential patterns they invoked. But philosophical appeals to these inferential norms were disconnected from the practices of theory acceptance and explanation or understanding that they supposedly governed. The difficulty was not that formal logic is not authoritative for what we say and do in science, but that the authority conferred by logical necessity turned out to make no binding claims upon what was at issue in theory confirmation or explanation.

My first chapter concludes by showing why we might not expect contemporary naturalists' appeals to a posteriori necessities to fare better than Carnap's or Husserl's appeals to the formal necessities of logical syntax or the essential structures of consciousness. Naturalistic responses to these earlier failures typically presume that their source was a misguided quest for a priori necessities to be uncovered by philosophical methods. Heidegger and Neurath challenged this assumption and instead located the problem in the attempt to ground normativity on necessity of any sort. This line of argument should be of particular concern to contemporary naturalists, who propose to "naturalize" normativity by situating it precisely in nomic necessities governing the concrete, historical-material world we inhabit. Of course Heidegger and Neurath each directed arguments against appeals to logical and transcendental necessity; however, their arguments can be generalized to apply to the nomic necessity invoked by contemporary naturalists. Moreover, grasping the relevance of their arguments for nomic necessity highlights scientific practices as the historical-material setting within which nomic necessity crucially lacks normative force. Both Heidegger and Neurath also appealed to the temporality of practical involvement in the world as an alternative way to make sense of the normativity of understanding. In later chapters, I will follow up on these suggestions as a way to reconceive scientific practices.

The conceptual problems within contemporary discussions of naturalism do not only concern the relation between necessity and the normative force it must explicate, however. In chapter 2, I explore another problematic dimension of the issues around naturalism. Not surprisingly, the rejection of formal-structural conceptions of meaning has led to efforts to locate meaning (and the understanding of meaning) within the world, as a phenomenon that could be investigated empirically. It might then seem appropriate to ask which domains of empirical science are relevant to understanding meaning, thought, and action, depending upon where within the world the phenomena of meaning and understanding are to be found. History and the social sciences, psychology, and neuroscience (or human evolutionary biology more generally) are common candidates for the successor scientific discipline to earlier philosophical characterizations of meaning as determined by formal or transcendental structural relations.

This second chapter begins with Charles Taylor's attempt to locate meaning specifically within the human sciences. Although Taylor's position within contemporary philosophy is in some respects idiosyncratic, his work draws attention to a difficulty confronting many other recent antinaturalists. Opponents of naturalism nowadays have mostly given up the aspiration to provide rational or transcendental grounds for the natural sciences. Their more limited aims are to exempt meaning, rationality, consciousness, ethical norms, or social life from natural scientific hegemony. Taylor's version of this project concerns the human sciences, which, he argues, ought to be sharply distinguished methodologically and ontologically from the natural sciences because they must account for how their objects of study (human beings) are self-interpreting in a way that "natural" objects are not. Human action is meaningful for human agents, whereas natural scientific phenomena are meaningless to the objects involved in them. Taylor's attempt to distinguish the human from the natural sciences on these grounds was sharply criticized by Thomas Kuhn, who insisted that the natural scientific disclosure of the world cannot be shorn in this way from the field of meaningful human activity. Kuhn's line of argument strongly suggests that Taylor's account of meaning must expand to encompass the supposedly meaningless behavior of natural objects. The parallel problem confronting other antinaturalists is how to block the expansion of their conceptions of meaning, thought, language, or other norms to encompass the scientific disclosure of nature. Their attempts to resist an all-encompassing metaphysical naturalism too easily slide into an equally all-encompassing rational or social humanism.

My point in juxtaposing Kuhn and Taylor is thus not to defend either position, but to explore the way the domains of meaning and nature have been conceptualized in their discussions. My central conclusion in chapter 2 is that Kuhn and Taylor are not alone in presupposing a dualism that conceives the natural world and the human world in ways that make it difficult to understand how they belong together. Wilfrid Sellars articulated similar concerns in contrasting the manifest and the scientific images of "man" and urging that these images be conjoined. But I would argue that such a conjunction (including Sellars's own conception) is blocked from the outset by the way in which nature and the meaningful world of human life have been conceived. I find compelling grounds for this argument in how the post-Kuhnian expansion of meaningful social life to encompass the natural world as understood scientifically is paralleled by the naturalistic expansion of the causal structures of the natural world to encompass whatever people say and do. The oscillation between a thoroughgoing metaphysical naturalism and an equally thoroughgoing social-conceptual constructivism shows how each position can be expanded to encompass everything—except the intelligibility of anything in terms of the other.

There is something deeply wrong in the conception and opposition of these positions. Even the best attempts to reconcile them, such as Donald Davidson's anomalous monism or its construal by Richard Rorty as a nonreductive physicalism, endorse the conceptual immiscibility of the natural and the normative and proclaim it as a virtue. Recent criticisms of Davidson, by John McDowell, Robert Brandom, and John Haugeland, begin to suggest the inadequacy of this solution, which I will elaborate upon in subsequent chapters. Scientific practices, then, provide the focal point in which this conceptual failure is most striking. My discussion of Kuhn and Taylor shows how scientific practices as presently conceived must balance precariously: they are meaningful, historically situated activities that belong to the human world; yet they are also that component of the human world that aspires to point beyond itself so as to make intelligible the supposed brute unintelligibility of the natural world.

W. V. O. Quine is perhaps the most prominent defender of an all-encompassing physicalist naturalism that is the counterpart to the Kuhnian expansion of Taylor's conception of meaning. Quine's official doctrine is that behavioristic psychology or neurophysiology must ultimately substitute for philosophical analysis of meaning and justification. The indispensable intermediary to this doctrine, however, is his insistence upon the indeterminacy of translation, which provides the initial focus of my discussion in chapter 3. On Quine's account, meaning is not a property of an utterance, but an interpretive theoretical construction governed by the concern to make comprehensible the utterer's subsequent behavior. Such interpretation is not merely underdetermined by what the utterer says and does (which provides the only data that could conceivably constrain it). Quine argues instead that meaning is radically indeterminate, that is, not an objective matter of fact at all. The ongoing activity of translation, of behavioral responses to behavior that engender further behavior, is sufficient to account for what happens without needing to postulate meanings to which translation is accountable. Neither philosophy nor social science should be construed as the analysis of meaning, therefore, for there are no meanings to analyze. Science shows us that there are natural facts, but no semantic facts.

I begin by defending Quine's doctrine of indeterminacy against Rorty's and Harman's familiar interpretations that mistakenly reduce it to a special case of the empirical underdetermination of theories. Yet semantic indeterminacy does not have the expected consequence of vindicating Quine's conception of naturalism. Ironically, his empiricism turns out to incorporate a last vestige of semantic realism, and his naturalism still holds onto a residual role for "first philosophy" in the form of a precommitment to a physicalist metaphysics. My argument for these claims emerges from within Quine's own conceptualization.

For Quine, utterances and actions can only be interpreted against a background of sustained interactions and shared material surroundings. Such triangulation is essential even in self-interpretation, for attributions of meaning must be accountable to publicly accessible standards—hence, "radical translation begins at home." The result is to recognize the irreducibility of language. Translation (or better, interpretation) into a language I already know how to use is as deep as understanding can go.[13] Quine concludes from such arguments that human utterances, actions, institutions, and cultures cannot constitute a distinctive intentional and intensional domain of inquiry. Talk about meanings is merely a quaint (and ultimately eliminable) way of saying that all facts must be part of a conceptual scheme, that is, an interconnected network of statements accepted as true. Quine, however, contrasts this absence of semantic facts with the

13. Davidson helpfully substitutes radical interpretation for Quine's notion of radical translation. The best reason for accepting his alternative is that 'translation' often connotes rephrasing in an already fixed idiom, whereas 'interpretation' rightly suggests that one's prior interpretive resources may need to be extended or transformed in the course of interpretation.

straightforwardness of physical or natural facts. Indeed, physical facts are *under*determined by sensory stimulations, the subset of physical facts that are available as evidence for our theoretical constructions.[14] But semantic facts are not even partially determined by these manifest physical events.

Quine is right to deny the existence of semantic facts as a separate domain of inquiry. But his contrasting of physical facts, which supposedly do constitute a distinct (indeed, exhaustive) domain, depends upon his commitment to what Davidson has criticized as the "third dogma of empiricism": the attempt to distinguish the empirical content of science from its linguistic or conceptual organization. Quine's characterization of the evidential base of science as physical events rather than sensory experiences removes some of the possible objections to this distinction, but not the crucial one. Any specification of the supposedly prelinguistic or preconceptual content to which language or theory must be held responsible must itself be semantically articulated. The attempt to distinguish the physical from the meaningful must take place *within language*. Meaning is thus not a property of some (possibly empty) subset of what there is, but an indispensable aspect of the manifestation of anything, including the natural world and its character as "physical."

The implications of similar internal criticisms of Quine's program have often been misunderstood as merely relocating his naturalized epistemology from empirical psychology to empirical social science, thereby oscillating back to the other side of the dualism articulated in chapter 2. The "Strong Programme" in the sociology of knowledge, for example, arose precisely on the basis of such misreadings of Quine and Wittgenstein.[15] However, most feminist scholars in science studies have grasped the point of such arguments (albeit usually not primarily through reading Quine). In chapter 4, I highlight some of the relevant contrasts between feminist science studies and the sociology of scientific knowledge in order to clarify the implications of these arguments for how to understand the sciences. Feminist science studies are often misread as attempting to find a precarious middle ground between sociological constructivism and realist or

14. It is crucial to Quine's view that the evidential basis for natural science is a set of publicly accessible physical events, that is, causally efficacious encounters between bodily surfaces and their environment, rather than intensionally or privately understood experiences.

15. For an insightful account of how the turn to a sociology of knowledge misread Wittgenstein, see Lynch 1992; Bloor 1992 is a thoughtful but unsuccessful response to Lynch's criticisms.

rationalist conceptions of scientific objectivity. I argue that they instead challenge the shared assumptions about semantic and epistemic normativity that govern the apparently opposed positions of realists, rationalists, and constructivists. Feminist work thereby points us in the direction of a more adequate conception of scientific practices.

'Practice' needs some crucial clarification, however, before one can understand the place of scientific practices in a more adequate philosophical naturalism. The term is often used in science studies and social theory to refer to social regularities in the doings of human agents. In chapter 5, after some initial clarifications of possible confusions about scientific practices, I develop an alternative way to understand practices more generally. I begin by reflecting upon Stephen Turner's recent criticisms of traditional accounts of practices as social regularities. Turner himself cannot imagine an alternative conception, but one emerges clearly with a slightly different perspective on his arguments. I argue that we can understand a practice through the *shared normative accountability of its constituent performances.* Turner mistakenly presumes that such accountability requires that a definite community at least implicitly agree upon or otherwise commit to definite norms or values. But various performances can be held mutually accountable, without requiring any underlying regularities among them, or any common formulation of "norms" or "values" at stake in that accountability.[16] To recognize this allows us to clarify frequently confused discussions of the role of language within supposedly social practices. It shows the need to replace substantive conceptions of linguistic meaning, knowledge, and power with dynamic, deflationary, constitutively normative conceptions. Indeed, the retention of a substantive conception of any one of language, knowledge, or power is enough to lead understanding of scientific practices astray. Chapter 5 concludes by setting the challenge for the remainder of the book: how to conceive scientific practices without reinscribing the dualism of nature and normativity, and thereby to vindicate the possibility of a coherent philosophical naturalism. The key to meeting this challenge is a conception of practices as held together by a

16. A central thesis of Davidson's essay "A Nice Derangement of Epitaphs" (1986) is that natural languages and the idiolects of particular speakers are coherent practices whose coherence does not depend upon underlying regularities of performance. Davidson concludes that "there is no such thing as a language" (446), in the familiar sense of rules, conventions, or behavioral regularities that enable communication between speakers who share such underlying commonalities.

normative accountability that outruns any supposed underlying regularities of performance or commitment.

I begin to take up this challenge with an extensive discussion of discursive practices in chapter 6. Robert Brandom's recent work has developed indispensable technical resources for articulating a constitutively normative conception of discursive practices. I endorse the central themes of Brandom's project: to explicate intentionality in practical-normative terms and to articulate semantics pragmatically in terms of proprieties of inference. I also thereby take over Brandom's detailed exposition of the linguistic aspects of discursive practices, which underscores the expressive role fulfilled by familiar logical and semantic vocabularies.

Brandom goes wrong at a crucial point, however, in his account of how intralinguistic performances are inferentially related to causal interaction with objects in perception and action. Brandom's approach reproduces the dualism of natural causality and rational normativity, and thereby fails to secure the objective accountability that he agrees is needed to secure the normativity of discursive practices. My arguments show how Brandom's account should be turned inside out. Instead of understanding intralinguistic practices as initially intelligible apart from "subsequent" inferential relations to perception and action, I show that discursive practices must be understood as patterns of practical/perceptual intra-action within the world. These arguments set the stage for later chapters, in which I develop a more adequate understanding of how such practical/perceptual intra-action is *causal,* thereby allowing my account of discursive practices to contribute to a more adequate philosophical naturalism. Before that part of my project can be undertaken, however, I need to follow up another consequence of my initial criticism of Brandom. My arguments about the intertwining of perception and action dissolve any clear boundaries between semantic/epistemic and practical/political normativity (a conclusion already foreshadowed by the discussion of feminist science studies in chapter 4). Brandom still seeks to maintain such boundaries, but his unsuccessful efforts to do so highlight another fundamental problem confronting explications of intentionality and normativity.

Brandom attempts to distinguish practical and moral from semantic and epistemic norms by contrasting the unity of truth and the plurality of goods. The latter, he argues, is explained by the underlying plurality of "bodies and desires" that serve as the grounds for those goods being taken as good. In chapter 7, I show how a second fundamental difficulty in Brandom's account emerges precisely from this plurality of bodies and desires. Brandom must be able to explicate how the commitments and entitlements that he takes to be instituted through discursive practices are both authoritative and binding. The concept of desire plays a crucial but largely unacknowledged role in his explication: the attribution of desires *expresses* the normative force through which discursive commitments and entitlements are binding upon practitioners. The first part of the chapter shows why 'desire' cannot fulfill that role, and hence why Brandom also fails in the end to explicate discursive normativity. Nor does it help to replace the deontic commitments and entitlements invoked by Brandom with John Haugeland's more resolute notion of existential commitments. Comparing Brandom and Haugeland highlights their problematic common ground: both of them seek the binding "force" of normative accountability in commitments undertaken and sustained by subjects (speakers or agents). My arguments show that voluntarist conceptions of normativity cannot succeed. Such accounts presume that there is no normative force except to the extent that subjects impose it upon themselves, but the intractable difficulty is to understand how such self-imposed commitments could ever be binding.

Further reflection upon Brandom's conception of discursive practitioners as subjects with divergent conceptual perspectives points toward a more adequate conception of normative force. Brandom thinks that perspectival differences arise from differences in bodies as well as desires. Bodies in the relevant sense, however, cannot simply be physical objects, but must instead be understood in terms of practical capacities and the correlative configuration of a practical situation. Taking seriously the role of practical-perceptual intra-action in discursive practices transforms Brandom's account of a subject's perspective *on* the world into a practical configuration *of* the world within which agents are differently situated. Moreover, one agent's situated environment and the possibilities it affords incorporate the activities of other agents as partially reconfiguring their shared surroundings. There is something at stake in intra-action with other agents, because its outcome shapes the intelligible possibilities for action and self-understanding by everyone involved.

Brandom and Haugeland were driven toward normative voluntarism by their prior assumption that the natural world is normatively inert. On their accounts, normative authority must be conferred upon objects, by us, because the natural world is composed of objects that *cannot* have or acquire normative significance otherwise. This assumption is familiar and hardly unique to Brandom and Haugeland; it is often taken to be the unquestionable legacy of modern science and its disenchantment of the world. Yet perhaps for that very reason, philosophers rarely undertake to show that this assumption is actually vindicated in scientific work.

My critical discussion of Brandom and Haugeland thus sets the stage for chapter 8, in which I take up the question of how nature manifests itself through scientific practices. I begin by asking how to reconceive traditional questions about the relation between experimental practice and theoretical conceptualization in the light of two decades of work on how scientific experiment has "a life of its own" (Hacking 1983). Experimental science does not merely aim to represent the world; it intervenes in the world to enhance its intelligibility. Such intervention is itself a causal transaction with what is to be understood, and my discussion begins with the question of how to conceive causality in this context. Many familiar conceptions of causal interactions or causal laws turn out to depend implicitly upon a "supernatural" or "God's-eye" standpoint from which causal interactions could be known despite the absence of any material interaction with them. Such conceptions should be unacceptable to naturalists, of course, because they render causal structure or causal relations opaque to experimental science. Yet many supposedly naturalistic conceptions of causality or laws depend upon precisely that mistake.

Contrary to Brandom and Haugeland, among many others, a conception of causality that avoids supernaturalism must assign causal significance to phenomena rather than to objects. The concept of a phenomenon is taken from Karen Barad's interpretation of Niels Bohr's reflections upon measurement. I extend her account to understand causal connections more generally, such that an effect is a "measure" of the capacities of the cause. Several examples from physics and biology show why the boundaries of causally intra-active systems are not properties of those systems in isolation, but are instead determined by the overall configuration of the phenomenon of their intra-action. Causal structure and causal capacities belong to the phenomenon as a whole rather than to its constituents understood as independent objects. Phenomena in this sense are not regularities, but differentially repeatable patterns involving norms of correct or incorrect repetition. Perhaps the most surprising result of the argument in chapter 8 is the discovery that conceptual practices and discursive norms are an ineliminable component of causally determinate phenomena. This conclusion is less surprising, however, in light of my earlier discussion of Quine. The "meanings" of the relevant concepts are not distinct "semantic facts," but simply articulate the phenomenon itself in its repeatability. By recognizing how the interpretation of meaning and the assessment of truth are themselves always causally intra-active phenomena, I show how this notion of scientific practices provides the basis for a naturalistic view of conceptual normativity. I conclude the chapter by showing how these arguments affect one's understanding of experimentation and its theoretical articulation, including the circumstances in which theoretical science can also come to have a "life of its own."

In this way, the familiar conceptions of natural causality that mark one side of the dualism of nature and normativity prove unacceptable on naturalistic grounds. Naturalists are not entitled to conceptions of causal relations or causal laws that presuppose a magical, God's-eye determination of the boundaries of causally intra-active systems or processes. A naturalistic conception of causal relations must instead be situated in the midst of the causal relations so conceived. My final chapter thus undertakes three tasks, whose collective fulfillment shows why this alternative account of scientific practices and the disclosure of natural phenomena should be understood as a more adequate philosophical naturalism. The first task is to recapitulate the booklong discussion of what is at stake in philosophical naturalism, and why naturalists should discard the familiar dualism of causality and normativity. The second task is to understand the appropriate place for laws, nomological necessity, and modality more generally within my view of scientific practices and causality. The third task is to consider how to understand the normative authority and normative force that govern scientific practices and the disclosure of the natural world. Unless I can make sense of the constitutive normativity of scientific practices and the scientific disclosure of nature, I cannot claim to have an adequate alternative to more traditional conceptions of naturalism.

My commitment to the fundamental role of causal intra-action in discursive practices, and scientific practices more specifically, is central to conceiving my project as naturalistic. I depart from traditional philosophical accounts of causality in some fundamental respects, however. In contrast to the familiar regularist or regulist conceptions of causality, I give explicative priority to causal phenomena themselves over the regularities or laws they supposedly instantiate. Philosophical objections to taking causality at face value in this way can be traced to Hume's doubts about causal connections between objects. Surprisingly, however, such objections take objects for granted, wondering only about how one object influences the behavior of another. My treatment proceeds in the opposite direction, beginning with causally intra-active phenomena and asking how objects and natural kinds are understandable through their practical-inferential role expressing the connections between various phenomena. This approach has obvious advantages for those inferential relations (such as wave-particle duality or the Bell inequalities) in which the systematic connections that can be established between phenomena cannot be expressed by assigning stable properties to constituent objects, but my discussion is primarily directed toward the inferential proprieties expressible by the use of natural kind terms.

I then consider how we should construe causal necessity as expressed nomologically. Many philosophers may conclude from my discussion of the normativity of causal intra-action that natural laws and nomological necessity must instead be the principal manifestations of the anormativity of the natural world. Drawing upon Nancy Cartwright's recent challenges to construals of laws as universally quantified generalizations, I argue that laws cannot serve this role, because the scope of nomological expressions must be determined in ways that are ineliminably normative. Cartwright herself does not explicitly acknowledge the constitutive normativity of natural laws, but it shows up implicitly in her antinaturalist commitment to empiricism: what she takes to determine the scope of scientific laws or theories is the availability of empirically adequate models that specify what the theory or law says about particular circumstances. I show why even her own epistemological and moral-political concerns are better served by an appropriate philosophical naturalism than by her sophisticated, pragmatic empiricism.

My discussion of Cartwright points toward one of my central claims in chapter 9: philosophical naturalists should take the normativity of causal intra-action as the basis for understanding alethic and nomological modalities rather than the reverse. I lay out my argument by juxtaposing Cartwright's implicit move in this direction to Brandom's and Haugeland's more explicit project of treating modalities as expressing normative commitments. However, Cartwright to some extent shares in the voluntarism of Brandom's and Haugeland's attempts to ground normativity on deontic or existential commitments, and I show how their common failure is connected to a misconstrual of the modality of scientific understanding. Brandom and Haugeland each take impossibility to be the crucial modality to explicate in normative terms, as expressing the joint incompatibility of various commitments or entitlements. I argue instead that we need to begin with a conception of possibility ("real possibility") that is more restrictive than those of alethic or nomological possibility (each of which can be defined in terms of negation and the corresponding form of necessity). This conception of real possibilities is indispensable to understanding scientific practices and their temporal orientation toward the future. Scientific research is always both grounded in capacities already attained and oriented toward possibilities that outrun any present conceptualization of their aims. Such possibilities are normative in a stronger sense than Brandom's and Haugeland's construals of the exclusion of impossibly conjoined commitments. Part of what distinguishes real possibilities from what is merely not-impossible is that real possibilities *matter*. They express a practical configuration of a situation such that there is something at stake in whether and how these possibilities are to be realized.

This shift from impossibility to real possibility as the fundamental modality involved in scientific understanding also compels a significant reconception of normativity itself. Philosophers have long been suspicious of normativity (and likewise, of causality); the common response is to reduce normativity (also like causality) to rules or regularities. Thus, it has commonly been assumed that norms can only be binding if they hold according to rule or principle ("necessarily"), or at least with regularity over time. The perhaps surprising conclusion of my arguments is that under such conditions, norms *cannot* be bindingly authoritative. Our normative reach always exceeds our grasp, and hence what is at stake in practices outruns any present articulation of those stakes. That is part of what it means to insist (as I began to do in chapter 5) that normativity cannot be reduced to regularities. We are accountable to what is at stake in our belonging (causally *and* normatively) to the material-discursive world: our fate is bound up with what is at issue and at stake in our practices, although those stakes are not yet definitively settled—indeed, that is part of what it is for them to be "at stake."

A failure to grasp this point is frequently expressed in demands to show the *source* of binding normativity, or its authoritative *grounds*. Sellars highlighted this point with his implicit recognition of the unavoidably mythical character of any retrospective disclosure of the source of normativity.[17] But Sellars did not recognize the need for a parallel treatment of any prospective (final) articulation of the norms to which practices are accountable. Normativity arises from practical involvement *in* a situation whose subsequent development is not yet determined: there are real possibilities for making a (significant) difference in how things *subsequently* turn out. The difference those possibilities can make transforms the situation, thereby changing what is at stake in responsiveness to it. That is why it is crucial that the normativity of practices not be reduced to underlying regularities of performance or belief. What is at stake in practices cannot be established with finality except at the cost of their normative authority and force. Or, to put it another way, the binding normativity (the normative force) of what is at stake in practices comes from always already belonging to patterns of *ongoing* causal intra-action. The world already has a (normative) grip upon us, through our belonging to a situation, understood as a field of real possibilities.

I conclude by showing how these considerations should inform an understanding of the temporality of scientific practices and of what is at stake in scientific investigation. Philosophers have often seriously misunderstood science by construing it retrospectively: they look to science to discover what is (already) known about what the world (already) is. They thereby overlook science as, in François Jacob's words, "a machine for making the future." Science is first and foremost research, an activity that transcends present understanding. Even such philosophers as Davidson, Rorty, Brandom, and Haugeland, whose criticisms of scientistic versions of naturalism bear close affinities to my own, still accept too readily a scientistic conception of natural science.

Such considerations have especially ironic significance for traditional conceptions of a naturalistic metaphysics, which insist that the sciences disclose nature as devoid of normative significance. Naturalists often eloquently insist that the anormativity of nature *matters:* the sciences free us from the illusions of theological, anthropomorphic, and other wishful projections upon the world. Such achievements are indeed the rightful legacy

17. Rebecca Kukla's brilliant analysis (2000b) of "Empiricism and the Philosophy of Mind" shows why readers must take seriously Sellars's aspiration "to kill a myth [of the Given] with a myth [of Jones]." Sellars's analysis, she argues, shows that any account of the non-normative source of normative authority involves a "constitutive misrecognition," since the alleged source must have *already* been normatively binding in order for its commands to be authoritative.

of philosophical naturalism. Yet that legacy has been claimed by metaphysical naturalists in a self-defeating way: they cannot adequately express how nature so conceived matters to us as scientists and as naturalists. The effort to "naturalize" semantic, epistemic, or practical/political norms thus illegitimately presumes that nature has already been "naturalized" in the same way. To allow the naturalistic legacy to claim *us* appropriately, I am arguing, we should recognize instead the normativity of nature, manifest in the possibilities expressible in part through scientific practices.

ONE

The Problem of Manifest Necessity

TWENTIETH-CENTURY PHILOSOPHY canonically began with Frege's and Husserl's extended criticisms of psychologism concerning logic and mathematics. Frege insisted that logic is a normative science. Husserl conceived of phenomenology as a rigorous science that would discover essential structures of meaning distinct from any empirical psychology. Each promised a new beginning for philosophical work at a time when the dramatic, rapid growth of the empirical and formal sciences seemed to leave behind traditional philosophical projects. The powerful tools of first-order quantificational polyadic logic and its metatheory, and to a lesser extent of the phenomenological and eidetic reductions, provided philosophers in both Europe and North America with new and promising resources. Their deployment quickly seemed to surpass the orthodoxies of the previously dominant neo-Kantian and neo-Hegelian schools, and they set the agenda for more or less continuous lines of subsequent philosophical development.

The familiar histories of these developments present Frege and Husserl as having effectively charted very different courses from their shared opposition to psychologistic accounts of the foundations of mathematics. Frege and his followers aligned the analytic tradition in philosophy with the formal sciences themselves, such that Frege's own foundational work in mathematics, Russell and Whitehead's *Principia Mathematica,* Wittgenstein's *Tractatus,* Hilbert's formalism, Carnap's *Logical Syntax,* Gödel's theorem, and Tarski's *Wahrheitsbegriff* were collectively perhaps even more integral to philosophy than to mathematics itself. Husserl looked elsewhere to secure the foundations of logic, turning inward toward the careful description of experience and the subjective constitution of meaning.[1] As a con-

1. Husserl's phenomenology was also very influential in the development of the intuitionist/constructivist tradition in the philosophy of mathematics, through his influence upon Luitzen Brouwer and Hermann Weyl.

sequence, the Fregean tradition continued to see itself as aligned with a scientific attitude and the pursuit of rationalist methods, whereas after Husserl, the phenomenological tradition turned toward introspective, hermeneutic, or existentialist styles of thought, and a more critical attitude toward the cultural hegemony of the sciences. Subsequent and far-reaching changes in each tradition, which defy characterization in terms of logical analysis or phenomenological description, have not significantly altered this fundamental difference in their orientation and style, at least according to their folk histories.

But accounts of the bipolar development of twentieth-century philosophy are deeply misleading, and not merely as caricatures of more complex histories. Taking them too seriously overlooks illuminating parallels in the development of problems and arguments on both sides. These parallels lead beyond a morass of argumentative details, however crucial for assessing specific claims and reasons, to uncover larger patterns in the posing of philosophical questions and responses to them. A larger pattern is in any case what I shall outline in this chapter. If Frege and Husserl mark one crucial turning point in the philosophical tradition, then another such turn occurs in the late 1920s and early 1930s, with the publication of *Sein und Zeit,* the protocol sentence debate within the Vienna Circle, and Wittgenstein's return to philosophy. Since my aim is primarily philosophical clarification rather than historical scholarship, I shall attend to perhaps the most striking and counterintuitive juxtaposition, which aligns Neurath's criticisms of Carnap's program for the rational reconstruction of scientific knowledge with Heidegger's criticisms of Husserl's transcendental phenomenology.

It may seem difficult to imagine a more bizarre philosophical juxtaposition than between Heidegger and Neurath. Neurath was a Marxist, internationalist, relentlessly secular and antimetaphysical exponent of the practical, "encyclopedic" unification of diverse scientific inquiries toward more effective prediction and control of our surroundings and ourselves. Heidegger was a conservative, provincial, neoromantic, theologically trained philosopher in lifelong pursuit of a thoroughly "impractical" way of thinking, which he nevertheless supposed to be more fundamental than any empirical or formal science. Their brief forays into revolutionary politics also stood in stark opposition: Neurath was the technocratic head of the Central Economic Administration of the Bavarian Soviet Republic in 1919; Heidegger was the visionary university rector who notoriously sought to implement the "Führer-principle" within Freiburg University for the sake of its spiritual renewal in accord with what he later called "the inner truth and greatness" of Hitler's National Socialism (1959, 166). Reading Neurath's terse manifestos for physicalism and a scientific world-conception in juxtaposition to Heidegger's circuitous and enigmatic contemporaneous lectures hardly dispels the strangeness of the comparison. What can such a confrontation show beyond their thoroughly alien orientations, which make readily comprehensible their evident mutual disdain?[2]

Certainly my aim is not to dismiss or diminish the philosophical, political, and personal significance of their very real differences, at a time and place when such differences urgently mattered. But my reasons for looking closely at patterns of argument in Heidegger and Neurath are not primarily historical. I do not discount the historical interest of the comparison; recognizing common patterns in the issues and arguments within the apparently divergent traditions of logical empiricism and phenomenology will enhance the recent interpretive reconfiguration of early-twentieth-century philosophy. My aim, however, is philosophically constructive and goes beyond the truism that our own philosophical thinking has been deeply shaped by its history, such that a better understanding of these internal criticisms of the analytic and phenomenological traditions will bring greater clarity to our own situation. Indeed, I think there is something importantly right in some convergent themes of Neurath and Heidegger that has not yet been adequately assimilated or developed. Thinking about their work can help clarify what is at issue in current debates over philosophical naturalism, removing some unfortunate confusions. Thinking about the issues they pose can also direct us toward a better response to those very issues.

At stake in the philosophical orientation shared by Frege, Husserl, early Wittgenstein, Carnap, and Cassirer (among many others) was the relation between facts and norms. In particular, they were united in thinking that the normativity (they would have spoken instead of "validity," "objectiv-

2. There is no reason to think that Neurath and Heidegger ever met, but they certainly knew of one another. Heidegger was teaching in Freiburg during Neurath's ministerial appointment and subsequent treason trial in neighboring Bavaria and at one point explicitly referred to the programmatic commitments of the Vienna Circle and *Erkenntnis*. Friedman (1996, 72 n. 18) convincingly suggests that in this passage, Heidegger's disdainful association of the Vienna Circle with Russian Communism was a reference to Neurath. Neurath's occasional references to Heidegger emphasize the latter's theological training and metaphysical orientation, both of which Neurath despised.

ity," or "rationality") that is *constitutive* of meaning and knowledge cannot be established by any contingent facts about the psychology, biology, or social history of speakers and knowers. What is merely contingently true about us can have no binding authority upon us, and hence cannot be correct or incorrect, appropriate or inappropriate, insightful or trivial. Moreover, without such binding normativity, there can be no factual claims in the first place; naturalism (of which both psychologism and historicism are variants) was thus apparently a radically self-defeating position. It was radically self-defeating because its truth would not be just self-contradictory, but self-deconstructive. If a naturalist thesis were really true, no statement, including the naturalist thesis, could be even an intelligible statement, a possible candidate for truth or falsity.

If contingent facts could not secure the normativity of meaning and knowledge, then it would have to be established by something *necessary;* the deepest disagreements among the most prominent early-twentieth-century philosophers thus concerned which necessities established normativity, and whether their necessity was manifest in their logical form, essential structures, transcendental conditions, or the like. These "forms," "structures," or "conditions," however, could admit of nothing contingent, and hence nothing material or historical. They provided the necessary form that enabled contingent facts to have content, to say something substantial about the world. Yet although these necessary structures of experience or meaning were not themselves contingent, they nevertheless had to be manifest within the contingent spatiotemporal world. Materially embodied and historically situated human beings had to be able to discern and respond to these necessities. Unless these necessary structures could make intelligible the actual occurrence of meaningful thoughts or expressions and justified beliefs, their alleged necessity would be idle.

This issue is what I call the problem of "manifest necessity." If rational reconstruction or phenomenological reflection were to discover necessary structures that successfully account for the normativity of thought, language, or knowledge, those structures and their necessity would have to become manifest within the concrete historical-material world. This problem is often misconstrued as epistemological: how do we know *of* these necessities, or *that* they are necessary? That epistemological misconception is reinforced when we consider where the problem of manifest necessity arose. Neurath developed the issue in the course of his disagreements with Carnap over the proper character of protocol sentences, the reconstructed observation reports that were to show the empirical basis for scientific knowledge. Heidegger likewise developed his fullest and clearest account of the difficulty in his 1925 Marburg lectures on Husserl's conception of phenomenological evidence, focused upon the sense of perception as a "fulfilling act." Yet both Heidegger and Neurath adamantly denied that the issues that concerned them were epistemological. Heidegger insisted that ontology rather than epistemology was at the center of his questioning. Neurath, in turn, repeatedly inveighed against the project of a theory of knowledge, and argued that his physicalism was a substitute for any such theory rather than a version of one. In both cases, their criticisms primarily concerned what Heidegger called the "being" of experience and language rather than the justificatory role of perceptual evidence.

Heidegger's and Neurath's turns from epistemology to ontology were also turns in the direction of naturalism, or so I shall argue.[3] Neurath is now widely recognized as a naturalized epistemologist *avant la Quinean lettre;*[4] Heidegger's engagement with naturalism has hardly been noticed at all. What I find significant, however, is not merely that they were members or fellow travelers of the party of philosophical naturalists, but that each anticipated a fundamental problem confronting later philosophical naturalists, and that each of their responses to that problem were initial moves in the right direction. The problem that they foresaw is a naturalistic parallel to the problem of manifest necessity that confronted antinaturalists. For the antinaturalists they criticized, the difficulty was to account for the "relation" between the formal structures of consciousness or language and the historical-material world. For many subsequent naturalists, there are comparable difficulties in making intelligible the relation between the causal or nomic necessity of nature and the concrete practice of science that discloses it. Heidegger's and Neurath's responses to these difficulties each involved reconceiving their philosophical work as taking a practical

3. Is it more strange to think of Neurath as doing ontology or of Heidegger as a naturalist? In the former case, the very word 'ontology' would surely have belonged on Neurath's *Index Verborum Prohibitarum*. But he would have prohibited the word only because it encouraged metaphysical misinterpretations that the more appropriate 'unified science' dispelled. 'Epistemology,' by contrast, needed to be abolished in deed as well as word: he saw the entire enterprise as irredeemably metaphysical. Heidegger's movement toward naturalism is more complicated; he was accused of naturalism by some critics (Husserl among them), but never quite embraced it himself. His conception of philosophy nevertheless moved in the direction of naturalism, although he then hesitated to identify his own thinking as philosophical.

4. Uebel 1992a provides the clearest and best defense of this interpretation.

rather than theoretical stance. Throughout this book, I shall be wrestling with these two themes emerging from Neurath and Heidegger: the shifts from epistemology to ontology, and from theoretical to practical philosophy.

The convergence between the two philosophers cannot be grasped without an examination of some common themes in Carnap's and Husserl's philosophical work, to which Neurath and Heidegger respectively directed their criticisms. I shall turn next to Heidegger, both because his criticisms of Husserl are more extensively articulated than Neurath's criticisms of Carnap, and because they are more familiar. Neurath's distinctive position and line of argument in the protocol sentence debates are not well known or understood even among philosophers who trace their lineage to the Vienna Circle. Only recently, in the work of Thomas Uebel, Nancy Cartwright, and a concurrent revival of German-language philosophical discussions of Neurath, has his philosophical stance become usefully articulated in the secondary literature.[5] After discussing the significance of Neurath's stance, I shall conclude by looking ahead to how this odd couple might help us grasp what is at stake in understanding philosophy naturalistically.

Husserl's and Carnap's Structural Antinaturalisms

The close affinities between Carnap's project from *Der Logische Aufbau der Welt* to *Logical Syntax of Language* and Husserl's phenomenology have been obscured both by the common misrepresentation of Carnap as an empiricist foundationalist and by his real but overemphasized differences with Husserl over the philosophical significance of modern polyadic logic.[6] To show how their projects confront similar philosophical difficulties, I therefore must first reconstruct their common features.

Both Husserl and Carnap sought to articulate and secure the conceptual and epistemological foundations of the sciences. Neither doubted the importance or adequacy of the sciences, but both detected widespread confusion over the nature and source of that authority. Both sought to overcome

5. Uebel 1991a, 1992a,b,c; Cartwright, Cat, and Chang 1996a; Cartwright, Cat, Fleck, and Uebel 1996; Uebel 1991b.

6. The widely held view of Carnap as an empiricist foundationalist has been decisively challenged in recent years by Friedman 1987, 1988, 1991, 1992; Richardson 1990, 1992; Coffa 1991; Ricketts 1994, 1996; Uebel 1992c; and others. My own discussion of Carnap has been strongly influenced by this work, especially that of Friedman.

the confusion by rejecting metaphysical interpretations of science. Carnap and the Vienna Circle are well known for their attacks on metaphysics, understood as "the field of alleged knowledge of the essence of things which transcends the realm of empirically founded, inductive science" (Carnap 1959a, 80). But Carnap also thought his rational reconstructionist program extracted from competing philosophical schools their common, legitimate core, while rejecting their illegitimate, metaphysical additions. Thus, Carnap insisted in the *Aufbau:*

> *The so-called epistemological schools of realism, idealism, and phenomenalism agree within the field of epistemology. Construction theory represents the neutral foundation which they have in common. They diverge only in the field of metaphysics, that is to say, only because of a transgression of their proper boundaries.* (1967, 286)

Husserl, in his own rejection of the contemporary schools through a return "zu den Sachen selbst" similarly argued that these programs sought the foundations of science within a philosophically illegitimate sphere, namely, the articulation or explanation of the real (whether reality was identified with material objects or with consciousness understood psychologically as a real constituent of the world). Philosophy, properly understood, had nothing to say about the actual existence or nonexistence of anything. Indeed, for Husserl, the confusion between the determination of what is real and the philosophical clarification of meaning and validity was the characteristic mistake of all forms of philosophical naturalism.[7]

For Husserl, this fundamental difference between metaphysical claims about reality and genuinely philosophical claims about the conceptual and evidential foundations of the sciences was revealed methodologically by the phenomenological reduction. Philosophical reflection began, for Husserl, with the commitment to put out of play any concern with what does or does not really exist. Thus, philosophy involved a deliberate changeover from the "natural attitude," characteristic of both empirical science and everyday life, to the "phenomenological attitude." The natural attitude is marked by a concern for the existence or nonexistence of real objects and

7. On the question of realism, idealism, and the philosophical standing of questions about reality, see Husserl 1982, sections 52, 88–90. For an especially clear discussion of Husserl's views about this issue, see Hall 1982. Husserl's most extended critical discussion of naturalism is in "Philosophy as Rigorous Science" (1981), although part 1 of his *Ideas Pertaining to a Pure Phenomenology* (1982) also takes up similar themes, as do many of his introductions to phenomenology.

their properties; the phenomenological attitude considers how these same matters are manifest as meaningful presentations, independent of any "existential" concern.

Carnap also insisted that philosophy properly construed puts traditional conceptions of reality out of bounds: "*The concept of reality (in the sense of independence from the cognizing consciousness) does not belong within (rational) science, but within metaphysics*" (1967, 282). More important, however, Carnap, too, eventually proposed to secure this exclusion through a *methodological* shift from a standpoint or mode of representation characteristic of science and everyday life to one more adequate for philosophical clarification. Appropriately, given his other commitments, Carnap saw this shift as linguistic and syntactical, rather than as an attitude or orientation on the part of the philosophical inquirer. Scientific and everyday discussions are typically conducted using "object-sentences," whose terms designate objects. This way of speaking is confusingly carried over into "the material mode of speech," in which properly syntactical matters of linguistic designation are spoken of using "pseudo–object sentences" that seem to designate objects when only syntactical forms are at issue. Such material mode sentences as "Five is a number" and "Babylon was discussed in yesterday's lecture" are more appropriately indicated by their genuinely syntactical correlates in the formal mode: " 'Five' is a number-word" and "The word 'Babylon' occurred in yesterday's lecture." The result is typically philosophical confusion:

> The habit of formulating in the material mode of speech causes us, in the first place, to deceive ourselves about the objects of our investigations: pseudo-object-sentences mislead us into thinking that we are dealing with extra-linguistic objects such as numbers, things, properties, experiences, states of affairs, space, time, and so on; and the fact that, in reality, it is a case of language and its connections (such as numerical expressions, thing-designations, spatial coordinates, etc.) is disguised from us by the material mode of speech. (1937, 298–99)

Both the formal mode of speech and the phenomenological reduction were methodological devices to focus investigation within the philosophical domain of linguistic or experiential manifestations, which were not to be confused with objects taken as transcendent to their forms of manifestation.

The respective turns to the phenomenological attitude and the formal mode of speech were intended to make possible a more rigorous explication of the meaning and epistemological grounding of the sciences. Carnap's conception of the epistemological aim of the *Aufbau* (an aim that he never abandoned despite other changes in his views) showed this clearly:

> Even though the subjective origin of all knowledge lies in the contents of experiences and their connections, it is still possible, as the constructional system will show, to advance to an intersubjective, objective world, which can be conceptually comprehended and which is identical for all observers. (1967, 7)

Recent scholarship has decisively challenged the familiar conception of the *Aufbau* as a phenomenalist tract.[8] Yet Carnap's choice of a phenomenalist construction among other acceptable concept construction systems was not arbitrary. For some purposes, a physical or other basis for construction could serve equally well, but the phenomenalist strategy crucially served the epistemological goal of displaying the objective validity of the conceptual structure of unified science. Epistemology was, for Carnap, a normative enterprise, which showed how the logical organization of private experience conferred the normative force and authority of objective knowledge. Empirical psychology, even an empirical psychology that discovered quite general regularities of human experience, could not make evident the *objectivity* of the concepts whose empirically confirmed interrelationships constitute unified science.

Carnap's continuing commitment throughout his career to a conception of epistemology as the rationally reconstructive explication of scientific knowledge showed his commitment to antinaturalism.[9] Husserl was more explicit in proclaiming the antinaturalist stance of phenomenology. In the 1911 essay "Philosophy as Rigorous Science," Husserl specifically posed phenomenology in opposition to both naturalism and historicism, understood as attempts to ground the normative role of meanings or validations upon empirically ascertainable facts. Phenomenology would ask questions such as these:

8. Friedman 1987, 1992; Uebel, 1992a,c; Richardson 1990.

9. We should not allow ourselves to be confused here by Carnap's commitment to seeing philosophy as part of unified science. The formal sciences of logic and mathematics, to which epistemology was to be assimilated, were crucially different in their grounds and authority from the empirical sciences for which they provided indispensable articulation. Where others might distinguish the normative task of philosophy from the descriptive and explanatory tasks of empirical science, Carnap distinguished mathematics and logic from empirical science. That no more amounts to philosophical naturalism than does Husserl's conception of philosophy as "rigorous science."

> How can experience as consciousness give or contact an object? How can experiences be mutually legitimated or corrected by means of each other, and not merely replace each other or confirm each other subjectively? How can the play of a consciousness whose logic is empirical make objectively valid statements, valid for things that are in and for themselves? (Husserl 1981, 172)

Husserl emphatically denied that any of these questions could be answered by an empirical science without begging the question. A different kind of scientific investigation was called for. It must rigorously show how experiential interrelationships *ought* to be understood, and how they could make authoritative claims upon us, not just provide descriptions (or predictions) of what we (will) in fact do.

Perhaps the least recognized parallel between Husserl and Carnap concerns their understanding of the source of semantic and epistemic normativity. Here is one of the points at which recent scholarship has most significantly transformed the received view of Carnap. The givenness or incorrigibility of sense experience could not secure the normative authority of the claims of unified science, for such experience could have no intersubjective authority.[10] Thus, Carnap noted in 1928:

> The series of experiences is different for each subject. If we want to achieve, in spite of this, agreement in the names for the entities which are constructed on the basis of these experiences, then this cannot be done by reference to the completely divergent content, but only through the formal description of the structure of these entities. For science wants to speak of what is objective, and whatever does not belong to the structure but to the material (i.e., anything that can be pointed out in a concrete ostensive definition) is, in the final analysis, subjective. (1967, 29)

There were, then, two stages involved in securing the objectivity of scientific claims on Carnap's view in the period from the *Aufbau* to *Logical Syntax*. The first was a formalization that removed any particular references from scientific claims whose standing is in question. In the *Aufbau*, for example, Carnap claimed that

10. "Normative authority" here has at least two dimensions, of meaning (how we ought to understand an expression) and justification (whether we ought to accept it). In neither case, we shall see, did Carnap locate such authority in the "givenness" of experience, his initial enthusiasm for the verificationist theory of meaning notwithstanding.

> each object name [but also the general names of concepts, classes, or relations] which appears in a scientific statement can in principle (if enough information is available) be replaced by a structural definite description of the object, together with an indication of the object domain to which the description refers. (1967, 28–29)

Contingent particular references were removed altogether when this claim was conjoined with "the fundamental thesis of construction theory" that there is only one object domain, the domain of unified science. The formal, polyadic relational structure of a scientific claim (exemplified by the five-dimensional structure expressed by claims about the visual modality of sense perception) was supposedly sufficient to locate it uniquely within the overall, unified domain of scientific knowledge, and to do so on purely *structural* grounds. Its objectivity (intersubjective validity) would be secured by its formal logical structure.[11] The role of Carnap's early phenomenalism was to provide a formulation of the constructional system of unified science specifically relevant to epistemological concerns. By showing how more complex levels of objective structure (physical objects, cultural institutions, etc.) were themselves formally constructable from structure statements corresponding to individual sensory experiences, his reconstruction would thereby show "how, through the application of uniform formal construction rules, entities result which have a structure that is the same for all subjects" (1967, 29).

Husserl, who argued that the objective validity of formal logic was as much in need of philosophical clarification as any other scientific domain,

11. In the *Aufbau* this second step, which removes all nonlogical primitives from the construction, was accomplished by introducing a variable over a restricted range of relations ("experiencable, 'natural' relations" [Carnap 1967, 236]), which under the empirical conditions of unified science, would allow for a unique, formal definition of the basic relations employed in the construction. As Friedman notes (1987, 531–34), this approach leads to a fundamental dilemma: either the relevant domain of experienceable relations is specified ostensively (thereby undermining the aim to produce an intersubjectively valid construction), or else it is accomplished formally (thereby collapsing the distinction between empirical and formal-logical truth). Carnap's eventual response in *Logical Syntax of Language* (1937) took the latter route, while attempting to preserve a distinction between logical and empirical truths within each particular formalized language through a syntactical difference expressed in a metalanguage. The second stage in securing objectivity was then no longer accomplished by the uniquely specifying structure of the one object domain of unified science, but was instead relativized to a chosen language whose formal (and hence, objective) structure could be specified in the metalanguage. Friedman (1988) points out that Gödel's theorem rendered this approach untenable, since logical and mathematical truth cannot be consistently identified formally by syntactical decidability.

could not avail himself of formal logic or its metatheory to secure the normative authority of scientific claims. Nevertheless, he followed a strategy that paralleled Carnap's in two crucially connected respects: he located the source of the normativity of both meaning and objective validity in structural relationships, and he saw the removal of any reference to particular objects or experiential subjects as the crucial feature of those relationships. Husserl's account proceeded in three distinct steps. The first, of course, was the phenomenological reduction, which examined conscious experience while putting out of play any reference to actual objects or the psychophysical reality of consciousness. What then remained for consideration was the structural correlation between a particular conscious act, or noesis, and its noematic sense, the object *as intended* through that act.[12] For example, Husserl suggested that each aspect of an act of perceiving has a structurally correlated aspect of the sense of its object *as perceived*. In the second step, the eidetic reduction, the particularity of such acts and their correlated meaning-contents was removed. One varied in imagination both the act and its intended sense, and noted when the correlated variant showed itself to be impossible (thus, if one tried to imagine an object visually presented as having no back side that could be explored by moving around it, one would fail in much the same way as when trying to imagine a square circle). It was these essential structures that circumscribed the possible forms that any particular conscious experience could take, and which supposedly marked the fundamental achievement of Husserl's phenomenology. Thus, having insisted that no empirical science can account for the normativity of its own meaning or validity, Husserl contrasted empirical sciences to phenomenology as a science of essences (essential structures) rather than as a science of norms.

The third step in Husserl's strategy was to perform a similar imaginative variation on the particular consciousness (transcendental ego) engaging in phenomenological clarification. Just as the eidetic reduction shifted from

12. The term 'act' is easily misleading here, since it suggests a psychophysical performance by an existing subject. Heidegger's lectures on Husserl's phenomenology in the 1920s speak especially clearly to this point: "What does act mean [in phenomenology]? Not activity, process, or some kind of power. No, act simply means *intentional relation*" ([1979, 47] 1985, 36; original emphasis). Heidegger had indicated the significance of that clarification earlier in the lectures: "What makes us blind to intentionality is the presumption that what we have here is a theory of the relation between physical and psychic, whereas what is really exhibited is simply a *structure* of the psychic itself" ([1979, 46] 1985, 35; my emphasis).

particular perceivings or imaginings to the essential structure of perception or imagination, so this third move accomplished the transition from a disclosure for "me" (this particular transcendental conscious field, not this particular psychophysical individual, whose existence was already put out of play by the phenomenological reduction) to a disclosure valid for any transcendental ego whatsoever. This third step paralleled the intended role for the unity of science in the *Aufbau*. The transformation of scientific statements into "structure statements" removed all particularity from them relative to their object domain; the unity and uniqueness of the object domain of unified science then removed that last contingent relation to something particular or contingent. Likewise, the eidetic reduction revealed the essential structures of *this* field of conscious experience, and its extension to the ego itself removed the last vestige of particularity within the "formal" constitution of the meaning and validity of any science.

The close relationship between Husserl's and Carnap's programmatic approaches for clarifying the normative authority of scientific knowledge emerges especially clearly in another way, through the efforts each made to incorporate the other's program within his own.[13] Husserl was well aware of the claims for formal logic as a systematic, normative science that would secure the grounds for all other scientific knowledge; he himself had insisted upon the irreducibility of logic to empirical science (1970b, vol. 1; 1969). He nevertheless argued for the epistemological priority of the "transcendental logic" introduced by phenomenological reflection, on the grounds that the objectivity of formal logic cannot be secured by itself. Instead, Husserl insisted, the understanding of logical relationships and their correctness or incorrectness had to be made evident as an intentional structure of transcendental consciousness:

> The formations with which logic is concerned and their universal forms are given at first in a *straightforward evidence;* and this comes first necessarily. But now a *thematizing reflection on this evidence* is demanded. . . . The formations and universal forms (formations belonging to a

13. There is no evidence that Husserl was directly aware of Carnap's work. The *Aufbau* was published just as Husserl retired from active university service, and Husserl was in any case notorious for ignoring the work of most other contemporary philosophers. Nevertheless, he was well aware that a prominent alternative to his own philosophical program was to seek the philosophical grounding of scientific knowledge in the purely formal structures of modern logic, and his *Formal and Transcendental Logic* (1969) was a systematic response both to that program and to its implicit criticism of phenomenology as still psychologistic.

higher level), which are "given" in the activity and are, at first, all that is "given," must now be "clarified" reflectively in order that, by clearing up the intentionality that aims at and actualizes its objective sense originaliter, we may rightly apprehend and delimit this sense and secure its identity against all the shiftings and disguisements that may occur when it is aimed at and produced naively. (1969, 176)

Carnap responded with a parallel attempt at the incorporation of Husserl's strategy within his own. After clearly summarizing Husserl's conception of intentionality, referring specifically to *Ideen I* (Husserl 1982), Carnap insisted that an adequate account of intentionality must itself be situated within a more general theory of relations:

> *The intention relation is not a relation of a unique kind* which can be found nowhere but between a psychological entity and that which is represented by it. . . . The relation between an element and a relational structure of a certain sort in which it has a place is one of the most important relations of the applied theory of relations. The *intention relation* is nothing but a sub-class of this relation, namely, the relation between an experience (or a constituent of an experience) and an order which has a real-typical structure. (1967, 262)

By the time of *Logical Syntax,* Carnap had reconceived the theory of relations as simply expressing the logical syntax of many-termed predicates (1937, 260), but there is no reason to think he had abandoned the commitment to construe intentionality in those terms. Indeed, the turn to logical syntax offered him a clearer response to Husserl's criticism that purely formal systems could not express their own formal structure, whereas phenomenology could account for its own grounding phenomenologically. Carnap began his specific syntactical constructions by noting:

> Up to the present, we have differentiated between the object-language and the syntax-language in which the syntax of the object-language is formulated. . . . Actually, it is possible to manage with one language only; not, however, by renouncing syntax, but by demonstrating that without the emergence of any contradictions the syntax of this language can be formulated within the language itself. (1937, 53)

Thus, Husserl and Carnap each defended his own general, structural explication of the normativity of any scientific or otherwise intelligible experience by trying to show its uniquely inclusive character.[14] It is not germane to my argument here to try to assess whether either was in a better position to make good on that claim, but only to show their common strategy.

When we turn to the question of how these formative structures and their normative authority were to become manifest, we do encounter a more fundamental difference between Husserl's and Carnap's programs. Carnap's project was articulated within the overlapping shadows of Frege's logicism, Hilbert's formalism, and Wittgenstein's combinatorial conception of logical form in the *Tractatus*.[15] He hoped to display purely formal structures that would explicate the normativity of meaning and justification without appealing to anything like an intuitive or rational faculty for recognizing the correctness of structural or logical determinations. For that reason, the concept of analytic truth (or, eventually, analytic-truth-in-*L* for variable *L*) played a central role for Carnap; moreover, the logical apparatus necessary to articulate the propositions of primitive recursive arithmetic was always assumed by Carnap to be unproblematic, and in need of no further philosophical exposition.[16] Husserl was instead happy to take as unproblematic the appeal to intuition embedded in what he once called the "principle of all principles" of his phenomenological philosophy:

> *That every originary presentive intuition is a legitimizing source of cognition,* that *everything originarily offered* to us *in "intuition" is to be accepted simply as what it is presented as being,* but also *only within the limits in which it is presented there.* (1982, 44)

Thus, whereas Carnap looked to the syntactic structure of a language to secure the validity of logical truths, Husserl called for careful attention to the evident intuitive grasp of appropriate demonstrations of those truths. These are important and far-reaching disagreements, both for consideration of the adequacy of each approach and for understanding the issues that arise in response to each, and the plausible ways to consider them.

14. By the time of *Logical Syntax,* Carnap had abandoned the idea of a single, uniquely inclusive formal structure of the world as a whole, recognizing the possibility of a plurality of languages. He nevertheless insisted upon the priority of the method of logical syntax for formulating the results of any logical analysis, and the substitution of the logical syntax of the language of science for the suspect metaphysical claims of traditional philosophy (1937, 332–33).

15. For an especially clear account of Carnap's evolving conception of analytic truth, and its central role in his program, see Friedman 1988.

16. For elaboration, see Friedman 1988.

The difference between the two programs must, however, be seen within a slightly larger frame. Carnap's insistence upon sharply distinguishing logical from empirical truths (at least within any particular language) reflected their distinct roles within his philosophical reconstructions of scientific knowledge. Analytic statements articulated the formal, inferential relations that enable any statement to be meaningful at all. Yet analytic statements themselves are without content.[17] They could articulate substantial content only by displaying the structural (syntactical) connections between other statements and the "protocol statements" that constituted the empirical basis for scientific knowledge. The nature and role of protocol statements was, however, the focus of Neurath's criticisms of Carnap's epistemological program for the rational reconstruction of scientific knowledge.

Not surprisingly, Husserl also recognized differences in the modes of "evidence" displayed by formal and empirical judgments, but he understood these differences within a unified conception of evidence that played a fundamental role within his phenomenological program. Phenomenological clarification was primarily directed toward the modes and aspects of the evidence with which various intentional acts showed themselves within the phenomenological and eidetic reductions. Husserl's conception of logical and mathematical truths as evidently fulfilled also encountered distinctive difficulties.[18] Yet a more fundamental problem afflicted Husserl's overall concept of evidence as givenness in a fulfilling intuition. This prob-

17. As both Friedman (1988) and Uebel (1992c) have noted, Carnap was deeply troubled by the related questions of the status of his own rational reconstructionist claims and of the distinction between analytic and synthetic truths. The latter problem, Friedman (1987) cogently argues, was not adequately resolved in the *Aufbau*. An apparently more satisfactory resolution was introduced in *Logical Syntax* by articulating the distinction between analytic and synthetic truths for any given language within the metalanguage in which one could express the syntactical structure of the original language. Friedman (1988) notes that this solution also foundered on Gödel's theorem, since maintaining the distinction for a given language depends upon being able to identify analytic truths within the finite, combinatorial resources of its metalanguage. The former problem had a somewhat different, but convergent history. For some time after the publication of the *Aufbau*, Uebel argues (1992c), Carnap was inclined to follow the early Wittgenstein by regarding the statements of the *Aufbau*'s "construction theory" as merely "elucidations," strictly meaningless expressions that nevertheless played a useful role in displaying the structures through which scientific statements were meaningful. Never fully satisfied with this proposal, he eventually proposed to resolve the difficulty by the dual move of understanding his own analyses as syntactical statements (in the formal mode) within a metalanguage, and then arguing that the physicalistic language of unified science could itself be its own metalanguage.

18. Husserl's treatment of mathematics had close affinities with Brouwer's and Weyl's intuitionism; a discussion of the difficulties of his program or its relation to theirs is beyond the scope of this chapter.

lem was at the center of Heidegger's critical reflections upon Husserlian phenomenology, showing up in his concerns about the "sense" of perceptual presentations as fulfilling intuitions, about "the being of consciousness," and about the inescapably "hermeneutic" character of all intelligibility. It is precisely this problem, and Heidegger's response to it, that I propose to consider in juxtaposition to Neurath's stance in the protocol sentence debates.

Heidegger on Phenomenological Evidence

During the period that concerns me in this chapter (from the early versions of *Sein und Zeit* that began to emerge in the mid-1920s, until the mid-1930s), Heidegger understood his project as phenomenological, which he conceived explicitly in the terms of Husserl's 1911 *Logos* essay. Philosophy was to be a rigorous science distinct from both the empirical and formal sciences, yet indispensable for securing the normative grounding of both. To this extent, of course, Heidegger's project stood in sharp contrast to Neurath's militantly antiphilosophical stance (in ways that Heidegger later came to question).[19] Yet Heidegger's phenomenology involved a distinctive and ultimately radically transformative focus: he proposed to begin with a phenomenological description of the natural attitude itself. Husserl had conceived the natural attitude as the characteristic intentional orientation shared by science and everyday life. The natural attitude was supposedly governed by everyday practical involvement, and more fundamentally by a naive acceptance of the world as simply there, as *existing*. The natural attitude involved both a *concern* for the existence or nonexistence of worldly beings (including human beings as psychophysically embodied, as well as the objects they encountered perceptually and practically) and an unreflective *acceptance* of one's familiar surroundings.[20] Husserl insisted that philosophy proper could only begin once the natural attitude had been suspended. Husserlian phenomenology did not doubt the existence of any-

19. In the end, Heidegger's identification with Husserlian phenomenology as a distinctively philosophical program may seem to be as misleading as is Neurath's identification with logical empiricism, if the latter is understood as committed to an antinaturalist conception of philosophy as the rational reconstruction of scientific knowledge.

20. Husserl only sharply distinguished these two aspects of the natural attitude (under the headings of the natural attitude and the theoretical attitude) in *Crisis in European Sciences and Transcendental Phenomenology* (1970a), undoubtedly as a response to *Sein und Zeit,* in order to insist upon the continuity in *theoretical* orientation between phenomenology and the empirical and mathematical sciences.

thing; it suspended its concern with existence in order to attend to something else that was systematically overlooked or misconstrued when philosophy was undertaken from this natural orientation. That something else was the phenomenon, the structurally correlated relation between conscious acts and their intentional sense.[21]

Husserl of course hoped to reveal the forms of intentional directedness at work in both science and everyday life, by suspending their characteristic involvement in the world. Heidegger's phenomenology of the natural attitude raised the possibility that what Husserl methodologically suspended was in fact essential to the phenomenon of intentional directedness. And that is just what Heidegger claimed: Husserl's conception of consciousness, the sphere of immanent, indubitable givenness distinct from any physical, psychological, or sociocultural reality, obscured rather than revealed the intentional directedness it hoped to clarify. Moreover, Heidegger criticized that conception in terms to which Neurath would have been broadly sympathetic. He argued that consciousness is a metaphysical prejudice taken over uncritically from the history of philosophy:[22]

> This idea, that *consciousness is to be the region of an absolute science,* is not simply invented; it is the idea which has occupied modern philoso-

21. Those readers unfamiliar with the phenomenological tradition can usefully understand Husserl's conception of the intended sense of a conscious act as a generalization of Frege's notion of the sense of a linguistic expression. Just as the sense of a definite description is its mode of presentation of its referent, the sense of a perceptual experience, a hallucination, or a judgment is its mode of presentation of its intended object as perceived (from this location, in these sensory modalities), as hallucinated, or as judged (in this specific content). Moreover, just as Fregean senses are ultimately correlated with linguistic types rather than tokens, Husserlian senses are also ideal structures that correlate with types of intending acts. When Husserl speaks of conscious acts, he means not a specific spatially and temporally located event, but an ideal meaning structure that could in principle be repeated as the same, not merely as similar (at least in many of its aspects). Even those aspects of a conscious act that are de facto unrepeatable, for example, its relative location within some extended temporal series of acts, is still an ideal, typical structure rather than a de facto occurrence.

22. In making this suggestion, I am using 'metaphysics' in something akin to Neurath's sense. I do not intend to imply that Neurath and Heidegger conceived of metaphysics in a similar way, that Heidegger explicitly objected to Husserl's project for being metaphysical, or that Neurath would have accepted (as a part of, or a substitute for, empirical scientific inquiry) Heidegger's account of how to avoid such prejudice by allowing the phenomenon to "show itself from itself." The parallel is only that Heidegger objected to Husserl's allowing traditional philosophical presuppositions about what there is, and about what philosophy is, to prevent him from attending to what actually shows itself experientially. In making this parallel, I thus abstract from profound differences in their conceptions of experiential manifestation.

> phy ever since Descartes. The elaboration of pure consciousness as the thematic field of phenomenology is *not derived phenomenologically by going back to the matters themselves* but by going back to a traditional idea of philosophy. ([1979, 147] 1985, 107)

In place of Husserl's phenomenological explication of consciousness as a field of immanent experience, Heidegger famously substituted his analytic of "Dasein." Leaving aside many complications about what or how Dasein is, the initial contrast to Husserl is striking. Dasein is *worldly:* its particular material and social involvements, and its concernful understanding of the "being" of its surroundings cannot be deliberately suspended, and if it could, the phenomena would be distorted. Moreover, Dasein is "in each case mine," such that there can be no analogue to Husserl's turn to transcendental consciousness in general. To shift from this particular locus of Dasein (whose own being is "at issue" for it) to what is essential to any "case" of Dasein would be to omit and obscure crucial features of "the being of the intentional."[23]

This fundamental shift of philosophical attention from immanent experience to practical involvement in the surrounding world will soon play an important role in my comparison of Neurath's and Heidegger's criticisms of their philosophical traditions. Nevertheless, it is even more important that, in the 1925 Marburg lectures in which Heidegger most extensively used an immanent criticism of Husserlian phenomenology to develop the themes of *Sein und Zeit,* his criticism began with the phenomenological description of what is manifest in perception rather than with the question of "the being of consciousness."[24] To understand the force and significance

23. The attentive reader will note that only two of Husserl's three methodological stages were thereby rejected: the initial phenomenological reduction that suspends our everyday and scientific concern with existence and nonexistence, and the final stage in which Husserl uses imaginative variation to turn away from any "particular" transcendental consciousness to the essential structures of transcendental consciousness in general. The intermediate stage, the eidetic reduction that displays essential structures of conscious experience, had as a prominent correlate in *Sein und Zeit* the essential structures of Dasein, which Heidegger characterized as "existentialia" rather than "categories" to acknowledge that these were not essential properties of a kind of thing, but instead characterized a way of being (in which "its own being is at issue for it"). Heidegger later came to question this remnant of traditional philosophizing, but the significance of that shift for Heidegger's own project will not be discussed here.

24. Friedman's insightful discussion (1996) of Heidegger's philosophical confrontation with Carnap in the 1930s rightly emphasizes Heidegger's shift from Husserl's commitment to the atemporal, immaterial structures of "pure logic" (which he of course construed as transcendental rather than formal logic) to the historically and materially concrete existence of Dasein, and his consequent rejection of the philosophical priority of logic. Friedman

of Heidegger's criticism, however, we need to say a little more about why perception was problematic for Husserl.

Husserl's phenomenology depended upon a basic distinction between "empty" and "fulfilled" intentional acts.[25] A signifying act such as a judgment that "the butler did it" presents its sense (that the butler committed the crime in question) as grasped in the absence of the intended event itself. Perception, by contrast, is a fulfilling act: to *see* the butler committing the crime is to have before you the sense of actual presentation of what is intended.[26] But how is one aware of this difference? Husserl argued that the primary intuitive presentation of an intentional content is always accompanied (usually quite unthematically and unreflectively) by "categorial intuitions" that directly manifest structures that previous philosophers sought through transcendental arguments or demonstrations from first principles.[27] Thus, the emptiness or fulfillment of particular intendings was

---

does not attend to the parallels discussed here between Heidegger and Neurath's differences with Carnap. Undoubtedly these parallels are less visible without attending specifically to the juxtaposition of Heidegger's and Neurath's accounts of perceptual manifestations (in the perceptual sense and the protocol sentence). Perhaps even more crucial to Friedman's historical reconstruction of the Heidegger/Carnap contretemps, these parallels were not readily visible to the participants. Neither Neurath nor Carnap were clearly aware of how deep their own differences were (Uebel 1992a,b,c), and their political, protoscientific, and antimetaphysical solidarity in opposition to Heidegger was far more salient than any antiphilosophical affinity between Neurath and Heidegger. It is even more doubtful that Heidegger would have ever read Neurath's key papers in *Erkenntnis,* let alone been well positioned to recognize their philosophical significance; moreover, the extent to which his own program moved in the direction of a historicist, antiphilosophical "naturalism" was not yet fully clear to him.

25. This was a distinction in degrees (on multiple "scales") rather than in kind. Intentions can vary in the certainty, the completeness, and the modality with which they are fulfilled. In the last case, for example, acts of imagination are empty intendings of the actuality of the intended object, but fulfilling (evident) presentations of their possibility. Husserl's central methodological moves turned on these points. The "principle of all principles" was that whatever is clearly presented to consciousness in the phenomenological reduction was fulfilled with certainty (I cannot be mistaken about what seems to be present in my experience when I clearly attend to it). The eidetic reduction depended upon a comparably certain fulfillment of the *possibility* of what is clearly manifest in imagination.

26. One can, of course, simultaneously engage in both acts. As I witness the scene, I might explicitly express the judgment by uttering, "My God, the butler *killed* him." But in this case, the perceptual presentation (together with the fulfilling sense of the identity between the sense of what is judged and the sense of what is perceived) fulfills the sense of what is judged, while the judgment itself remains an empty intending. Strictly speaking, what is identical is only the "core" content of the senses of what is perceived and what is judged, since these two presentations clearly differ in the mode of evidence (as empty or fulfilled) with which they present that identical content.

27. Readers unfamiliar with the phenomenological tradition should not be misled by the word 'intuition.' As Heidegger noted in the Marburg lectures, "The term 'intuition'

itself supposed to be directly given. For Husserl, 'truth' denoted any directly presentative categorial act of identification of intention and intended object;[28] its correlated intended sense was what is meant by 'being,' the *sense* of what is presented categorially as directly present to consciousness in the appropriate way. Hence, Husserl recognized different modes or regions of being, corresponding to the different ways in which an intended sense could be fulfilled: the being of natural objects that can only be presented incompletely in perception is different from the being of numbers, which in turn differs from the being of conscious presentations. Each is fulfilled differently in experience, and hence the corresponding categorial intuition of the sense of fulfillment (the being of what is presented) is different.

The problem that Husserl confronted is what it could possibly mean for the categorial sense of fulfillment (being) itself to be fulfilled. Consider cases of perception as fulfilling acts. In seeing an object before me, I not only have the sense of the object as perceived with a definite shape, color, and orientation, but I also have the categorial sense of the object's actually being there, shaped, colored, oriented, and so forth in that way.[29] Is the sense of the being of the object a sense that could be intended emptily? Husserl confronted a dilemma. If the sense of being presented in a categorial intuition could itself be intended emptily, then one would need another categorial intuition of fulfillment that would identify the empty sense of being with its actual presence (the being of being, we might say). The resulting regress would be vicious. No intuition could ever actually be fulfilled (and hence, the very distinction between empty and fulfilled in-

corresponds in its meaning to . . . 'seeing' in the broad sense of that word. *Intuition* means: simple apprehension of what is itself bodily found just as it shows itself. . . . Intuition in the phenomenological sense implies no special capacity, no exceptional way of transposing oneself into otherwise closed domains and depths of the world" ([1979, 64] 1985, 47)

28. 'Truth' was thus not essentially propositional for Husserl. Propositional truth was the intuitive grasp of the identity of what was said in the proposition with what is directly presented in a fulfilling intuition (whether the appropriate form of fulfillment was perceptual, imaginative in the case of fulfilling a judgment that some meaning-complex is possible, demonstrative in the case of a mathematical proposition, etc.). There could also be nonpropositional truths (as when I return to my boyhood home and *see* its façade just as I had remembered it to be). Even when the sense that is fulfilled is propositional, strictly speaking, it is not the proposition itself that is true, but the conscious recognition of its being-true.

29. Perceptual presentations differ from other fulfilling acts in their uncertainty. Although the perceived thing is presented as actually being there as perceived, that presentation is not given with the sense of indubitably being so. Indeed, there are differences in degree of evident "certainty" even among perceptual presentations, as one has a more or less optimal distance, illumination, orientation, or context for what is seen.

tendings would be phenomenologically illegitimate, because it could not be made evident in intuition).[30]

So perhaps we should assume that the categorial sense of fulfillment (i.e., being) is always essentially fulfilled: it can only be intended in its presence.[31] That may not be a problem when what is fulfilled is the being of ideal objects (numbers, conscious acts, the senses of such acts, propositions, and so forth) because the being of these kinds of objects might be compatible with their direct presence to an immaterial, ahistorical, nonparticular transcendental consciousness.[32] Husserl took them to be "ideal" in the sense of having no spatiotemporal instances; each evident intending of the number 2 is directed toward one and the same identical number. Matters are altogether different for perception. On any plausible account, perception gives its objects as materially present, existing in space and time. Moreover, their presence is typically indicated (perhaps unthematically) as given through bodily interaction with them. So the material existence of the perceived object and the perceiving body belongs to the sense of what is perceived. If perception is a fulfilling act, and the fulfillment of its correlated categorial intuition of fulfillment is essential to its being a fulfilling act, then the material existence of perceived objects and perceiving bodies is an essential component of any perceptual presentation whatsoever.[33]

30. Heidegger took pains to point out that it was not the regress itself that undercut the conception of the sense of what is perceived as a possibly empty representation of the object, but rather that such a conception is belied by the perceptual phenomenon itself, and hence is a prejudicial philosophical imposition: "It is not because we fall into an infinite regress, and so explain nothing, that the infrastructure of the consciousness of a picture for the apprehension of an object is to be rejected. . . . It is rather because this is already contrary to every phenomenological finding" ([1979, 56] 1985, 43).

31. Note that, on this hypothesis, one could still intend 'being' emptily in judgments, musings, memories, nonpropositional significations (for example, I could indicate being by intentionally writing the sequence of marks b-e-i-n-g for that purpose), and so forth.

32. What cannot be made evident, however, is the presence of such ideal objects to anything actually existing, through which they would be intelligibly manifest within the spatiotemporal world as what they are (namely, as ideal objects within the atemporal, immaterial space of "pure logic"). Friedman (1996) trenchantly shows how this problem bedeviled virtually all of the participants in what he calls the pure logic tradition: the Marburg and Southwest neo-Kantians (including Cassirer among the former), Husserl, Schlick, and Carnap. Discussions of how Schlick and Carnap failed to resolve the problem, couched in terms of how to distinguish logical from empirical truths, show up more extensively in Friedman 1983, 1987, 1988, 1991, 1992.

33. Perception is of course fallible. But in the case of mistake or illusion, the categorial intuition of the material being of the perceived is not fundamentally affected. The perceived object is discovered to be other than it seemed, but not as not being at all, or as being immaterial.

And that means that the phenomenological reduction, which purported to suspend any concern with actual spatiotemporal existence (or any other mode of "actual" existence or nonexistence) in order to reveal the intending and intended sense of perception, instead abolished its perceptual character.

This line of immanent criticism of Husserl's conception of the perceptual noema was clearly articulated almost thirty years ago by Hubert Dreyfus (1972). What Dreyfus could not have known, however, was that the same line of argument had been explicitly developed in Heidegger's then unpublished 1925 Marburg lectures, as a prolegomenon to a recognizable preliminary draft of *Sein und Zeit*. Heidegger's response to the dilemma was to give up any sense of perception as representational, in favor of taking "perception, with its kind of givenness, [as] a superlative case of intentional fulfillment" ([1979, 59] 1985, 44):

> The authentic moment in the perceivedness of the perceived is that *in perception the perceived entity is bodily there*. In addition to this feature, another moment of every concrete perception of a thing in regard to its perceivedness is that the perceived thing is always presumed in its *thing-totality*. . . . When I go into a room and see a cupboard, I do not see the door of the cupboard or a mere surface. Rather the very sense of perception implies that I see the cupboard. ([1979, 57–58] 1985, 43)

Heidegger took up for comparison the perception of a picture to emphasize that the sense of what is perceived is *not* related to the perceived object in anything like the way a picture is related to the object it depicts:

> It goes against all the plain and simple findings about the simple apprehension of an object to interpret them as if I first perceive a picture in my consciousness when I see that house there, as if a picture-thing were first given and thereupon apprehended as picturing that house out there. There would thus be a subjective picture within and that which is pictured outside, transcendent. Nothing of the sort is to be found. Rather, in the simple sense of perception I see the house itself. ([1979, 56] 1985, 42)

Heidegger consequently abandoned the phenomenological reduction, which for Husserl had secured the difference between philosophical reflection and the natural attitude. As the Marburg lectures clearly show, Heidegger simply omitted any mention at all of the phenomenological reduction in his own initial presentation of the task and achievements of phenomenology! He began instead with intentionality, and through a careful consideration of the intentional directedness of perceiving, arrived at the claim that perceiving is directed toward and discloses the "bodily presence" *(Leibhaftigkeit)* of its object, not an intended sense that could conceivably be separated from the object itself.

The published version of *Sein und Zeit* omitted this extended discussion of the intentionality of perceiving, but its conclusion and its order of inference (from the sense of what is perceived to the worldliness of Dasein) still figured quite centrally in the book's argument:

> It requires a very artificial and complicated attitude to "hear" a "pure noise." The fact that we initially hear motorcycles and wagons is, however, the phenomenal proof that Dasein, as being-in-the-world, always already maintains itself *together with* innerworldly things at hand and initially not at all with "sensations" whose chaos would first have to be formed to provide the springboard from which the subject jumps off finally to land in a "world." ([1953, 164] 1996, 153)[34]

Moreover, as is well known, everyday perception for Heidegger was not even a thematic appearance of an object to a subject, but a concernful attending to and noticing what is significant within the practical nexus of one's surroundings—not a disengaged, immanent *Sicht,* but an actively engaged, worldly *Umsicht:*

> The less we just stare at the thing called hammer, the more actively we use it, the more original our relation to it becomes and the more undisguisedly it is encountered as what it is, as a useful thing. . . . No matter how keenly we just *look* at the "outward appearance" of things constituted in one way or another, we cannot discover handiness. ([1953, 69] 1996, 65)

In contrast to Husserl's (and Carnap's) attempts to express and account for the objectivity of the world in terms of its formal structure, Heidegger insisted this interrelated nexus of practical significance could not be

34. Better translation would express the point more strikingly. Dasein is not merely "together with" (Stambaugh) or "alongside" (Macquarrie and Robinson) equipment in the world. The German *bei* being translated here, like the French *chez,* suggests a kind of "being at home among" the interrelated paraphernalia of everyday life (their "home," not Dasein's!).

adequately understood as a formal structure abstractable from its actual material interaction:

> The referential context that constitutes worldliness as significance can be formally understood in the sense of a system of relations. But we must realize that such formalizations level down the phenomena to the extent that the true phenomenal content gets lost. . . . These "relations" and "relata" of the in-order-to, for-the-sake-of, the with-what of relevance resist any kind of mathematical functionalization in accordance with their phenomenal content. Nor are they something thought, something first posited in "thinking," but rather relations in which heedful circumspection as such already dwells. ([1953, 88] 1996, 82)[35]

I have been arguing that Heidegger's criticism of Husserl on the sense of what is perceived, and his insistence upon the worldly entanglement of perceiving, played a fundamental role in the conceptualization of *Sein und Zeit*. In the Marburg lectures, it was only after his discussion of the sense of the perceived and the categorial intuition of being that Heidegger went on to raise as a fundamental problem "the being of the intentional" and insisted that the locus of intentional directedness was not an immanent consciousness, but an active, involved worldly "person."[36] My argument does not require a detailed exposition of Heidegger's conception of Dasein, as the way of being that we ourselves are. For our purposes, organized by the juxtaposition of Heidegger and Neurath, only a few of its prominent characteristics matter.

First, as I have already shown, Dasein was not understood in terms of the immanence or immediate incorrigibility of consciousness, or a difference in kind between my self-awareness and my awareness of others:

> Others are not encountered by grasping and previously discriminating one's own subject, initially objectively present, from other subjects also

35. A more extensive discussion of this passage would be necessary to show how Heidegger's later philosophical development moved him closer in the direction of philosophical naturalism. Heidegger's continuing attempts to articulate just what is crucially omitted or covered over by such a structural formalization of the world were at the center of his extended engagement with naturalism. Such later work as "Der Ursprung des Kunstwerkes" and *Die Frage nach dem Ding* are concerned in part to show how and why it is that the world "resists any kind of mathematical functionalization." These works further developed Heidegger's distinctive interpretation of what Neurath characterized as physical.

36. In both his exposition of phenomenology in the Marburg lectures, and in *Sein und Zeit* itself, Heidegger appealed to Dilthey's conception of human beings as persons, as preparatory to his own conception of Dasein.

> present. They are not encountered by first looking at oneself and then ascertaining the opposite pole of a distinction. They are encountered from the *world* in which Dasein, heedful and circumspect, essentially dwells. . . . This nearest and elemental way of Dasein of being encountered in the world goes so far that even one's *own* Dasein *initially* becomes "discoverable" by *looking away* from its "experiences" and the "center of its actions" or by not yet "seeing" them all. Dasein initially finds "itself" in *what* it does, needs, expects, has charge of, in the things at hand which it initially *takes care of* in the surrounding world. ([1953, 119] 1996, 112)

Second, although Heidegger still insisted upon an ontological difference between Dasein's way of being *(Sein)* and that of equipment or objects, there were no sharp bounds to Dasein as an entity *(Seiend)*. Heidegger insisted that Dasein "is" where its concerns are located, out among significantly interrelated equipment and its uses. Nor was any particular case of Dasein sharply individuated from others, at least in its everyday understanding. Heidegger insisted that "*initially,* 'I' 'am' not in the sense of my own self, but I am the others in the mode of the anyone" ([1953, 129] 1996, 121).

Third, the lack of sharply bounded identity to "cases" of Dasein was also characteristic of Dasein's surroundings as the locus of its everyday, practical activity. Heidegger made this point especially clearly about the spatiotemporality of everyday Dasein:

> In the calculative sense, these estimations may be imprecise and variable, but they have their *own* thoroughly intelligible *definiteness* in the everydayness of Dasein. We say to go over there is a good walk, a stone's throw, as long as it takes to smoke a pipe. These measures express the fact that they not only do not intend to "measure," but that the estimated remoteness belongs to a being which one approaches in a circumspect, heedful way. But even when we use more exact measures and say "it takes half an hour to get to the house," this measure must be understood as an estimation. "Half an hour" is not thirty minutes, but a duration which does not have any "length" in the sense of a quantitative stretch. This duration is always interpreted in terms of familiar, everyday "activities." ([1953, 105–6] 1996, 98)

This imprecision was neither an eliminable imperfection in us nor a characteristic of our understanding of the world in contrast to the world itself. Heidegger insisted:

> When there is a prior orientation toward "nature" and the "objectively" measured distances of things, one is inclined to consider such interpretations and estimates of remoteness "subjective." However, that is a "subjectivity" which perhaps discovers what is most real about the "reality" of the world, which has nothing to do with "subjective" arbitrariness and the subjectivistic "conceptions" of beings which are in themselves otherwise. ([1953, 106] 1996, 99)

The world we inhabit is shot through with vagueness and imprecision that cannot be supplanted by more definite measurements, either in practice or as a postulated reality to which we can only approximate.

The next point of Heidegger's critique of Husserl to consider concerns the role of language within phenomenology. Husserl had sharply distinguished the ways in which linguistic or other materially instantiated signs could "indicate" some object or state of affairs from the ways in which consciousness could "expressively" articulate meanings for itself. Moreover, Husserl insisted upon the philosophical priority of expression. Mere indications can only mean something for a consciousness that could grasp their indicative sense expressively, in an evident, intuitive presence before itself.

> The concrete phenomenon of the sense-informed expression breaks up, on the one hand, into the *physical phenomenon* forming the physical side of the expression, and, on the other hand, into the *acts* which give it *meaning* and possibly also *intuitive fulness,* in virtue of which its relation to an expressed object is constituted. In virtue of such acts, the expression is more than a merely sounded word. It *means* something, and in so far as it means something, it relates to what is objective. (Husserl 1970b, 1:280)

On Husserl's account, language is only derivatively intentional, dependent for its meaningfulness upon the immanently expressive activity of consciousness.

The role ascribed to language in *Sein und Zeit* is contested; this is not the place to untangle and assess the competing interpretations.[37] However, two points are important here. First, Heidegger reversed the priority Hus-

37. For a clear overview of the issues, albeit one that imports some of Brandom's own semantic inferentialism into his reading of Heidegger, see Brandom 1999. For a more extensive, well-informed discussion of Heidegger on language, see Boedeker 1998. Lafont (2000) provocatively situates Heidegger's account of language primarily against the background of the German romantic tradition of Hamann and Humboldt.

serl assigned to expression over indication. The more basic intentional phenomenon for Heidegger was the indicative relation among entities in the world. For Heidegger, this indicative relatedness is not primarily linguistic, but practical and normative. The way in which hammers "refer" to nails, boards, the practice of carpentry, and the broader range of purposes for which carpentry makes sense is the genus incorporating the species in which signs indicate something signified. Indeed, signs, including linguistic signs, are themselves encountered as "equipment for indicating." In the end, what they indicate is not fundamentally different from the referential whole toward which hammers or other equipment are directed:

> Signs always indicate primarily 'wherein' we live, what our heedfulness is concerned with, what the relevance is. . . . *Signs are something ontically at hand which as this definite useful thing functions at the same time as something which indicates the ontological structure of handiness, referential totality, and worldliness.* ([1953, 80, 82] 1996, 74, 77)

The second crucial point is that indications are grasped not expressively in an intuitive presentation to consciousness, but interpretively, by taking them up and using them appropriately in context. Whether what is to be interpreted is a hammer or a word, one interprets it by using it in a way that accords with and further articulates its practical involvement within the world. In the special case of assertions, their practical involvements extend to those things that we can assertively "point out" in our immediate surroundings, other assertions to which they are inferentially connected (along with what these other assertions disclose and discover), and a communicative directedness toward other speakers and listeners. What sustains the possibility of such practical orientations is not the formal structure of a language, but its actual discursive use by speakers and listeners who are already involved with one another in a significantly articulated world. It was not language *(Sprache)* but discourse or articulation *(Rede)* and its constant falling into idle talk *(Gerede)* that Heidegger took to be an essential structure of Dasein. Even the ideal outcome of such interpretive disclosures could never be complete clarity or adequate representation of a whole, but only a better focused orientation toward further interpretation.

Despite this rejection of the very ideal of a systematic clarification of complete and completely clear theoretical representation, the wholeness of Dasein and the interpretive unification of its self-understanding were crucial concerns for Heidegger. Dasein in its everyday way of being is absorbed in and dispersed among its various dealings with things. It has also abdicated responsibility for what it says and does, not merely doing what "one" does, but also deferring to this undifferentiated sense of what its activities are ultimately "for the sake of." Yet this everyday dispersal among the many things that "have to be done" conceals its own underlying possible unity.[38] What unifies Dasein, according to early Heidegger, is the practical grasp of its own temporal finitude, and the responsibility it can thereby take over for who and what it makes of itself. This unification is always from within. We cannot stand back and survey our lives as a sequence from birth to death and plan accordingly; we can only act at any given time in anticipation of the indefinite but always impending possibility of death. Such "authentic" ways of comportment resolutely take up some of the possibilities available within the surrounding world so as to take responsibility for the finite whole of one's life: one takes a stand upon who one has been by taking up available possibilities, for the sake of their becoming one's own.

Neurath's Physicalism

The comparison of Carnap and Husserl required a shift of reference frame. Carnap talked about unified science rather than Husserlian transcendental consciousness, and he analyzed the logical structure of scientific language rather than the essential structure of conscious acts and their corresponding modes of presentation. Protocol sentences rather than perceptual intuitions were then the form through which claims about the world became manifest as objective knowledge. One must see through these manifest differences between Carnap and Husserl in order to grasp that each accounted for the meaning and justification of scientific claims by situating them within an atemporal, immaterial logical structure. The empirical basis of science was itself part of that structure, and its role in verification or fulfillment was itself structural. For our purposes, it matters less that for Carnap that role was expressed by structural or syntactic relations within the language of unified science, whereas Husserl saw it in terms of essential structures of the transcendental consciousness of an eidos Ego.

Understanding the parallels between Neurath and Heidegger involves

38. Heidegger famously argued that this concealment was itself an essential dimension of Dasein. Dasein fearfully flees into its absorption in everyday activities for the sake of social conformity in order to cover up its complete and ungrounded responsibility for who and what it becomes.

getting past another hurdle. Heidegger saw that his criticism of Husserl was fundamental, and he introduced an unfamiliar terminology and a novel methodological conception to express their differences. Dasein's being-in-the-world replaced consciousness, circumspective interpretation replaced perception, and the "idealism" invoked by the phenomenological reduction was supplanted by fundamental ontology. Neurath, by contrast, was concerned to minimize his differences from Carnap, out of solidarity in their common commitment to science and the *neue Sachlichkeit,* a philosophical commitment to collaborative work, and to some extent, a genuine misunderstanding of their differences. Neurath and Carnap both spoke of unified science, physicalist language, protocol sentences, and logical syntax, but in Neurath's work these concepts were significantly transformed. Neurath did not seek a rational reconstruction of scientific knowledge claims to secure their objectivity, either through their formal structure or their empirical foundations. Instead, he was committed to a program of scientific, political, and philosophical reform, from the midst of historically and socially situated activities. Language for Neurath was not a formal system, but a part of the social, material world. Logical analysis was intended to clarify and reform the actual use of language, not to serve philosophical ends. Establishing connections across the very different language and orientation of Heidegger and Neurath thus requires a prior grasp of underlying divergences masked by the apparently similar language and philosophical approaches of Neurath and Carnap.

Neurath's differences from Carnap were articulated through an ongoing debate over the nature and role of protocol sentences. Carnap had originally introduced the phrase to characterize the empirical basis of science:

> Verification is based upon 'protocol statements,' . . . statements belonging to the basic protocol or direct record of a scientist's (say a physicist's or a psychologist's) experience. Implied in this notion is a simplification of actual scientific procedure as if all experiences, perceptions, and feelings, thoughts, etc., in everyday life as well as in the laboratory, were first recorded in writing as 'protocol' to provide the raw material for a subsequent organization. (Carnap 1934, 42–43)

Moreover, in this original formulation, Carnap restricted its application so as "to exclude all statements obtained indirectly by induction or otherwise and [to] postulate therefore a sharp (theoretical) distinction between the raw material of scientific investigation and its organization" (1934,

43). Carnap then identified "the language to which such statements belong [as] the *protocol-language* (sometimes also termed 'language of direct experience' or 'phenomenal language' [although] the neutral term 'primary language' is less objectionable)," even while acknowledging that "it is not [now] possible to characterize this language with greater precision, i.e., to specify its vocabulary, syntactical forms and rules" (1934, 44–45). Carnap was initially agnostic as to whether protocol sentences, the "raw material of science," should be couched in terms of isolated sensations, sensory gestalts, or material objects. When he eventually settled upon the priority of a physicalistic protocol language, prompted in part by Neurath, he did so only because a physicalist language offered some pragmatic advantages over the useful alternative choice of a phenomenalistic language.[39] The phenomenalist language nevertheless still played an indispensable epistemological role in making available the evidential base of science:

> A person S *tests* (verifies) a system-sentence [in science] by deducing from it sentences of his own protocol language, and comparing these sentences with those of his actual protocol. The possibility of such a deduction of protocol sentences constitutes the *content* of the sentence. (Carnap 1959b, 166)

Articulating such empirical content in "one's own" protocol language provided a reconstruction of the system of science that displayed each person's access to its meaning and justification.

Carnap thus had clearly interpreted Neurath's defense of a physicalist language for unified science as opting for one among many possible linguistic forms for the protocol language. On Carnap's interpretation of Neurath, taking "protocol statements [to] have approximately the same kind of form as 'A red cube is on the table' " expressed an epistemological commitment. As most commonly expressed in the material mode, the claim

39. In his immediate response to Neurath's paper on protocol sentences, Carnap (1987) formulated this contrast as between situating the protocol sentences outside the system language of science and situating them within that language. By then, he had concluded (for reasons discussed below) that science must actually *use* physicalist language to be intersubjectively communicable and objectively valid, but it can be reconstructed in a phenomenalist idiom for purposes of understanding how this intersubjective system becomes subjectively manifest. Hence, the physicalist protocol language is part of the system language, while a phenomenalist one is external to this publicly accessible system. Carnap's conception of the universality of the physicalist language hinged upon the availability of a physicalist translation for every expression in a phenomenalist language (but not vice versa).

was that "material things are elements of the given: a three dimensional body is perceived as such directly and not as a series of successive two-dimensional projections" (Carnap 1934, 47). That interpretation was seriously mistaken, however.[40] Neurath was not merely criticizing the specific form taken by a protocol language for articulating the empirical basis of science. His target was the very conception of a protocol language as articulating an internally coherent sphere of subjective appearances, and thus also the epistemological role that Carnap still attributed to such a (linguistic) sphere of appearances. Neurath's proposed reformulation of protocol sentences was thus a radical and thoroughgoing reformulation of the project of the Vienna Circle. Here we can already begin to see the outlines of my comparison to Heidegger. Heidegger had similarly argued that "the perceived as such" was a concrete material object, not a separable perceptual sense.[41] Yet Heidegger's discussion of the perceived as such was not simply an alternative account of the perceptual noema, but a thoroughgoing reformulation of phenomenology and its turn to immanent presentations to consciousness.

Neurath rejected both the intelligibility and the utility of a phenomenalist protocol language. A phenomenalist language was supposed to display the epistemic accessibility of the intersubjective system language of science. Yet Neurath argued that the supposedly intrasubjective domain of phenomenal experience was itself beset by the same problem of accessibility:

> If someone makes predictions and wants to check them himself, he must count on changes in the system of his senses, he must use clocks and rulers; in short, the man who supposedly is in isolation already makes use of the intersensual and "intersubjective" language. The forecaster

40. The parallel mistake would be to interpret Heidegger as taking the perceived as such to be the *sense* of the concrete object as a totality. Heidegger rejected the entire Husserlian domain of ideal senses, and Neurath likewise rejected Carnap's conception of a protocol language. A comparable mistake would be to construe Quinean radical translation as a mode of epistemic access to "ideas" expressed by utterances. In all three cases, the notions of formally structured languages, senses, or ideas was the target of the criticism.

41. Husserl took the perceptual noema to be an articulable meaning, whereas many of the Vienna Circle were inclined toward a phenomenalism in which the "given" was a complex of sensations. Yet both were understood to be incorrigibly presented in experience. Moreover, Carnap noted (1934, 47–48) that even the phenomenalists in the Vienna Circle had increasingly turned against an atomistic conception of sensation in favor of a more complex conception influenced by gestalt psychology; at roughly the same time, Aron Gurwitsch was developing his influential interpretation of Husserl's perceptual noema as a perceptual gestalt (Gurwitsch 1966). Dreyfus 1972 offers an illuminating discussion of Gurwitsch's interpretation.

> of today and the controller of today are, so to speak, two persons. (Neurath 1983, 55)

We do not have direct access to our own subjective experiences (or their expression in a phenomenalistic protocol language), at least not if there is to be any coherence and intelligibility to what is thereby encountered. Public, material engagement with shared surroundings, through an intersubjective, physicalist language that enables oneself or others to correct one's claims, is indispensable—even for making the most minimal sense of what is presented in the most basic "protocols." Instead of needing a phenomenalist language to reconstruct subjective access to science, we need scientific language and practice to make intelligible the content of experience: "If Robinson [Crusoe] wants to make use of a language at all, he must make use of the '*inter-subjective*' language" (Neurath 1983, 96). Neurath thus rejected both the reasons for appealing to a phenomenalistic protocol language and the possibility that such an appeal could succeed.

Neurath also rejected claims for the incorrigibility and epistemic priority of protocol sentences, however they are expressed. Carnap's own commitment to the incorrigibility of protocol sentences was always somewhat ambivalent. Contrary to many subsequent portrayals, Carnap had always sought the objective validity of science in its formal structure rather than in its empirical foundation. Moreover, during the course of the protocol sentence debate in the early 1930s, Carnap's commitment to incorrigibility gradually wavered.[42] Yet Carnap retained a distinctive epistemological role for sentences in the protocol language. Neurath rejected any such role for protocol sentences:

> *Each law and each physicalist statement of unified science or of one of its factual sciences is subject to [replacement by other statements]. The same is true for each protocol statement.* In unified science we try to create a *consistent system* of protocol statements and non-protocol statements (including laws). . . . The fate of being discarded [to protect the consistency of the system] may befall even a protocol statement. (1983, 94–95)

Moreover, he continued, no privilege attached to one's own experience over that of another: "That a man clings more obstinately to his own pro-

42. Uebel (1992c, esp. 112–16) gives a clear and convincing account of this gradual shift in Carnap's views.

tocol statements than to those of another, in general, is a historical fact—without any fundamental significance for our discussion" (1983, 97). What matters to science is the ongoing, corrigible construction of an intersubjectively consistent system of statements tested by making and checking predictions.

So far, we have seen Neurath reject the autonomous intelligibility of a private, experiential protocol language as the epistemological basis upon which scientific knowledge was to be reconstructed. The meaning and objectivity of scientific knowledge were secured by public, intersubjective correction rather than private, autopsychological reconstruction. One might say that Heidegger had similarly replaced the immanent phenomenological constitution of meaning with Dasein's ongoing concern to secure conformity to public norms of comportment. In both cases, moreover, self-correction and correction by others are not significantly differentiated. Yet my comparison of Neurath and Heidegger does not rest upon such easy parallels (which could also be made to later Wittgenstein, Quine, or Davidson, among others who accounted for meaning in terms of public corrigibility rather than private transparency). It is more interesting to see how their arguments developed further along parallel lines. Heidegger's shift from perception as intuitive fulfillment to the public normativity of practical, "circumspective" comportment led him to reconceive "the being of the intentional." Husserl's conception of consciousness as the locus of manifestation was replaced by the worldliness of Dasein's disclosedness. Neurath's rejection of a phenomenalist protocol language likewise led him to challenge Carnap's implicit understanding of what Heidegger might have called "the being of language."

When Carnap proposed that physicalistic language was the universal language of science, he took himself to be in basic agreement with Neurath's claims that "the unified language of unified science is the language of physics" or that "unified science contains only physicalist formulations" (1983, 62, 54). Their formulations were only superficially similar, however, because they understood language quite differently. For Carnap, a language was a formal structure defined by a vocabulary and a syntax. Actual utterances were at most instances of its structure, and often accorded with it only approximately. When Carnap spoke of a phenomenalistic protocol language outside of the system language of science, or a physicalistic protocol language specified as internal to that system language, all three "languages" were formal reconstructions. When Carnap said that the physicalistic language was universal, he meant that all meaningful expressions in other languages could be systematically translated into physicalistic language. When he identified a distinct protocol language, he meant only that one could in principle construct a closed vocabulary and well-defined syntax in which protocol sentences could be expressed without using additional vocabulary whose application could only be introduced inferentially.

In contrast, Neurath conceived of "a language" as something more akin to a natural language, a pattern of actual word and sentence use by some group of speakers who thereby successfully communicated: "Protocol statements could [never] be anything else but real events. . . . The terms 'statement,' 'language,' etc., must be defined historio-sociologically" (1983, 108). When Neurath spoke of physicalist language, he thus was not referring to a formal reconstruction, but to ordinary discursive practice: "The view advocated here starts from a given state of everyday language, which in the beginning is essentially physicalist" (66). The language of physics was not a distinct formal structure into which other formally specified languages might be translated, but the shared communicative field in which predictions are formulated to guide action and coordinate with others. Thus, Neurath objected to Carnap's conception of a primitive protocol language not merely because its content, subjective appearances to a single knower, cannot cohere as an autonomously intelligible domain, but because he objected to this conception of a language as a formalized reconstruction:

> The fiction of an ideal language composed of neat atomic statements is as metaphysical as the fiction of Laplace's spirit. Scientific language, with its ever growing equipment of systematic symbol formations, can by no means be regarded as an approximation to such an ideal language. (91)

Neurath's conception of "the language of physics as the universal language of unified science" did not concern the choice of the proper formal structure for rational reconstruction, but the recognition of the actual discursive context in which scientific work is done. When Neurath then went on to talk about developing a "physicalist language of advanced science that we can design to be free of metaphysics from the very start" (91), this was intended not as a formal reconstruction for philosophical purposes of conceptual clarification, but as a concrete proposal for the improvement and clarification of actual speech, for example, forbidding the use of certain words whose metaphysical misuse was strongly entrenched.

Neurath clearly reinterpreted Carnap's whole program for articulating the logical syntax of scientific language as aimed toward such linguistic reform rather than toward formalist reconstruction. He emphatically rejected any understanding of conceptual clarification as an enterprise distinct from the ongoing work of unified science. Instead, it was simply part of the work of doing science effectively in the one and only language we have available for that purpose:

> The Vienna Circle is making particularly vigorous efforts to give unified science a solid framework through 'syntax,' along the lines of the logicians. . . . The defects of syntax reveal the standard of scientific study. An unblemished syntax is the foundation of an unblemished unified science. Language is essential for science; within language all transformations of science take place. (1983, 54)

Carnap would not have said in this way that scientific work itself actually takes place within a protocol language, or even within the "system-language of unified science"; both languages were conceived as formal reconstructions of the much messier flow of actual scientific talk. The syntactical rules that identify such formal systems could be instantiated by discrete physical marks or utterances, but the language itself was an abstract formalism.

Neurath understood language to be itself part of what physicalist unified science describes:

> Scientific language itself is a physical formation whose structure, as physical arrangement (ornament) can be discussed by means of the very same language without contradiction. . . . The study of language can perfectly well be combined with the study of physical processes; for one always stays in the same field. (1983, 53)

That is why Neurath always insisted that "it is impossible to separate the 'clarification of concepts' from the 'pursuit of science' to which it belongs" (59). Conceptual analysis was not a formal or metascientific reconstruction, but a change in which statements are made and accepted. Like any other development within the sciences, clarification effected an alteration in "the mass of [scientific] statements, by comparing new statements with statements taken over from the past, thus creating a consistent system of unified science"; no other standpoint was possible, he went on to say, because "as makers of statements, we cannot, so to speak, take up a position outside the making of statements and then be prosecutor, defendant and judge at the same time" (61).

This understanding of statements as physical events accomplished by physical beings was carried over into Neurath's own proposal for the proper form of protocol statements. Statements could not be understood as internal to a private domain of experience, but only as publicly accessible events in the world, located in relation to a publicly recognizable speaker.[43] That is why Neurath insisted:

> It is *essential* for a complete protocol statement that the name of a person occur in it. 'Now joy' or 'now red circle' or 'a red cube is lying on the table' are not complete protocol statements.[44] Even as expressions within the innermost brackets they are not acceptable, [but must] feature a personal name and a term that belongs to the sphere of perception terms. (1983, 94)[45]

Here we find the fundamental agreement between Heidegger and Neurath on what the former called "the being of the intentional": perceptual appearances and other intentional contents accrue to a concrete, physically and socially situated entity, not to an immanent field of experience. Moreover, just as Heidegger argued that those entities whose way of being is Dasein could not be sharply differentiated from their surrounding environment, so Neurath insisted that this indispensable reference to actual speakers and perceivers introduced an ineliminable imprecision into unified science:

> The statement, 'Otto observes an angry man' is less precise than the statement, 'Otto observes a thermometer registering 24 degrees' inasmuch as "angry man" must be less precisely defined than a 'thermometer

43. Obviously, it is possible to entertain a statement in thought without uttering it publicly, but Neurath also thought that such "speech-thinking" was itself physical, and more important, that it must be made in an intersubjectively physicalist language in order to be accessible to the speaker herself over time, even just the time it takes to formulate the sentence. That latter point was implicitly made by the clock-referenced shifts in time-coordinates in his example of a complete protocol sentence: "Otto's protocol at 3:17 o'clock: [Otto's speech-thinking at 3:16 o'clock was: (at 3:15 o'clock there was a table in the room perceived by Otto)]" (Neurath 1983, 93).

44. Neurath refers here to Carnap 1934, 46, 47.

45. To say that these examples cannot go in the innermost brackets is to say that they cannot even be the content of a person's inner thoughts to which their outer speech might refer; the most minimally contentful expression of a perception must include a reference to the perceiver as herself an embodied, worldly, socially recognized person.

> reading of 24 degrees'; but 'Otto' itself is in many respects an imprecise term; the statement 'Otto observes' can be replaced by the statement, 'The man whose carefully taken photo is No. 16 in the file observes'; but the term 'photo No. 16 in the file' has not yet been replaced by a system of mathematical formulas that is unambiguously coordinated to another system of mathematical formulas that takes the place of 'Otto.' (1983, 91)

In Neurath's view, such clarifications could not eliminate the "multitude of imprecise, unanalyzed terms" *(Ballungen)* that belong to ordinary language from the achievement of unified science:

> If we want to embrace the entire unified science of our age, we must combine terms of ordinary and advanced scientific languages, since in practice, the terms of both languages overlap. . . . If imprecision is diminished at one place, it may well re-appear at another place to a stronger degree. (92)

Unified science is not a philosophical reconstruction of an abstract system of symbols outside physical space and historical time, but a concrete project within the world, partaking of its vague boundaries and imprecision.[46]

Neurath's conception of the unity of science provides another point of comparison to Heidegger, perhaps the most surprising of all. Of all the members of the Vienna Circle, Neurath was the most insistent upon identifying their shared standpoint under the banner "unity of science." Yet Neurath's conception of what that meant was at odds with the orientation of others who accepted the slogan. For Carnap in the *Aufbau,* "unified science" denoted the systematic representational schema within which particular sense modalities could be uniquely specified by their formal relational structure. A multitude of logically possible relations were excluded from consideration so that the objectivity of particular subjective appearances could be uniquely secured by their intersubjective formal structure. The significance of "unified science" was that it constituted the sole object domain to which scientific investigation was responsible. All particularity could then be removed from science, to allow each scientific statement to be construed as "nothing but a structure statement" (Carnap 1967, 29).

46. Cartwright's interpretation of Neurath places illuminating emphasis upon the ineliminable role of *Ballungen* (imprecise, "congested" cluster concepts) in Neurath's view of science and language. Her most extended discussion of the importance of *Ballungen* is in Cartwright, Cat, Fleck, and Uebel 1996, part 3.

Neurath rejected the very idea of a unified representational scheme for organizing the entirety of science, and with it Carnap's aspiration to conceive of scientific knowledge shorn of all particularity:

> [The comprehensive scientific conception] recognizes no 'world' as a whole, it does not aim at comprehending a mighty world-picture in its totality, at a world-view. If one speaks of a scientific world-'conception' in contradistinction to a philosophical world-'view,' 'world' is not to indicate a definitive whole, but the daily growing sphere of science. This conception is deduced from individual scientific work, which one wants to incorporate into a unified science. (1983, 32)[47]

"Unified science" for Neurath thus meant the practical task of bringing together the methods and results of various scientific disciplines in order to formulate better predictions to guide action in concrete situations. The limited domains, idealizing assumptions, and specialized vocabulary that enabled a scientific discipline to develop a more extensive and refined understanding and formulate specific, testable laws could be barriers to understanding the particular situations that matter to us. Such theoretical orientation was indispensable to scientific understanding, but it then needed to be surpassed by a unified science that would bring its results together with the predictive capacities of other sciences. Thus, Neurath insisted:

> In order to formulate the individual prediction: "This forest fire will soon be extinguished," we combine biological statements (concerning trees, etc.), chemical statements (concerning fire, etc.), sociological statements (concerning fire service, etc.) and statements of other disciplines. The situation is different for theories, they can be restricted to definite terms. On the other hand they cannot be controlled in isolation, but only in connection with the concrete predictions that we have just characterized. (1983, 132–33)

Neurath put at the center of his conception of science these concrete unifications of multiple disciplines, not just because they allowed for empirical control, but because they enable people to anticipate and respond appropriately to their surroundings. Here especially, Neurath emphasized that

47. Although my interpretation does not turn on this point, it is instructive to note that, like Heidegger, Neurath takes the "world" to be an open-ended practical field of activity rather than the totality of entities, and that he is also suspicious (for different reasons) of the conception of science as aiming toward a total "world-picture."

science should be grasped as a concrete, material project rather than an idealized formal reconstruction:

> 'Making predictions' is what all of science is about. At the beginning of the process are observation statements, which, of course, contain measurements of time and space, if only in an approximate manner. There are always spatiotemporal formulations behind which we cannot reach at all without saying something meaningless. 'Saying' itself is a spatiotemporal arrangement. With the help of observation statements we formulate laws, [which] are not to be seen as proper statements but as directives for finding predictions of individual courses of events. (53)

Unified science is the ongoing task of combining the efforts of many different scientists into a collective, encyclopedic project. The aim is not systematic purification or methodological grounding, both of which Neurath ridiculed as "pseudorationalist." Scientific work is oriented toward the future, and it always involves a decision to accept or reject particular statements for inclusion within the body of those statements currently accepted and used. Such decisions cannot be grounded methodologically or epistemologically, for any grounds to which one might appeal are simply themselves part of the body of current science about which one must make a decision. There is no further or higher check upon such decisions than the ongoing work of the sciences.

It would be a mistake to look to see whether Heidegger also emphasized the concrete unification of diverse scientific disciplines in their response to particular situations. In fact, Heidegger did at one point in his career emphasize a need to get beyond the centrifugal tendencies of the scientific disciplines, but it is not "unified science" in the narrow sense that provides interesting or substantial connections to Neurath's philosophical orientation.[48] Rather, the interesting point is that, having abandoned views that located the normativity of (scientific) understanding in the formal structure of knowledge, both Heidegger and Neurath turned to the concrete temporality of ongoing scientific practice. Heidegger aimed at "an *existential concept of science* [that] is distinct from the 'logical' concept that understands science with regard to its results. . . . The existential concept understands science as a mode of existence and thus as a mode of being-in-the-world"

48. Most notably in "What Is Metaphysics?" (Heidegger 1998, 82–96) and the notorious Rectoratsrede of 1933, "The Self-Assertion of the German University" (Heidegger 1990).

([1953, 357] 1996, 327). Scientific understanding was both "thrown" into a historically concrete situation and articulated as a practical competence for taking up specific possibilities that that situation presented. The possibilities that the sciences afford are public possibilities, available to "anyone" prepared to take them up; the decision to proceed in a particular way can be taken responsibly or irresponsibly (Heidegger says "resolutely"), but in either case what is required is a "decision," a taking up of one possibility rather than another.[49] That decision is then open to further assessment, but only in light of its outcome in relation to the situation as previously understood.

Although Neurath did not have anything like Heidegger's conceptions of "thrown projection" or the contrast of "authentic resoluteness" to the curiosity, ambiguity, and idle talk of falling into conformity to what one does, he likewise emphasized the decisive character of doing science, and the prospective rather than retrospective character of its assessment:

> 'Induction' that leads to laws is a matter of 'decision,' it cannot be deduced. The attempts to give 'induction' a logical foundation are therefore bound to fail. If a statement is made, it is to be confronted with the totality of existing statements. If it agrees with them, it is joined to them; if it does not agree, it is called 'untrue' and rejected; or the existing complex of statements of science is modified so that the new statement can be incorporated; the latter decision is mostly taken with hesitation. *There can be no other concept of 'truth' for science.* (1983, 53)

Once one recognizes that "statements" are actual events or groups of events rather than propositional contents, one sees here the striking affinity between Heidegger and Neurath. Both account for the correctness or incorrectness of a statement in terms of its actual, occasioned use rather than its place in a formal structure or its relation to "reality." Both locate the truthfulness of its utterance in its contribution to a larger, open-ended pattern of practical comportment. Heidegger's account of truth as disclosedness admittedly applies not just to statements, but to all practical comportments. Neurath's account of how statements are true, however, was no less ambitious in scope. He simply thought that any meaningful comportment had to be analyzed by making explicit what it implicitly stated: "In staying within the closed area of language one can express

49. For a clear and insightful articulation of what an "authentically resolute" engagement in scientific research might look like, see Haugeland 2000.

everything" (1983, 53).[50] He offered no theory of truth, because he thought that extensive talk about truth encouraged metaphysical confusion (135, 161) Truth is not a property of statements, or a relation between statements and things, but instead concerns how statements are used in everyday and scientific practice.[51] Truth should be appropriately applied to the ways statements or comportments, as "physical" or "innerworldly" entities, function together to provide a coherent, concrete setting for ongoing engagement with the world.

Neurath, Heidegger, and the Question of Naturalism

I can now offer a clearer formulation of why I think Neurath and Heidegger matter to subsequent philosophical debates about naturalism. Hei-

50. An important part of the practical program of Neurath's physicalism was a commitment to the reform of scientific and cultural life through clarification. He was a relentless opponent of contemporary conceptions of the human sciences as *Geisteswissenschaften,* in substantial part because he thought such conceptions involved appeal to ineffable meanings that could not be articulated in publicly accessible, debatable speech. His focus upon language expressed a practical commitment to remain within the domain of public, critical discussion, and to reject as meaningless whatever could not be articulated in statements with a clear use in relation to everyday comportment. Thus, in line with my earlier discussion of Neurath's conception of physicalist language, the rejection of metaphysical claims as meaningless had a different sense for Neurath than for others in the Vienna Circle (especially those most heavily influenced by Wittgenstein's *Tractatus*). He offered no a priori conception of what can meaningfully be said, either by its logical form or its relation to some generally specifiable domain of protocol sentences (protocol sentences had no special status for Neurath). The rejection of metaphysics as nonsense was instead a practical commitment to articulate whatever one had to say as part of an exchange of statements that could be made continuous with everyday, discursive/predictive interaction with one's surroundings, that is, as part of a total system of statements that function together in people's coping with the world. Consider this characteristic expression of Neurath's conception of the rejection of metaphysics: "It is also possible to formulate the terms of scientific language with the help of terms of ordinary language. . . . Einstein can somehow be expressed by means of Bantu language, but not Heidegger, unless one first introduces misuses into the Bantu language which have been adapted to those of the German. A physicist should, in principle, be able to satisfy the witty thinker's demand: 'It must be possible to make the main features of any strictly scientific theory plain to a hackney-coach-driver in his own language' " (1983, 92).

51. Neurath's few remarks about truth, and his insistence that statements can only be justified in relation to other statements, have led to his frequent identification as a "coherence theorist" about truth. That seems to me to misread his position in a fundamental way. Neurath saw the accommodation of one claim within "the mass of statements already accepted" as a practical matter requiring decision (whether and how to make adjustments to accommodate the claim) rather than as the discovery of a logical relation of "coherence" that could be a criterion for or an interpretation of a preexisting conception of truth. When he said that "no other concept of 'truth' is possible," I think he is best read as rejecting *any* substantive theory of truth (including coherence theories) rather than opting for one such theory among others.

degger and Neurath each turned toward philosophical naturalism as a response to what I have called the problem of manifest necessity. This problem concerns how an allegedly necessary, immaterial, and atemporal structure of language, consciousness, or representation is actually instantiated in the concrete, historical-material world, in such a way that its supposed necessity is bindingly authoritative. Each articulated a particular version of this problem in the course of criticizing the structural antinaturalisms of Husserl and Carnap. Their lines of argument can nevertheless be generalized. Indeed, an important reason for comparing such apparently divergent philosophers is to see how the more general line of argument emerged in these markedly different contexts. Upon recognizing (and accepting) the more general line, however, we can also ask how far the criticism legitimately extends. In asking that question, we may then discover that Heidegger and Neurath also pose a fundamental challenge to many subsequent philosophical naturalists.

Naturalism today, after all, is not quite the position that Frege or Husserl criticized as radically self-defeating. Three subsequent developments have made the philosophical world safer for thoughtful naturalists. First, naturalists have recognized that a coherent philosophical naturalism needs to account for the normativity of cognition, linguistic meaning, and scientific knowledge in naturalistically respectable terms. Naturalists today thus typically acknowledge as a criterion of philosophical adequacy the commitment that John Haugeland (1998, chap. 7) attributed to the infielders among his "Intentionality All-Stars": to offer a concrete account of how the normativity of intentionally directed thought or action can be accounted for as a "nonaccidental larger pattern" within a "vapidly materialist" nature. Second, they have acknowledged that the natural world is not intelligible without appeal to the natural necessity of causes or laws, and hence that naturalists cannot provide an adequate account of normativity without the use of modal locutions. The development of possible-worlds semantics has provided technical resources that make modality more philosophically respectable; hence, one might say, Frege, Husserl, or Carnap could not then have envisioned the a posteriori necessity now attributed to causal relations or scientific laws (discovered empirically rather than by logical analysis or transcendental clarification), which supposedly makes intelligible a naturalistic understanding of normativity. Third, as any good naturalist would expect, these technical and philosophical developments have been coordinated with philosophical attention to important changes in the sciences. The development of the cognitive sciences and neuroscience, and of neo-Darwinian evolutionary theory, is supposed to provide the material for naturalistic accounts of thought and language that far surpass the crude psychologisms and historicisms against which Frege and Husserl rightly inveighed.

Neurath's and Heidegger's lines of argument, however, strongly suggest that contemporary naturalists have accepted too much from the tradition. The problem with Husserl and Carnap, more recent naturalists might say, was that they sought to understand the normativity of thought and meaning through an a priori necessity discovered by philosophical analyses of language or consciousness. But Neurath's and Heidegger's challenges cut deeper; it was not the apriority of rational reconstruction or phenomenological analysis that mattered for their criticisms,[52] but the alleged necessity of the structures of language or consciousness. The problem of manifest necessity poses a difficulty for any attempt to account for normativity on the basis of necessity, whether a priori or not. To see this, recall how and why the problem of manifest necessity arose.

The initial concern was to understand the normativity of meaning and knowledge. Philosophers sought necessary structures of language or consciousness as grounds for normativity, because the contingencies of actual thinking, speaking, and acting could provide no compelling grounds for criticism of other contingencies. Only by locating our actual, contingent practices in relation to a structure of atemporal, immaterial necessity could we supposedly make sense of their failings. In order to ground the normativity of meaning and justification, however, these structures had to be manifest within the contingent, historical-material world. Actual thinking and speaking beings had to be at least implicitly responsive to the difference between accord and failure to accord with necessity (between being meaningful and meaningless, justified and groundless, etc.), and such responsiveness required the capacity to recognize these differences (again, perhaps

52. Neurath, of course, did object to any attempts at a priori philosophizing. He even anticipated Quine in calling for an interpretive-behavioral conception of logical truth (see Neurath 1983, 92, 104). But what he most objected to was metaphysics, including any substantive notion of necessity, whether logical or physical. Heidegger's Marburg lectures also included a reformulation of the "original sense of the a priori" among the three fundamental issues in phenomenology, alongside others I have discussed above. But a crucial part of his reconception was to shift the meaning of the a priori away from questions about epistemological grounds: "The 'earlier' [i.e., the 'prior' in *a priori*] is not a feature in the ordered sequence of knowing, but . . . a *feature . . . in the ontological structure of being*" ([1979, 102] 1985, 74).

only implicitly). For this purpose, Husserl introduced his conception of "evidence" (expressed in the "principle of all principles" that whatever was "evidently" given to consciousness was indubitable). Carnap proposed a dual solution corresponding to the difference between logical and empirical truths: logical truths were determined procedurally by combinatorial rules of syntax, while empirical truths were grounded in the "certainty" (at least autopsychologically so) of protocol sentences.

The resulting problem of manifest necessity is not the epistemic problem of whether protocol sentences or evident intuitions really are certain, or whether they really could justify more complex constructions. The problem is ontological. It concerns whether the manifestation of necessary structures within the material-historical world is itself a necessary structure or a material-historical contingency. Husserl and Carnap were driven toward the former alternative. Husserlian fulfilling intuitions were themselves ideal, atemporal meanings. Carnapian protocol sentences were to be pure structure statements within a rationally reconstructed protocol language. Only through such ideal structures could fulfillment or empirical grounding actually have any binding normative authority. Heidegger and Neurath saw that this putative normative authority was purchased at the cost of their normative force. Husserlian fulfilling intuitions could only present the unfulfilled sense of fulfillment. Carnapian autopsychological reconstructions of perceptual manifestations dissipated in a solipsism of the moment that could never be accessible to materially situated, temporally extended persons. Moreover, both Heidegger and Neurath saw that their criticisms of these structural accounts of perceptual manifestations then extended more generally to apply to "consciousness" or "language" as the presumed locus of any manifestation. Husserlian transcendental consciousness could have no psychophysical or social reality. Carnapian languages could not actually be spoken or understood. So Heidegger turned to the analysis of Dasein as being-in-the-world, and Neurath to understanding statements in the language of unified science as "real events" that are part of a currently accepted "mass of statements." The locus of the normativity of such concrete, historical-material activity then shifted from the necessity displayed by their atemporal structure to their contingently situated futural temporality. Unified science was conceived as a historically situated task oriented toward making predictions; authentically resolute Dasein was a thrown project of being-toward-death. Normativity was then to be located not in a structure outside the concrete historical-material world, but *in medias res*.

Contemporary naturalists may seem to have avoided these problems by rejecting the epistemological projects embedded in Carnap's and Husserl's idealized reconstructions of knowing. They are the heirs of postempiricist philosophy of science, which similarly challenged the relevance (i.e., the normative force) of logicized theories of confirmation or explanation for scientific understanding. They have consequently opted to locate the sources of normativity in the world (specifically, the natural world) rather than in supposedly a priori necessities of consciousness or language. Epistemology is to be naturalized, such that the account of how the world becomes manifest to beings like us is to be an a posteriori scientific explanation of our cognitive capacities and causal interactions with nature. Yet like Carnap and Husserl, many contemporary naturalists would still account for the normativity of meaning and knowledge through its place in a structure of necessities, now identified with the causal or nomological necessities of nature. These necessities then likewise have to become evident in the concrete, historical-material world. Such manifestations of natural necessity occur at two levels. First, physical events, including all of our actions, exemplify physical possibilities; some intended actions fail because the conditions needed for their success are not possible. Second, and more important, scientific practices articulate and make intelligible the necessary structures of the natural world, and thereby reveal the significance of the successes and failures of the first kind.

We can then see how contemporary naturalists might encounter their own distinctive versions of the problem of manifest necessity. The locus of manifestation is now the public, material, and institutional practice of scientific research, rather than individual conscious experience or the reconstructed language of science. The problem for naturalists in the philosophy of science is then likewise ontological rather than epistemological. The problem has nothing to do with skepticism, relativism, or other possible doubts about the objectivity or reliability of scientific work. Rather, it concerns how to make intelligible the normative force and authority of scientific practices within the world disclosed by the sciences. The issue becomes clearer in asking what we are talking about when we talk about science. Are we talking about actual, concrete historical-material activities, or are we talking about an idealized, rational reconstruction of scientific norms?

Similarly, ambiguities arise in naturalist conceptions of nature. Are we talking about the causal or nomological structure of possible worlds that the actual world exemplifies? Or are we only talking about what actually exists historically and materially?

Contemporary naturalists often avoid confronting such questions through an academic division of labor. Among philosophers of science (including many in the philosophy of biology or cognitive science), "naturalism" has meant that philosophy must be closely engaged with actual scientific practice. Strongly influenced by the demise of post–World War II logical empiricism, whose collective portrait of the norms of scientific inquiry came to seem utterly divorced from what scientists actually say and do, these philosophical naturalists have insisted that philosophical reflections on science, life, or mind take into account the work being done in the relevant scientific fields, or in empirical studies of the history or sociology of science. A different strain of naturalists (principally in philosophy of mind, epistemology, and ethics) have interpreted the naturalist dictum of continuity between philosophy and science to require instead that philosophical explications of intentionality, knowledge, or morality must ultimately refer to nature as disclosed by the sciences. If intentional contentfulness, epistemic authority, or moral norms are to be philosophically respectable, it is argued, they must be explicable in the non-normative (but often modal) terms of natural laws, causal powers, selective mechanisms, or predictive enhancements.

When push comes to shove, virtually everyone acknowledges the need to connect these two poles of philosophical naturalism. After all, the only plausible reason to insist that normativity must be explicable in terms of laws, mechanisms, or powers would be that the sciences (as actually practiced) tell us that that is what the world (or its intelligibility) consists in. And the obvious reason to take philosophical inquiry to be continuous with what scientists do is because the sciences (understood normatively) are the best way we have to figure out what there is or can be in the world, or how we ought to understand it. Yet we do not have, even in outline, a plausibly developed account of how the normative accountability of scientific practices might be intelligible as part of the world that those practices disclose. Alexander Rosenberg's insightful "field guide" (1996) to the current state of the art, for example, offers compelling reasons to believe that none of the currently favored Darwinian approaches to a naturalized philosophy of science succeed in making the normativity of scientific practices naturalistically explicable, even as credible promissory notes.[53]

The demand for such a detailed "naturalization" of the scientific disclosure of the world might, of course, be premature. Naturalists especially might expect that better science will have to come before better philosophy. Perhaps the functional evolution and development of scientific practices is far too intricate and historically embedded to be yet accessible to currently articulated explanatory models in evolutionary biology or cognitive science. Yet I think a different moral should be drawn once we recognize the possibility of extending Neurath's and Heidegger's criticisms of Carnap and Husserl. The difficulty confronting contemporary naturalists is not merely a need for patience until scientific progress catches up to our philosophical criteria of understanding. The difficulty, I suspect, instead lies in the very idea of making sense of normativity in terms of necessary structures; natural necessity and causal-functional roles in the end fare no better than the logical or transcendental necessities sought by Carnap or Husserl. We cannot make internally coherent sense of the manifestation of the causal or nomological structure of the world through the concrete, historical-material practice of scientific research, any more than they could make sense of the concrete, historical-material manifestation of the logical structure of knowledge or the transcendental structure of consciousness.

Neurath and Heidegger each offer a promising, and more radically naturalistic, response to the problem of manifest necessity. Each abandoned the effort to discover a supposedly necessary structure governing the normativity of our scientific and practical engagement with the world. They instead sought to understand how one's practical situatedness in the world, in all its contingency and concreteness, could nevertheless make intelligible a practical normative accountability to something at stake in that situation. What was "at stake" in Dasein's existence or the task of unifying science was (and is) not something already fully determinate that only needs to be

53. Rosenberg unfortunately still construes the problem epistemologically as the need to defend the naturalist's reliance upon science (as "objective") against the doubts of skeptics and relativists. His arguments could nevertheless be readily adapted to apply to the related ontological issue. Those of us who have other reasons to reject skepticism or relativism as serious concerns nevertheless need to make intelligible to ourselves *how* the normative accountability of scientific practices fits into our broadly scientific understanding of the world. Rosenberg's arguments show clearly why the currently available Darwinian programs fail to do so, in ways that are not readily remediable.

discovered. The articulation of what we are accountable to is inseparable from the practical process of holding ourselves accountable to it. The normative authority of what is at stake in our practices is not readily separable from how it sustains its normative force, that is, how it actually makes binding claims. The normativity of scientific and everyday practice thus comes from being in the open-ended contingencies of a historical-material situation rather than in a relation to something "outside" or "beyond" it.

My aim in the remainder of this book is to articulate further such a "naturalistic" understanding of the normativity of meaning and understanding. Normative accountability, I shall argue, is constituted practically, through belonging to an already meaningfully configured situation with something at stake. But what one's situation is, how one "belongs" to it, and what is thereby at stake, only become fully determinate through actual responsiveness to that situation, and its "outcome" (including its possible failure). The reader who has followed me this far should not be surprised at how I shall proceed. My discussion of these issues itself emerges from a particular philosophical, scientific, and political situation, which I have already begun to articulate through my discussion of Heidegger and Neurath. Understanding what my response to that situation is, including what I mean by calling that response "naturalistic," is not readily separable from developing a more complete understanding of the situation. In particular, I need to develop a clearer account of how nature, science, and naturalism have been conceived in philosophy, and what has been understood to be at issue in discussions of naturalism. The success (or failure) of my response, in turn, involves enabling others (you, for example) to recognize this situation as our common situation, and the response as indeed appropriate to "where we're at." Historical interpretation, constructive philosophical work, and programmatic suggestions for where we should go from here are therefore going to be closely intertwined.

TWO

The Dualism of Nature and Normativity

HEIDEGGER'S AND NEURATH'S constructive responses to the problems they identified in phenomenology and logical analysis have been mostly ignored in subsequent discussions of philosophical naturalism. Brief consideration of the alternative lessons that have been drawn from the failures of transcendental phenomenology and the logicist programs of Frege, Russell, early Wittgenstein, or Carnap should help explain why this has been so. To understand these alternative responses, it is helpful to recognize that naturalism can be challenged at two different levels. The two levels are perhaps most clear in Husserl, who explicitly rejected naturalism in both respects. First, and fundamentally, Husserl argued that philosophical claims could not be grounded upon the empirical facts of nature or history, because the sense and the validity of such facts then became unintelligible. He instead grounded meaning and justification in essential structures of transcendental consciousness, which he conceived as altogether separate from the natural world. Second, Husserl argued that within the world made accessible through conscious experience, naturalism collapses a crucial distinction. Nature and the "lifeworld" (the meaningful realm of human psychology and culture) are differently constituted "regions of being." Their manifestation in experience is different, and any claims about how matters stand in each domain are adjudicated on quite different grounds.

Within what became "analytic philosophy," the relation between these two levels of antinaturalism played out slightly differently. For Frege and Carnap, logic and mathematics rather than consciousness served as the "constitutive" source of semantic and epistemic norms. The parallel distinction at the level of the "constituted" was thus not between the experiential lifeworld and nature, but between mathematical and nonmathematical domains. The most influential criticisms of Frege's and Carnap's programs marked that divide in two distinct ways, however. For Quine and the logical empiricists, heirs to the Vienna Circle and its programmatic commitment to the unity of science, there was no reason to anticipate sharp methodological or epistemological differences between the natural and the human sciences. Relations between mathematics and the empirical sciences in general marked a more philosophically interesting boundary. For later Wittgenstein and many of his followers, however, the interesting "nonmathematical" domain was not empirical science, but everyday life and the ordinary uses of linguistic expressions.

Despite these differences, a striking common feature of the contemporary philosophical scene is the widespread rejection of antinaturalism at the constitutive, "transcendental" level. In Husserl's more fundamental sense, philosophers are almost all naturalists nowadays. As thinking, acting, speaking beings accountable to norms, we humans are worldly beings who understand ourselves empirically (in the broadest sense of that term). That commitment holds not merely for the followers of Quine, but also for most postphenomenological Continental philosophers, and for those analytic philosophers more influenced by Wittgenstein or Sellars.[1] Debates over naturalism are thus now situated almost entirely at Husserl's second level, which only concerns the place of mind and action within nature and history. Most contemporary antinaturalists defend not the constitutive priority of logic, consciousness, or rationality over empirical facts, but merely their autonomy from determination by natural scientific causes or laws. That is why today's antinaturalists almost invariably pay homage to natural science, which usually remains outside of their primary concerns. They may argue that mathematics, psychology, social science, or ethics should not be naturalized, but nature is almost invariably conceded to the natural sciences. Indeed, many antinaturalists (Thomas Nagel, John Searle, Charles Taylor, Bernard Williams, and Hubert Dreyfus are prominent examples) endorse rather conservative conceptions of natural scientific understanding as foils for their own commitment to the autonomy of human understanding and action from natural scientific determination.

There is a latent instability within such commitments to the mutual

1. Thus, for example, Heidegger's and Merleau-Ponty's early existential phenomenologies situated meaning and truth within the concrete, historical world inhabited by acculturated bodily agents. Even Heidegger's later emphasis upon the world-disclosing importance of poetry located such disclosure within historically emergent natural languages. Post-Wittgensteinian analytic philosophy also looked to natural languages, or to human psychology and social life, as the worldly locus of meaning and understanding.

independence of nature and human understanding and action, however. The natural sciences themselves have an ambiguous role in such conceptions. Scientific inquiry and understanding must surely provide central cases for any adequate conception of human rationality and action. Yet to account for the normativity of science, the meaning and validation of natural scientific claims must also be accountable to the natural world. The difficulty for contemporary antinaturalists is to maintain their precarious balance between two competing attractors. If natural scientific practice and understanding are naturalistically explicable, then it is hard to see why other aspects of thought and action cannot likewise be accommodated. If the objectivity or rationality of scientific understanding is instead governed by autonomous principles or norms, it is unclear why natural science and the world it discloses should not be subsumed under a more encompassing rational, psychological, or social-cultural humanism. Shorn of any constitutive, transcendental aspirations, the antinaturalistic center may not hold.

In this chapter, I argue that this latent instability within nontranscendental antinaturalisms manifests a deeper conceptual problem that troubles contemporary naturalists and antinaturalists alike. For what generates the instability is a conceptual dichotomy between the normativity of human action and understanding and the anormativity of nature. The problem for naturalists is how to account for the normativity of understanding and action as thoroughly natural phenomena ("to bake a normative cake out of non-normative ingredients" in Fred Dretske's memorable phrase). The problem for antinaturalists is to avoid the arbitrary projection of human norms onto nature. The worry I shall begin to articulate in this chapter is that both tasks have been rendered impracticable by shared preconceptions of the issue.

In the first part of the chapter, I explore the instability in contemporary antinaturalism by examining Charles Taylor's conception of the autonomy of meaning, understanding, and action, as criticized by Thomas Kuhn. This choice may seem idiosyncratic. Taylor's work stems from a brief convergence between postphenomenological and Wittgensteinian criticisms of the positivist legacy in the philosophy of history, psychology, and the social sciences.[2] Within Anglo-American philosophy, this rapprochement has

2. The Anglo-American heyday of historicism in the philosophy of the social sciences as a philosophical program of broad import extends roughly from the publication of Peter Winch's *The Idea of a Social Science* (1958), to the immediate aftermath of Taylor's 1979 article "Interpretation and the Sciences of Man" (in 1985b, 15–57), Stephen Toulmin's

been largely supplanted by the cognitivist challenge to behaviorism and the reorientation of philosophical semantics toward reference and truth instead of meaning. Naturalistic themes in the work of Quine, Sellars, Putnam, Kripke, and Davidson have come to overshadow the hermeneutic historicism of Peter Winch, Taylor, Stephen Toulmin, or Alasdair MacIntyre. Why give philosophical prominence to history and social science if cognitive psychology and rational decision theories provide more fundamental accounts of human understanding and action? Moreover, concurrent developments in the philosophy of natural science ironically undercut the significance of the historicist turn in the human sciences. Why be impressed by the critique of positivist methodology for history and the human sciences if *no* sciences answer to positivist models? The most prominent contemporary antinaturalists thus commonly focus upon the irreducibility of consciousness and mind rather than history and social life.

I look initially at Taylor rather than at Nagel, Searle, McGinn, or other antinaturalists about the mind for three related reasons. First, there are distracting ambiguities of scope concerning antinaturalism in the philosophy of mind: does human understanding in general resist natural-scientific explication, or is such resistance confined to its qualitative manifestation in consciousness? For the purposes of a naturalistic account of scientific understanding, it is somewhat beside the point whether there could be a scientific explanation of the qualitative experience of "what it is like" to understand something scientifically; those philosophers of mind for whom only the qualitative aspects of consciousness resist naturalistic explication have sided with naturalists on how to understand the content and justification of scientific work. Second, grasping the difficulties that arise in Taylor's attempt to situate human understanding within social practices rather than rational consciousness gives stereoscopic depth to the more familiar problems raised by the normativity of cognition. Third, and most important, however, is my conviction that the cognitivist turn toward mental representation is a dead end—the way out of the conceptual difficulties confronting all of these approaches will be found through a practical-discursive rather than a mentalist account of scientific understanding. I shall not argue for this conviction now; the best case I can make for it is the constructive program that I develop in the latter half of the book.

Human Understanding, volume 1 (1972), and Bryan Wilson's influential anthology *Rationality* (1970).

In the final section of the chapter, I articulate the more general conceptual difficulty that the instability of Taylor's antinaturalism exemplifies. The problem is the irreconcilability of the normativity of thought and action with the anormativity of nature, a problem that afflicts the more generic sense of naturalism shared by professed naturalists and antinaturalists alike. A fuller grasp of the problem places John McDowell's recent criticisms of Donald Davidson in a new light. Davidson may seem to have constructed his philosophical program in response to the difficulty that worries me. He conjoins a thoroughgoing physicalism with an insistence upon the irreducible autonomy of human thought and action, by treating thoughts and utterances as token-identical with physical states of bodies or sign-tokens. What McDowell's objections show is that Davidson's proposed approach to the reconciliation of nature and normativity merely names the underlying problem rather than solving it. Moreover, understanding the problem from my broader perspective shows why McDowell's proposed response is no more promising. The difficulties that emerge in these discussions then set the issues for the remainder of the book.

Kuhn and Taylor on Natural and Human Science

Thomas Kuhn's challenge to Charles Taylor's insistence upon fundamental metaphysical and epistemological differences between the natural and the human sciences thus provides the setting for my initial consideration of the instability of recent philosophical antinaturalisms. Taylor's arguments transcend the demise of logical empiricism and the deductive-nomological model of explanation, which had provided the original context for his emphasis on the autonomy of the human sciences. His strategy turns upon the central issue in debates over philosophical naturalism, namely, the relationship between natural causality (or laws) and semantic, epistemic, or moral normativity. Yet Taylor's proposed distinction between the natural and human sciences is still challenged by Kuhn's reinterpretation of the natural sciences. Not only does Kuhn's work threaten to reassimilate the two kinds of science that Taylor so carefully distinguished; Taylor's account of the human sciences provides an ironically instructive way to read Kuhn, and a challenge to some residual differentiations of natural from human science in Kuhnian terms.

The question of the methodological unity or duality of the sciences eventually recedes from the forefront of their disagreement, however. What is at issue in the juxtaposition of Kuhn and Taylor turns out to be not method but metaphysics, the "relation" between nature and the normativity of action and understanding. The specter that Taylor and Kuhn raise is thus not a cleavage between two kinds of science, but an irreconcilable dualism between the "world" of human action and understanding (including scientific understanding), and the natural world in its brute indifference to how it is described or understood. In this context, I shall argue, what the juxtaposition of Kuhn and Taylor reveals is the pivotal significance of scientific practices for making sense of naturalism and its philosophical prospects.

Taylor's work belongs to an influential tradition in the philosophy of social science, going back at least to Wilhelm Dilthey and Max Weber, which insists that human action requires interpretation in some methodologically distinctive sense. Interpretation is necessary, according to this tradition, because actions (including speech acts) are meaningful, and their identity and significance as actions can be disclosed only by uncovering or clarifying the meaning they express. The meaningfulness of action has been variously located in actors' intentions, rules constitutive of their behavior (including the rules that structure languages), background practices as the context within which action is intelligible, and so forth. Such views typically conclude, however, that the need to interpret action introduces important methodological and epistemological differences between the human and the natural sciences, perhaps grounded in ontological differences between nature and the social world.

Taylor argues for such distinctions on the grounds that the objects of the human sciences are essentially self-interpreting agents, whereas the natural sciences deal with objects that need not and should not be understood as self-interpreting. Taylor's version of this claim bypasses familiar debates about naturalizing consciousness, the mind, or propositional attitudes as functionally significant mental states. Self-interpretation is not to be construed in terms of fully explicit beliefs or attitudes, but instead emerges as a situated bodily engagement with the world. Taylor thus follows Merleau-Ponty in taking the subject of meaningful action to be an embodied agent, one whose interaction with the world is already structured by its bodily orientation and the capacities it thereby discloses. Agents are also situated within a field of "intersubjective meanings," which belie any monological interpretation of meaningful action. Raising one's hand can only be voting, and one's behavior toward others can only be bargaining if they are situated among appropriate institutions and practices.[3] Finally, agents are affectively engaged by their situations. Taylor notes:

> An emotion term like "shame," for instance, essentially refers us to a certain kind of situation, the "shameful," or "humiliating," and to a certain mode of response, that of hiding oneself, of covering up, or else "wiping out" the blot. . . . But this situation in its turn can only be identified in relation to the feelings it provokes; and the disposition is to a goal that can similarly not be understood without reference to the feelings experienced: the "hiding" in question is one which will cover up my shame; it is not the same as hiding from an armed pursuer. (1985b, 23)

On Taylor's account, then, understanding self-interpreting agents is itself a mode of self-interpreting agency, a way of interacting with others whom I thereby take to belong to a more or less shared field of meaning. Thus, even if one wanted to say that the natural sciences are also interpretive activities in some sense, one would then have to conclude that the human sciences were doubly interpretive, interpretations of interpretations.

Thomas Kuhn's investigations of the historical development of the natural sciences led him to question Taylor's way of drawing this distinction. Kuhn has figured prominently among those philosophers and historians who insist that scientists never encounter the natural world unmediated by previous understanding. Knowledge of nature is a response to prior knowledge and the successes and failures it engendered. The result is that there are conflicts of interpretation over such things as voltaic cells, classical resonators/quantum oscillators, and chemical reactions that parallel the interpretive issues that Taylor thinks are characteristic of human activity alone. The reason is straightforward: cells, oscillators, and reactions are not just objects, but aspects of a meaningful field of human activity, with a history and a significance that cannot be disentangled from that history. Interpretation of the voltaic cell is inseparable from interpretation of practices for identifying and coping with it.

The dispute over differences between the natural and the human sciences might thus seem to turn on the question of realism. Take realism to

3. These examples and others are discussed at some length in Taylor's classic criticism of behaviorist and cognitivist social science in "Interpretation and the Sciences of Man" (1985b, 15–57).

be the claim that beliefs within some domain are true or false depending upon the real properties of objects in that domain, that is, properties they possess regardless of anyone's beliefs, practices, or standards. Taylor wants to be a realist in this sense about the objects of natural science, even while insisting that a realist construal of "social reality" is unintelligible because the interpreters' beliefs, practices, and standards are part of what is to be interpreted. Kuhn, of course, notoriously rejected realism in the natural sciences, with his insistence upon the incommensurability of competing scientific practices and standards, and his denial that sense can be made of what the world is "really" like apart from a history of (scientific) interpretation.

Yet the juxtaposition of Kuhn and Taylor relocates what is at issue in debates over realism. Kuhn's concern was not whether there "really" are objects, natural kinds, or causal powers independent of scientific practices. Rather, his point was that what scientific practices are normatively accountable to are objects and phenomena manifest through scientists' engaged, meaningfully situated practices of experimentation and theorizing. Muon-detecting is in this respect no less "intersubjectively meaningful" than voting or bargaining, and grasping the significance of an experimental result need be no less affective than recognizing one's humiliation (scientists' accounts of their own work often highlight its affective significance, but philosophical discussions rarely give sufficient credence to this aspect of their accounts). Nor does it matter that scientific practices work with equipment and objects that are not themselves usually regarded as agents. Taylor's work reminds us that shame likewise incorporates the shameful situation, while collective bargaining is typically concerned with the organization of a material workplace.

Kuhn's challenge to Taylor's distinction between two kinds of science, then, is that the scientific interpretation of nature belongs squarely among the self-interpreting practical engagements with the world that Taylor takes as objects of the interpretive human sciences. The force of this challenge is ironically revealed by the extent to which Taylor's account of the human sciences usefully informs a reading of *The Structure of Scientific Revolutions*. Many readers of that book have misunderstood Kuhn to be saying that research communities in the natural sciences are, and perhaps ought to be, united by an enforced consensus about fundamental issues of ontology, method, and value. Taylor, however, highlights the difference between "consensus," an agreement in belief between individuals who each hold that belief separately, and what he once called the sharing of "intersubjective meanings."[4] He had in mind the way in which understanding a common language and participating in the social practices, institutions, and norms associated with that language shape the intelligible possibilities for individual belief, and thereby make either consensus or dissensus possible. Taylor thus offers an illuminating reading of Kuhnian paradigms as "ways of experiencing action in society which are expressed in the language and descriptions constitutive of institutions and practices" (1985b, 38). Paradigms, that is, are shared fields of meaningful activity rather than shared beliefs.

Taylor is also helpful in dispelling a still common but mistaken reading of Kuhn as a linguistic idealist for whom language is somehow constitutive of the world. Taylor rightly emphasized the inseparability of language and social practice. Linguistic distinctions and ways of employing them are embedded in larger contexts of practical activity, which in turn could not exist without the appropriate linguistic resources. Taylor took this to mean that one could not characterize "social reality" apart from the language and practices that constitute it, but the reverse is also true: the language is only meaningful as part of the practically configured reality it articulates. In the case of scientific inquiry, of course, the relevant practices include ways of encountering, responding to, and being resisted by what scientists are dealing with. The intertwining of language, social practice, and reality cannot be neatly bounded at the points where they run up against the natural world, for encounters with the world, and indeed the very boundaries between self and world, belong to interpretive human practices.

Finally, Taylor's work ironically dispels at least two ways of establishing a (perhaps contingent) difference between the natural and the human sciences in specifically Kuhnian terms. The first case turns on the internalism Kuhn ascribed to science guided by paradigms. He insisted that a characteristic feature of normal science was its effective isolation from criteria of social relevance. Notwithstanding examples such as the role of calendar reform in the downfall of Ptolemaic astronomy, Kuhn noted with approval:

4. Taylor has subsequently noted in conversation that this term may be misleading, since it suggests that subjects are independently constituted as individuals, and intersubjective meanings only then arise through their subsequent interactions; part of Taylor's point, however, was that such a field of meanings is constitutive of agency and understanding.

> A paradigm can insulate the community from those socially important problems that are not reducible to the puzzle form, because they cannot be stated in terms of the conceptual and instrumental tools the paradigm supplies. Such problems can be a distraction. . . . One of the reasons why normal science seems to progress so rapidly is that its practitioners concentrate on problems that only their own lack of ingenuity should keep them from solving. (1970, 37)

Some commentators have seen in such passages a crucial distinction between natural and social science, the latter being constitutively unable to free itself from demands for social relevance; others have seen here Kuhn's allegiance to the contemporary "end of ideology" movement in the United States.[5] But Taylor's own criticism of "end of ideology" behavioral science can help set the issue straight.

Taylor argued that empiricist political science seriously misunderstands some politically important phenomena when it limits itself to considering only political action that is identifiable *within* a given configuration of "intersubjective meanings" and "common meanings." Political conflict can occur over the shaping of that field of meaningful action, and the presence or significance of conflict can be suppressed by taking the dominant intersubjective meanings for granted. As one example, Taylor considered the 1960s countercultural challenge to "the vision of society as a large-scale enterprise of production" (1985b, 46) and to the politics of negotiation, which is "bound up with the distinct identity and autonomy of the parties and with the willed nature of their relations" (32). Such conflicts are especially tricky, because the very language in which to characterize what is at issue in the conflict is part of what is at issue. The result is that the conflict is easily misrepresented or trivialized by those who occupy the position of the dominant field of action, including empiricist political scientists.

Similar conflicts over the meaningful configuration of a practical field are at issue in various feminist, postmodernist, animal-rights, or neo-Romantic criticisms of the dominant culture of "technoscience," and in different ways in the resurgence of "fundamentalist" movements in several religious traditions.[6] There is an understanding of natural science, and of

5. Dreyfus 1984 exemplifies the former interpretation, Fuller 1992b and 2000 the latter.

6. Bruno Latour (1987) introduced the term 'technoscience' to recognize the artificial character of any boundaries between science and technology or science and society.

nature as the object of scientific knowledge, that represents not just a consensus of many people within the industrialized nations and elsewhere; it shapes what they—we—understand as intelligible possibilities for action. Indeed, the ideals of objectivity and rationality that inform that understanding of science are part of the common meaning through which a globally dominant community identifies itself to itself. Conflicts over such ideals are telling examples of the sort of dispute that Taylor thought would be obscured unless the practices at issue were understood hermeneutically. Proclamations of the political neutrality of the natural sciences are thus part of a contestable social reality in need of interpretation, rather than exemptions of scientific practices from cultural or political conflict.

A different way of excluding most contemporary natural sciences from the supposedly interpretive disciplines came from Kuhn himself. Kuhn insisted that there remains a distinctive activity of interpretation often required of historians, ethnographers, or literary critics, which physicists or economists can eschew. Interpretation in his sense is needed to get inside an alien culture or text, learn one's way about, and eventually perhaps explicate what is going on there. Such deliberate and explicit interpretation is fundamentally different, he argued, from the ways in which people live more or less unproblematically within their own cultural milieu, including particular scientific cultures (Kuhn 2000, chap. 10).

Kuhn and Taylor obviously did not draw the line between interpretive and noninterpretive activities in quite the same place, and they certainly did not do so in the same way. But Taylor helps show what was wrong with Kuhn's distinction. He insisted that "already to be a living agent is to experience one's situation in terms of certain meanings; and this in a sense can be thought of as a sort of proto-'interpretation'" (1985b, 27). The reason is twofold: first, making sense of one's own culture and even clarifying one's own sense of a situation are more reflective and interpretive than a sharp distinction between understanding familiar and alien meanings would suggest; second, there is less homogeneity and more room for construal and misconstrual even within a familiar and shared field of activity than is often assumed. After all, Thomas Kuhn (1970, chap. 5) himself recognized that even the shared practices of normal science incorporate partially divergent interpretations of that practice. Scientists usually "read" one another's work in ways that largely manage to smooth over those differences, and it is not clear why these readings are different in kind from making sense of practices one explicitly recognizes to be puzzling or alien. Kuhn is right that there are times when it is important to be attentive to the possibility that others are using words or things in ways that do not neatly fit one's own assumptions, and other times when one can safely make those assumptions for the time being. This difference does not neatly divide disciplines, however, for the need to be attentive in this way can sometimes arise in the midst of what one takes to be familiar surroundings.

Interpreting Nature

Taylor or his defenders might respond to such assimilations of the natural sciences to his accounts of the human sciences by instead distinguishing natural scientific practices from nature. Scientific practices and the language they invoke may themselves belong to the social world, but they also disclose a natural world as transcendent to its meaningful, scientific articulation. The crucial boundary to secure would then be the divide between the meaningful human lifeworld, including scientific practices, and the natural world.[7] Any methodological distinctiveness to the natural sciences would arise from their orientation toward a world devoid of human significance. Taylor himself attempts such a strategy by insisting upon the importance of what he calls an "absolute" conception of the world, freed from "the meanings it might have for human subjects, or of how it figures in their experience" (1980, 31).

There are important sources of instability, however, in the supposed boundary between the natural and social "worlds." Kuhn indicated a first source of instability in the way that scientific practices incorporate their object of study. Taylor's descriptions of the way in which social reality cannot be identified "in abstraction from the language we use to describe, invoke, or carry out" various social practices (1985b, 33) also aptly characterize physical reality as the object of natural scientific research. Kuhn reiterates that scientists do not and could not achieve knowledge of things except within particular historically contingent categories and practices. They always encounter nature as a configuration of significant possibilities for research. That is why Kuhn first identified "normal science" as a kind of social practice, and why he said that "after discovering oxygen Lavoisier

7. Husserl (1970a) prefigures this response, although unlike Taylor, Husserl regards the meaningful configuration of a shared lifeworld to be constituted by the activity of transcendental consciousness rather than by the practical interaction of embodied, situated agents.

worked in a different world" (Kuhn 1970, 118). Taylor thought that the human sciences had to be interpretive because what they were trying to understand was a field of meaningful human activity, constituted by a shared understanding of what could be intelligible and significant action within that context. Kuhn insisted that learning how to practice research within a scientific field is very much like being socialized into a culture, while trying as an outsider to understand that research activity and the world it reveals is very much like the work of an ethnographer. If it still seems odd to you to think of electrons or tectonic plates as belonging inescapably to social reality in Taylor's sense, remember that the social world on Taylor's account certainly includes the various sorts of equipment people use in their activities. Kuhn's argument is that a field of natural scientific research is a work world in which the objects of research are understood as equipment and make sense only through the ways we deal with them in research and development. Hence the appropriateness of his famous analogy between scientists who too readily abandon a paradigm in the face of anomaly and carpenters who blame their tools.

Kuhn's resistance to any separation between the natural world and its meaningful interpretation is reinforced by a recognition of how deeply their intertwining affects any attempt to identify what the natural world is. Mary Hesse once noted (responding not to Taylor but to Jürgen Habermas): "It is impossible in studying theories of evolution, ecology, or genetics, to separate a mode of knowledge relating to technical control from a mode relating to the self-understanding of man [since] the very categories of these theories, such as functionality, selection, survival, are infected by man's view of himself" (1980, 186). Even the most basic ontological distinctions, for example, between an organism and its environment in evolutionary theory are fundamentally engaged with conceptions of individuation and agency.[8] Nor are these difficulties confined to the biological sciences. Evelyn Fox Keller (1985, chap. 8) has suggested how the notion that chemical processes are regulated or controlled by "master molecules" implicates a gendered conception of human agency. The attempt to characterize the state of a physical system in quantum mechanics and to interpret what it means for a system to have a definite state

8. For useful discussion of this point, see Levins and Lewontin 1985, part 1; Lewontin 2000. In chapter 8 below, I return to the issues raised by relations between organisms and their selective environment.

is likewise difficult to disentangle from involvement with human self-conception.[9] The point is not to suggest that scientific theories are debilitated by cultural or ideological preconceptions, although that does sometimes happen. More commonly, however, the opposite point is apropos: the natural sciences have proceeded quite well without needing to eliminate all aspects of human self-understanding from the interpretation of the natural world. The world those sciences aim to disclose is within the scope of meaningful human practices, not transcendent to them.

Taylor recognizes these affinities between the mediation of the natural world by prior scientific understanding and his own account of interpretation in the human sciences, but he still insists there is an important residual difference. He responded to Richard Rorty's assimilation of the natural and human sciences thus:

> The understanding which is relevant to the sciences of man is something more than this implicit grasp on things. . . . We are talking here of what you could call human understanding, understanding what makes someone tick, or how he feels or acts as a human being. (Taylor 1980, 30)

I will not take up here the difficulties confronting Taylor's attempts to explicate this supposedly distinctive sense of human understanding in terms of "desirability-characterizations." My concern, which makes up my second line of argument, is that however Taylor tries to work out the notion of understanding a human being, it will not support his point about the natural and human sciences unless there can be an extensional correspondence between two distinctions that in fact do not match neatly at all. The boundary between natural and social reality, which also supposedly demarcates the boundary between the objects of natural and human science, needs to match up with the distinction between those areas of inquiry in which our self-understanding as human agents is at issue and those areas

9. Keller (1985, chap. 7) offers one account of how the issues surrounding the interpretation of the Schrödinger equation display the entanglement of self-understanding and the interpretation of nature. Barad (1996) goes much further, arguing that reflection upon quantum mechanics shows that what the natural sciences can understand *are* the phenomena of practical, agential involvement in nature, and not objects independent of that involvement. I discuss Barad's arguments more extensively in later chapters. A third example is the notorious invocation, by some physicists and other theorists, of "consciousness" as an ineliminable aspect of any quantum measurement. I find such appeals to consciousness indefensible, and bordering upon unintelligible; yet ironically, that such a strange view would be even a conceivable temptation perversely illustrates the extent to which the interpretation of nature has been bound up with the categories of human self-understanding.

in which it is not. Any such match would at best be historically contingent, but I want to argue that it does not occur at all.

Remember that drawing the relevant line between the human sciences and the natural sciences is not just a simple matter of ascertaining the objects of the various sciences. Taylor certainly excludes human physiology or genetics from the interpretive human sciences, and he may well want to include some aspects of primatology or other studies of animal behavior among them. Hence the line depends upon there being a fairly sharp split between the objects that are somehow intrinsically meaningful in human terms and a natural world of physical objects that are not intrinsically meaning-bearing. The difficulty is that this distinction is itself a contestable point of human self-understanding. Whether there is such a distinction, and where it lies, is inescapably part of a sense of who we are and what it means for us to act. It is no accident that anthropologists study cosmology as an ineliminable part of culture. Human beings understand themselves in significant part through their understanding of nature. The disenchantment of the natural world, Darwinian accounts of natural selection, or the turn from qualitative and perfectionist to compositional accounts of substances are now integral to many people's self-understanding as agents (even for those people, like Taylor, who work to expand what can be accommodated within a disenchanted, compositionalist, selectionist conception of embodied human agency).

An illuminating connection can be made here between Taylor's views about the human sciences as the interpretation of the self-understanding of human agents and his communitarian political arguments against liberal individual conceptions of the self. In the latter case, Taylor argues that it is a serious mistake to draw the boundaries of the self at a point short of our social relations with others. There are irreducibly social dimensions to self-understanding, just as there are social goods that cannot be reduced to individual preferences.[10] Taylor objects, rightly, to those approaches to the human sciences that rule out on methodological grounds the possibility of an irreducible social dimension to human life, and that regard all of our social relations instrumentally: "The ontology of mainstream social science," he argues, "lacks the notion of meaning as not simply for an individual subject; of a subject who can be a 'we' as well as an 'I' " (1985b,

10. These points are central themes throughout Taylor's writing, but are best illustrated by his papers "Atomism" (1985b, 187–210), and "Irreducibly Social Goods" (1995, 127–45).

40). Yet Taylor similarly rules out on methodological grounds the idea that the boundaries of self-understanding extend beyond social relations in the narrowest sense to encompass dealings with the natural world.

The result is a remarkably Whiggish interpretation of the history of science since the seventeenth century. On Taylor's view, it was unequivocally progressive to interpret nature as dead and noncommunicative, such that interaction with it can be of only instrumental value (with respect to social "preferences," to maintain the analogy to his critique of liberal individualism). Taylor believes that the further extension of atomistic and mechanistic understanding to a liberal utilitarianism was a mistake, but he presumes that that mistake has now been exposed. He hopes to forge a more adequate distinction between social practice (as the realm of meaningful interpretation) and physical nature, to replace the more familiar dualism between the mental as the realm of meaning and physical nature as what is unaffected by how it is meant. He would then have to show how the disastrous political history of distinctions between nature and society, which has been used to justify social hierarchies of race, gender, culture, and class, can be avoided if only one draws the boundary in just the right place. I remain suspicious, however, of the history of science that Taylor's distinction may compel us to accept.

My suspicions are heightened when I cast about for examples where it would make a difference to follow Taylor in denying that our understanding of nature involves interpretation in his strong sense. I cannot resist beginning with the example Taylor himself offers in defense of his view. Taylor says of traditional Japanese society that "they live under the same heavens as we do, only understand it differently, but it is not true that they have the same kind of bargaining as we do" (1985b, 33). Perhaps it is just a remediable slip that he takes "living under the heavens" to capture a meaning-independent reality identifiable apart from human self-understanding, but I do not think so. I suspect that Taylor's phrase reflects a complacent identification with the ascendancy of a contentious interpretation of science and nature comparable to the 1950s understanding of the ascendancy of liberal pluralism as the "end of ideology."

The parallel emerges clearly in the way Taylor's distinction between the natural and human sciences bears upon some recent intellectual and political controversies. Feminist interpretations of the gendering of scientific knowledge and of nature as the object of such knowledge offer one important group of examples. Taylor's view can accommodate those feminist philosophers and scientists for whom the discovery of gender in science is the discovery of a bias to be eliminated in accord with an unchanged notion of scientific objectivity. But his view rules out in advance, on methodological grounds, any stronger feminist account of natural scientific knowledge as gendered. Thus I believe Taylor must reject out of hand the idea, suggested by Nancy Chodorow and Evelyn Fox Keller, that the separation between self and nature, which he thinks essentially characterizes natural science, is the product of male psychodynamics in societies where women nurture infants. He must similarly reject the poststructuralist feminisms of Donna Haraway, Judith Butler, or Karen Barad, which locate meaning and agency within historical configurations of material-discursive practices. Indeed, Taylor cannot countenance the idea, put forward in rather different ways by von Wright and Keller, that causality is not conceptually separable from human self-understanding as agents.[11] For it is important to Taylor that efficient causality be a real relation in nature, disconnected from the intersubjective meanings of a particular historical culture. If scientists like Barbara McClintock or the pseudonymous Anna Brito find it appropriate to interpret their understanding of maize or lymphocytes in terms of friendship or love (Keller 1983; Goodfield 1981), this *must* at best be metaphorical or merely psychological. These are not the sorts of interactions that can really be reciprocal relations, so they must be reduced to something else in much the same way the individualist reduces social relations to individual subjective values or preferences. On Taylor's view (1985b, 134–51), such accounts *must* concern a kind of attunement to the world that is separable from theoretical understanding, even if the scientists themselves somehow fail to see this.

A related example is suggested by the various attempts to create an ecological politics that would take noninstrumental relations to the natural world as irreducible goods. Social ecologists, deep ecologists, ecological feminists, Marcusean critical theorists, and perhaps some defenders of animals are engaged in precisely the sort of challenge to the common and intersubjective meanings of our culture that Taylor recognized in the counterculture of the 1960s. To insist that they and their political concerns must be understood within the parameters of the dominant conceptions of scientific rationality seems to me comparable to the attempt, which Taylor

11. In chapters 8 and 9, I develop a different account of how causality and agency are intertwined, one closer to the work of Barad and Haraway than to that of von Wright or Keller.

deplores, to impose the vision of a society of work and the social practices of negotiation as the horizon within which we must make sense of the protests of the 1960s.

My point in raising these examples is thus not to endorse any or all of these views, any more than Taylor wanted unequivocally to endorse the counterculture protests. Indeed, the parallel is a strong one. Taylor thought that a social science and social world premised upon political and methodological individualism was incapable of understanding what was at issue in those debates. The problem was not simply that an individualist science and political culture could not recognize where countercultural criticisms raised seriously troubling concerns. Individualists were also bereft of an adequate response to those aspects of the countercultural challenges that Taylor found unacceptable. I similarly want to argue that a conception of natural science and the natural world constituted by a sharp separation between the social and the natural is incapable of grasping much of our current cultural and political situation, and is thus incapable of responding adequately to the various criticisms of technoscientific cultures. My first line of argument suggested that Taylor's distinction between the human and natural sciences could not accommodate what we have learned from Thomas Kuhn and others about natural science. Now this second line of argument suggests that the distinction also cannot accommodate the kind of political insight that I have long associated with the work of Charles Taylor.

There is a third line of argument against Taylor's distinction, which I have developed elsewhere and which should be mentioned here because it accounts for much of the intuitive appeal that distinction continues to possess. Taylor sees a deep connection between his view of the human sciences and his views about what it is to be a human agent. The human sciences must be interpretive in a stronger sense than is true of the natural sciences, because they have as their objects human agents who are in an important sense constituted by their interpretive activities. A human science that left out of account the self-interpretive dimension of human beings would be radically incomplete or misguided about its own object, while this would not be true of the natural sciences. Unfortunately, this argument involves a mistaken inference.[12] From the fact that all human understand-

12. Okrent 1984 first developed this argument, and my discussion (Rouse 1987, chap. 6) draws extensively upon his account.

ing takes place within a field of intersubjective meanings, as Taylor rightly insists, it does not follow that this is a fact about human beings that must always be taken into account in the human sciences. The point parallels Kant's Paralogisms of the Soul: conditions on the possibility of knowledge are not facts about knowers as objects of knowledge. The argument from human beings as self-interpreting subjects to the human sciences as interpretive in Taylor's strong sense is a non sequitur.

Causal Laws and Rational Norms

What conclusions follow from my critical juxtaposition of Kuhn's treatment of the natural sciences and Taylor's account of the human sciences as interpretive? It makes a crucial difference whether one is considering the methodological distinction between two kinds of science or the ontological distinction between nature and the meaningful social world. Rejecting the first distinction seems to me unproblematic.[13] Taylor's defense of a strongly interpretive account of the human sciences was directed against the logical positivists' insistence upon the unity of science, and against the construction of mainstream American social science to accord with a naively empiricist reading of the logical empiricist program. It would be a mistake to identify the rejection of these views with an alternative methodological unification of the sciences, now modeled upon the human sciences rather than physics. The appropriate conclusion is instead that neither natural nor human science forms a natural kind. There are many interesting and important differences between the various scientific disciplines. Paleontology or meteorology may be as interestingly different from one another and from high-energy physics as they are from macroeconomics or political science. All have been shaped by a history of internal development and interaction with other scientific fields and other social practices. We should follow Nietzsche in his insistence that "only that which has no history is definable" (1967, 80); successful practice in the various sciences always has and will continue to escape the constraints placed upon them by meth-

13. Taylor (1994, 233–36) now concedes this first point that there need be no sharp methodological divide between natural and human science, but he still insists upon the second, which he calls "the [neo-Diltheyan] conviction that different domains of study yield different answers to the question of what makes a science really successful, what makes accounts insightful and illuminating" (234). I question whether one can usefully separate any answer to "the question of what makes a science really successful" in its domain from the findings of the science in question.

odologists armed with the latest account of what a science must be like in order to be successful.

The relation between nature and the social world is a different matter altogether. The tenor of my arguments in this chapter so far has been to assimilate nature, the world as manifest through scientific practices, into what Taylor would have described as the social world, the meaningful field of human activity. Yet this disappearance of any boundary between the human and the natural also has a naturalistic counterpart. Such a thoroughgoing metaphysical naturalism would treat the sayings and doings of human agents themselves as causally efficacious events in the natural world. Taylor's own resolute insistence upon the embodiment of human agency makes this point especially clear, for bodies and bodily performances seem fair game for causal understanding. If Taylor's field of meaningful human practices seems to expand to engulf their material setting and objective referents, one could equally well argue against any artificial restriction upon the boundaries of the natural world and its causal unfolding. Taylor's meaningful field of embodied activity would then be encompassed within the causality of nature.

Such naturalistic unifications of the world have in fact been prominent within contemporary Anglo-American philosophy. Causal theories of reference and possible worlds semantics have provided powerful tools for situating the meaningfulness of people's sayings and doings within the natural world. The supposedly normative character of semantically significant performances might then be accounted for by causal interaction with their surrounding circumstances (including, perhaps, their prior adaptive-evolutionary circumstances). Not just what an expression refers to, but also what an utterance means, or what action someone performed, might be understood to be determined by the causal chains or adaptive/functional roles in which it is situated. Similar treatments have been offered for epistemic norms, which would be replaced either by the de facto interactions of psychological states or by the causal reliability of responsive dispositions. Taylor's strategy in response to these projects has been to argue that the sense of meaning or rationality that can be accounted for in naturalistic terms falls short of the richness and complexity of social life.[14] Yet his strategy simply invites naturalists to appropriate any such articulation of a field of "meaning" that supposedly resists their explanations. This appro-

14. See especially Taylor 1985a, chap. 10, and 1995, chaps. 1 and 4.

priation is a counterpart to the strategy I have been exploring so far in this chapter, which would display any construal of nature and its transcendence of human significance as itself a meaningful interpretation within a field of human practices.

Taylor's attempt to confine nature and meaningful social practices each within their supposed proper bounds would thus fail on both sides, as each voraciously encompasses whatever is taken to remain alien to it. This dialectic is, of course, familiar; Foucault articulated a similar conundrum in his account of the inevitable "doubling" of the modern episteme ([1966] 1970, chap. 9). Like Foucault, I take this doubling of the world to be an artifact of an insistent philosophical misconception, although my diagnosis differs noticeably from his. I call attention not so much to a pointless dialectical oscillation between the ascendancy of a thoroughgoing naturalism and that of its humanist counterpart, as to the unintelligibility of each view from the standpoint of the other. What the failure of Taylor's distinction shows is that natural causality and the normativity of meaning and justification have been conceived in ways that make their relations to one another unintelligible. Taylor and Kuhn are especially illuminating, because the difficulty that emerges from their encounter is widespread within contemporary philosophy and science studies. The proponents of an unremitting naturalism and of a thoroughly humanist social constructivism can each plausibly claim to have left nothing out of their accounts—nothing, that is, except the entire world conceived in the opposing terms.

Robert Brandom has perceptively identified such mutual unintelligibility as the mark of a debilitating dualism, for which he offers Descartes's dualism of mind and body as the exemplary case (1994, 615). Scientific practices would then be the pineal gland of the contemporary dualism between nature and the normativity of meaning and justification, the site that must magically reconcile what has been conceived irreconcilably. Scientific practices, after all, are meaningful activities undertaken by human beings to fulfill aims that make sense within a particular sociocultural setting. Yet they are also supposed to be that human activity that discovers and explicates a natural causality that transcends all human significance.

Wilfrid Sellars once presented a similar diagnosis of the philosophical situation at midcentury as a clash between the manifest and scientific images of humanity. Sellars, however, thought that when appropriately understood, these two images could be conjoined in a kind of binocular vision:

> To complete the scientific image we need to enrich it *not* with more ways of saying what is the case, but with the language of community and individual intentions, so that by construing the actions we intend to do and the circumstances in which we intend to do them in scientific terms, we *directly* relate the world as conceived by scientific theory to our purposes, and make it *our* world and no longer an alien appendage to the world in which we do our living. (1963, 40)

I share something of Sellars's vision in this passage, but I think that he was too sanguine about the compatibility of the familiar conceptualizations on each side. Sellars still envisioned a complementary conjunction of two distinct conceptions; natural science would then no longer seem like an alien appendage to the meaningful human world, but it would still be an appendage. I see a need for a more fundamental reconception of both scientific understanding and human agency, one that more closely integrates saying what is the case and belonging to "what is the case" as a practically significant situation.

Donald Davidson and Richard Rorty claim to have achieved such an integration of the rational-discursive and natural orders in their conceptions of "anomalous monism" or "nonreductive physicalism."[15] On their views, the world is thoroughly physical and causal, yet many of these physical states or processes are caught up within autonomously explicable discursive practices.[16] Semantically articulated performances are token-identical with causally efficacious physical events, but there are no type-type correlations between causal relations and the semantic interrelations within which the same events are understood as meaningful. This view

15. See especially Davidson 1980, essay 11, and Rorty 1991, 113–25. For a long time, I accepted Davidson's and Rorty's responses to this question. I thought that their conceptions of causality needed significant revision (see note 17 below), but I presumed that semantic normativity and a more adequate account of causality would provide compatible but nonoverlapping ways of conceiving the world. My gradually coming to see the inadequacy of this conceptual segregation was an important step in the process of conceiving this book.

16. Davidson uses the concepts 'causal' and 'law-governed' interchangably, which enables him to characterize the absence of intelligible connections between physical and rational descriptions as "anomalousness," an absence of any lawlike connections between them. This characterization is doubly mistaken, in light of my arguments in chapters 8 and 9. To characterize physical events as law-governed is already to bring them within the space of reasons, while doing so also mischaracterizes scientific understanding in terms of subsumption under laws. To this extent, Davidson's view is thereby doubly dualistic: he makes unintelligible the relation between physical events and the "same" events in the space of reasons, but he also takes for granted the overcoming of a dualism between physical events and their supposed scientific understanding in terms of laws.

supposedly sustains the autonomous irreducibility of the semantic/psychological and physical domains, without positing a metaphysical dualism that should offend the scruples of metaphysical naturalists. There are no nonphysical events, only irreducibly different descriptions of physical events. These descriptions in turn, of course, always take the form of psychological states or overt performances, which are themselves token-identical to other physical events.

Recent arguments by John McDowell, Robert Brandom, and John Haugeland show that Davidson's and Rorty's approaches merely restate rather than resolve the unacceptable dualism of natural causality and semantic or epistemic normativity.[17] McDowell prominently claimed that in Davidson's interpretive reconstruction of thought and talk, "he . . . helps himself to the idea of a body of beliefs" (1994, 68). The implicit objection is that interpretations of a speaker's performances only acquire semantic significance through an intelligible connection to (in the form of a rational constraint by) something external to language. Although he doesn't put the point this way, McDowell is arguing that Davidsonian truth theories really only amount to uninterpreted formal systems, with a syntax but no semantics, unless their normative accountability is more than intralinguistic. Davidson claims to connect the sentences in the interpreted speaker's idiolect to their extralinguistic truth conditions, via the description of those conditions in an antecedently understood metalanguage. McDowell's objection is that Davidson has only postponed the difficulty: his truth theories merely provide the formal means to connect sentences within the interpreted language to other sentences in the metalanguage. It cannot express

17. McDowell's argument against Davidson is developed throughout his *Mind and World* (1994), but the key passages are 10–23, and 66–70. Haugeland's argument is primarily developed in his "Truth and Rule-Following" (1998, 305–61). Brandom's criterial insistence upon the genuine objectivity of conceptual norms is expressed briefly (1994, 52–55), but chapters 8 and 4 are entirely devoted to his attempt to satisfy this criterion by making conceptual norms genuinely accountable to objects (including their causal interface with discursive practice through perception and action), where Davidson would argue that beliefs are only normatively accountable to other beliefs. Brandom goes on at the beginning of chapter 9 (614–18) to characterize the underlying problems to which these arguments respond as a threefold dualism in Kant's account of concepts, which he takes to have been inherited by most of the ensuing philosophical tradition. Although Brandom does not specify where more recent philosophers run afoul of these dualisms, at least the third dualism (between spontaneity and receptivity) plausibly applies to Davidson's characterization of the disconnection between the rational and causal orders of explanation. Brandom's more recent discussion (2000b) of this issue, however, suggests that he still takes his own account to be consistent with a broadly Davidsonian position.

the semantic contentfulness of those sentences (their significance beyond the substitution of one intralinguistic sign for another).[18]

McDowell's criticism of Davidson thus recapitulates what I called in chapter 1 the problem of manifest necessity. He is arguing that the formal relations internal to a Davidsonian truth theory are akin to the logical syntax that defines a Carnapian language.[19] They specify a rationally reconstructed set of formal relations that are not manifest in the concrete, historical-material world, and hence neither answer to nor constrain worldly events. Davidson wants to say that the utterances and thoughts incorporated on the right side of his Tarskian truth theorems (p in the formula "S is true iff p") are token-identical to physical events (and therefore simply are manifestations of these formal relations in the world), but his theoretical reconstruction cannot express that claim until their connection to events outside the formal system of the truth theory is rendered intelligible within his own overall conception.[20] Davidson thinks this

18. A fully adequate reconstruction of how McDowell's argument is telling against Davidson requires an intervening argument that I do not present until chapter 6. Davidson's official view is that his truth theoretic models are interpretations of an object language in a metalanguage (typically, the object language is the idiolect of an individual speaker, but it could also be the language of a community, or a portion of a speaker's or community's idiolect). Following a line of argument implicit in Brandom 1994, I argue in chapter 6 that what is actually being interpreted in a Davidsonian truth theory is the discursive interaction that incorporates *both* the ostensible object language and the language used in the interpretation. This conclusion is crucial to the success of McDowell's argument, because it blocks the response that the needed external constraint upon the sentences of the object language is provided by successful interpretation in the metalanguage, whose relation to the world is what is understood in "understanding a language."

19. The object language of a Davidsonian truth theory is specified by a recursive axiomatization of the set of sentences that would be assented to by the speaker(s) to be interpreted, rather than by a Carnapian logical syntax. Davidson nevertheless here remains on Carnap's side of the latter's implicit disagreement with Neurath over what is meant by "a language." The object language is a set of formally interrelated marks or noises, whose status as a language is only vindicated by their interpretation in a metalanguage that Davidson implicitly takes to be a language in something more like Neurath's sense. Davidson (1986) begins to recognize the underlying difficulty when he notices that the ordinary sense of what it is to "have a language" (a sense that he still must take for granted as applicable to the metalanguage in which the truth theory is constructed) diverges from the sense in which an object language is specified by a recursive Tarskian truth theory. He therefore acknowledges that his truth theoretic reconstructions are at best models of a "prior theory" that an interpreter must always supplement with a "passing theory" to close the gap between this formal structure and the linguistic competence of the speaker being interpreted. But the passing theory, if construed as a formal system, has the same underlying problem.

20. I remind the reader once again that this argument is not complete until I show (in chapter 6) that what is being interpreted in a Davidsonian truth theory is not just the idiolect of a speaker, but the discursive practice that incorporates the sayings and doings of both speaker and interpreter.

problem can be avoided because he does indeed "take for granted" that his own utterances are appropriately occasioned by their worldly circumstances (because he already speaks a language in something like Neurath's sense). He clearly believes that the interpreter's own unquestioned status as a language user is sufficient to secure the semantic contentfulness of the sentences used in the interpretation. McDowell's point, I argue,[21] is that Davidson's own account of the relation between rational interpretation and causal explication (anomalous monism) undercuts the status he implicitly assigns to such interpretations, by preventing any rationally explicable relation between circumstances and the utterances they occasion, and hence their capacity to sustain genuinely semantic content. Another way to put the point is that, while Davidson must insist that "radical interpretation begins at home" so that the interpreter's metalanguage has no privileged status, the possibility of interpreting the metalanguage in turn does not actually account for its semantic contentfulness; the language that one can interpret in this way is, like Carnap's formal languages, a language that no one speaks [22]

Brandom (2000b, 167–68) has recently tried to defend Rorty's nonreductive physicalism against an analogue to the charge I have been making against Davidson. Specifically, he denies that Rorty's use of 'vocabulary' (his analogue to Davidsonian languages) "commits that great foe of dualisms to a dualism of norm and cause" (167). Brandom's defense fails, how-

21. McDowell himself does not explicitly say why Davidson *cannot* legitimately "help himself" to the concept of a belief in the language the radical interpreter uses. My suggested interpolation to complete his argument is that it is precisely anomalous monism that is supposed to render even the interpreter's own utterances opaque to any connections sufficient to express how semantic contentfulness is intelligible.

22. McDowell thinks this issue has been overlooked, because the question of how expressions acquire semantic significance through their normative accountability to something other than relations to other signs has been conflated with the question of epistemological skepticism. McDowell does not doubt Davidson's ability to counter the epistemological skeptic once semantic contentfulness has been secured; the question is instead the normative status of thoughts and utterances as contentful: "Davidson manages to be comfortable with his coherentism, which dispenses with rational constraint on thinking from outside it, only because he does not see that emptiness is the threat. He thinks the only point of wanting a rational connection between intuitions and thoughts is reassurance that we are justified in endorsing the thoughts, as if we could take it for granted in any case that they are thoughts, that they possess content" (McDowell 1994, 68). McDowell's distinction highlights the connection between his argument and Heidegger's and Neurath's criticisms of Husserl and Carnap. Their concern was also not epistemological, but ontological, a concern for what it is for an expression or thought to be intentionally directed toward, and normatively accountable to and binding upon, worldly agents.

ever, in two respects. First, he acknowledges that for Rorty, the normative side of the distinction takes pride of place, so that in the end, Rorty reverts to the humanist side of the dualism.[23] Specifically, Brandom notes that for Rorty. "the vocabulary of causes is a vocabulary [that] we can discuss in the metavocabulary of vocabularies" (167). The more important difficulty, however, concerns the relations Rorty can recognize between norms and causes. Brandom concludes:

> The fact that we can use the vocabulary metavocabulary to discuss the causal vocabulary (its emergence, peculiarities, practical virtues and vices, and so on), and the causal metavocabulary to discuss vocabularies (the role of reliable differential responsive dispositions in empirical vocabularies, the practical capacities they enable, and so on) shows that the distinction between the vocabulary of causes and the vocabulary of vocabularies is not drawn in terms that make relations between them unintelligible. So it is not playing the functional expressive role of a dualism. (167)

But these uses remain compatible only so long as each vocabulary remains comfortably domesticated and is not allowed to incorporate the other wholesale. What neither can accommodate in its own terms is the intelligibility of its own assimilation within the other vocabulary. That the Rortyian conversation of mankind is a causal nexus while a thoroughgoing causal materialism is a contingently emerging moment in a historically shifting play of vocabularies can both be asserted, but they cannot be coherently understood together in these terms. Like Davidson, Rorty simply proclaims the mutual opacity of normative and causal vocabularies to be a virtue.

McDowell diagnoses the source of these problems in Davidson's and Rorty's conception of natural causality as impervious to reason.[24] His own

23. In the final section of chapter 3, I show in more detail how Rorty's construal erases the causal dimension of scientific practices and scientific understanding. In chapters 6 and 7, I show in turn how the priority Brandom ascribes to intralinguistic norms over causal interaction with the world in perception and action also turns out to be dualistic.

24. McDowell expresses the point this way: "According to [Davidson's] ontological thesis [anomalous monism], the items that satisfy the *sui generis* spontaneity-related concepts have a location in the realm of [natural] law. But the concepts are *sui generis* precisely in that it is not by virtue of their location in the realm of law that things instantiate those concepts. So if we go on equating something's place in nature with its location in the realm of law, we are debarred from holding that an experience has its conceptual content precisely as whatever natural phenomenon it is" (1994, 75–76).

response to these difficulties is to locate the connection between spontaneous conceptualization and natural causality in our own animal natures: "To see exercises of spontaneity as natural, we do not need to integrate spontaneity-related concepts into the structure of the realm of law; we need to stress their role in capturing patterns in a way of living" (1994, 78). McDowell proposes to accomplish this "naturalization" of reason by distinguishing nature as the realm of law from the "second nature" that inculcates the demands of reason through the acculturation of human organisms whose "first nature" is presumably found in the realm of natural law.

Unfortunately, McDowell's account simply reinstates the dualism of nature and normativity. Consider his own presentation of the relation between first and second nature:

> We cannot credit appreciation of [rational norms] to human nature as it figures in a naturalism of disenchanted nature, because disenchanted nature does not embrace the space of reasons. But human beings are intelligibly initiated into this stretch of the space of reasons by ethical upbringing, which instils the appropriate shape into their lives [as second nature]. (1994, 84)

McDowell then owes an account of how second nature intelligibly emerges from the activities of organisms that are also understandable in terms of biological first nature, such that first and second natures can coexist in the same organism. But such an account has been ruled out in advance by his insistence that first nature has no place for the space of reasons. McDowell in the end differs from Davidson only in the locus of his analogue to anomalous monism: it is not in mental events (or the pineal gland) that McDowell postulates the inexplicable coincidence of the natural and the normative, but in the "patterns of living" of the acculturated human organism as a whole. For McDowell no less than Davidson has to recognize, but cannot render intelligible, a metaphysical identity between human beings as rational animals and as biological animals. Like Davidson in the end, he finds philosophical virtue in this necessity:

> Human life, our natural way of being, is already shaped by meaning. We need not connect this natural history to nature as the realm of law any more tightly than by simply affirming our right to the notion of second nature. (McDowell 1994, 95)

Davidson might well retort that anomalous monism similarly does no more than affirm our right to the notion of rationality. What McDowell has shown is that neither is yet entitled to his preferred affirmation.

Davidson's and McDowell's shared conundrum can be readily traced back to its source in the version I extracted from the encounter between Taylor and Kuhn. Both Davidson and McDowell aspire to accommodate the normativity of understanding and action within human self-understanding as natural, and thus to satisfy the anti-supernaturalist commitment of philosophical naturalism. Yet they have taken for granted their compliance with naturalism's concomitant commitment to impose no arbitrary constraints upon scientific inquiry. That science discloses nature as the realm of law is supposedly taken from science rather than imposed upon it. If they acknowledged an obligation to vindicate that assumption, however, Davidson's and McDowell's dualisms would be superimposed upon the dualistic treatment of scientific inquiry. Like rationally interpreted mental events, second nature, or Sellars's manifest image in its other articulations, scientific practices are difficult to accommodate within what McDowell dubbed a "baldly naturalistic" conception of nature. But scientific practices provide whatever rationally compelling access we have to this "first nature," whose supposed inhospitality to rational normativity posed McDowell's, Davidson's, and everyone else's difficulty in the first place.

The dualism of nature and normativity thus calls for further reflection upon scientific *practices*. I emphasize practice in this way, because what needs articulation is precisely the connection between the representations of nature that philosophers attribute to science and the activities through which those representations are actually rendered accountable to the world they represent. Such reflection might then usefully indicate how to reconceive nature and normativity in ways that makes intelligible how they belong together more intimately than Sellars had in mind. Davidson and McDowell are not alone among contemporary philosophical naturalists in taking scientific practices for granted; the sciences have presumably already secured the basic outline of the intelligibility of nature, such that the philosophically interesting issues arise in accounting for more problematic phenomena in those terms. In the remainder of this book, I shall instead look more closely at scientific practices and what they disclose, not because I doubt the accomplishments of the sciences, but because I doubt the adequacy of familiar conceptualizations of those accomplishments. The problem that concerns me is more conceptual than empirical, although Quine has rightly inspired suspicion of any sharp division between matters of fact and their conceptual-discursive articulation. Recent scholarship in the history and social studies of science are certainly relevant to my argument (and vice versa, I hope!), but my concern is to reconceive and clarify 'practices' and 'nature.' In the next chapter, I begin with reflections on Quine, whose work is the apparent antithesis to my Kuhnian extension of Taylor, and who is in any case a pivotal figure in the contemporary revival of philosophical naturalism.

THREE

Quinean Indeterminacy and Its Implications for Naturalism

ANY SERIOUS DISCUSSION of naturalism in twentieth-century philosophy must contend with the work of W. V. O. Quine. Quine's "Two Dogmas of Empiricism" developed one of the most influential programmatic criticisms of philosophical appeals to the logical analysis of meanings. The constructive program for philosophy of language and epistemology developed in *Word and Object* and "Epistemology Naturalized" placed naturalism at the forefront of possible alternatives to the empiricist-formalist dogmas. My discussion of Quine is focused upon his notorious insistence upon the indeterminacy of translation. In the end, I take Quine's claim to be fundamentally correct, but its philosophical implications and its relations to Quine's other commitments (including his naturalism) have been widely misunderstood. Hence, I must begin by considering why translation is supposed to be indeterminate, and what should follow from acceptance of those arguments.

Indeterminacy of Translation and Theoretical Underdetermination

Quine sharply distinguished his claim that translation is indeterminate from his concomitant claim that theories are always underdetermined by empirical evidence:

> The indeterminacy of translation withstands the totality of truths of nature, known and unknown, observable and unobservable, past and future. Where indeterminacy of translation applies, there is no real question of right choice; there is no fact of the matter even to *within* the acknowledged underdetermination of a theory of nature. (Davidson and Hintikka 1969, 303)

Quine's insistence upon this distinction ironically complements the methodological distinction Charles Taylor and other hermeneutically inclined philosophers have wanted to draw between the natural sciences and the human sciences. For Taylor, the objects of the human sciences are meaningful thoughts, actions, or institutions, which are inaccessible to the empirical methods exemplified in natural science. For Quine, however, the crucial difference between meanings and natural facts is that there are no meanings to be right or wrong about, apart from the ability to predict what people say or do.

Philosophical readers of Quine have often found this distinction perplexing, however, in the light of his other claims, in particular, holism, epistemological naturalism, and his commitments to empiricism, behaviorism, and physicalism. If there is anything like a received view on these matters, it is the claim—advanced in different ways by Gilbert Harman (1967), Hilary Putnam (1975b), and Richard Rorty (1972, 1979)—that the indeterminacy thesis is a consequence of Quine's prior commitments to empiricism, naturalized epistemology, and holism. This line of interpretation has led Putnam and Rorty, among others, to claim that the indeterminacy of translation turns out (contrary to Quine's own insistence) to be only a special case of the underdetermination of all theories by empirical evidence. Adopting the "intentional stance" by ascribing meanings is just a different strategy for theoretical explanation. Evidence alone (apart from pragmatic considerations) is insufficient to determine which translation is correct, but this insufficiency equally characterizes physical theories. There are no "facts of the matter" of the sort available to undergird physical theory but absent for semantics, and meanings are therefore no more ontologically suspect in principle than are electrons or quarks. Just as with physical theory, "the reasons normally given for translating languages one way rather than another are justified simply by their internal coherence, and such practices as translation and ascription of intentional states are justified by their social utility" (Rorty 1979, 194).

These reasons for assimilating Quinean indeterminacy to theoretical underdetermination are familiar and widely accepted. For a long time, I also found them compelling, but I am now convinced that they are mistaken.[1] Quine's refusal to assimilate indeterminacy of meaning to the empirical underdetermination of theories does not depend upon a problematic methodological distinction between translations and natural scientific the-

1. Since writing this chapter, I discovered that Rorty (2000) now also recognizes, in response to Bjorn Ramberg (2000), that it is a mistake to assimilate indeterminacy of translation to the underdetermination of theories.

ories. Instead, the radically distinctive character of the indeterminacy of translation follows directly from Quine's holism and the indispensability of intersubjectively available cues for language learning.[2] The consequences of this vindication of Quinean indeterminacy are nevertheless subtle. In particular, two tempting but mistaken inferences must be resisted. First, this defense of indeterminacy does not support Quine's commitments to empiricism and to epistemological naturalism.[3] The indeterminacy of translation properly understood, far from sustaining "the last mooring of empiricism" (Roth 1987, 53), shows all the more clearly why Quine's conception of empiricism is indefensible. Second, this reading of indeterminacy does not restore, on Quinean grounds, a fundamental methodological or epistemological distinction between the natural and the human sciences. This latter inference is especially tempting, because Rorty drew explicitly upon his assimilation of indeterminacy to underdetermination to challenge Taylor's and other defenses of the distinctiveness of the human sciences. If that assimilation fails, one might wonder, why not restore Taylor's distinction? This suggestion rests upon a crucial mistake, however. Indeterminacy does not solely or specially affect those disciplines whose "objects" (e.g., utterances, actions, institutions, or practices) are meaningful, but instead characterizes any understanding of anything.

In this chapter, I consider first why indeterminacy of translation should not be construed as the underdetermination of interpretive theories. Then I can show why indeterminacy does not vindicate Quine's nonfoundationalist empiricism. An important consequence of my conclusions is that indeterminist rejection of "meaning realism" is as much an issue for understanding the natural sciences as the human sciences. This argument for the pervasiveness of meaning and indeterminacy will highlight the parallel mistakes underlying Rorty's rejection of the methodological distinctiveness of the human sciences and Quine's anti-intensionalist defense of naturalized epistemology. Ironically, Quine's version of naturalism turns out to conflict with his own rejection of epistemology as "first philosophy," a theme I shall continue to explore in subsequent chapters. If I am

2. This way of interpreting Quinean indeterminacy was developed by Paul Roth (1987). My own presentation closely follows Roth's line of argument, although we disagree about its implications for Quine's philosophical views more generally.

3. Roth (1987) himself failed to recognize these consequences of his interpretation, claiming instead that it vindicated Quine's insistence that an adequate pragmatism is ineliminably empiricist (Quine 1981), against Rorty's thoroughly pragmatist revisionism.

correct, "naturalized epistemology" turns out to be a philosophical oxymoron.

To begin, how are we to interpret the indeterminacy thesis, and why should it be accepted? On familiar interpretations, indeterminacy results from methodological difficulties in choosing between alternative manuals of translation. If these interpretations were correct, indeterminacy would indeed be on a par with (and perhaps only a special case of) the underdetermination of all scientific theories. There would be no compelling reason to deny that there is something (the "meaning" of an utterance or expression) for translation manuals to be correct or incorrect about, but only a reason to deny that one could ever uniquely determine the correct translation of that meaning into another idiom.

The indispensable premise for an alternative construal of Quine's argument for indeterminacy is holism, the claim that theoretical sentences acquire meaning and evidence only as part of a theory. Holism by itself, however, leads to what Roth called the "paradox of language learning":

> If we maintain that the sentences accepted as true in both natural language and natural science have their meaning and their evidence only as a related block, then someone with no knowledge of a language (e.g., an infant) should find any single sentence to be meaningless, and so incomprehensible. But children initially learn to speak by learning single sentences. So [it seems] Quine must be wrong in claiming that sentences have significance only within a theoretical context. (Roth 1987, 21)

The appropriate response to this paradox, however, is not to abandon linguistic holism, but to find a way to make holism compatible with the evident fact that natural languages can be learned through extensive social interaction in a shared environment.

Quine makes holism compatible with the need for languages to be publicly learnable by positing that any natural language must include some "observation sentences," whose "usage conforms to concurrent observation in about the way that it would if everyone had learned it ostensively" (Quine 1970, 4). The paradox of language learning then only shows that there must be "shared stimuli to which we respond and around which our use of language is initially coordinated" (Roth 1987, 19). Quine's insistence that language is a social art, coordinated by shared surroundings, is thereby restored to its central role; only publicly accessible cues could make language learning possible.

Invoking observation sentences in this way does not contravene the holist claim that sentences confront observation only collectively, because *which* sentences turn out to be observation sentences under what observable circumstances are not specifiable independently of an assumed theory. That a (learnable) language always contains some such sentences is only known by reflection upon indispensable conditions for language learning. (As I shall show, the argument for observation sentences is nonconstructive; it provides no criteria for identifying either the observation sentences in a language or the circumstances that cue them.)[4] No such argument can be constructed, however, to show that the existence of "semantic facts," or of any specifically intentional evidence, is needed for language learning. Hence, the epistemological parallel that Rorty and others draw between theories about the publicly available world and theories about meanings is undercut. There is strong reason to believe that there is a fact of the matter (however underdetermined) underlying the one, and no such reasons in the case of the other. Indeed, Roth gave this claim a stronger rendition, claiming that there is an

> "in principle" cast to the argument provided by the fact that, given our understanding of how language is learned, no additional evidence suffices to make meaning realism plausible. Rather, meaning realism requires wholesale revisions in our beliefs about language acquisition. (Roth 1987, 20)

The indispensable role that publicly accessible circumstances play in the acquisition and maintenance of language, and the absence of any comparable role for shared meanings, vindicates Quine's claim that the indeterminacy of translation is independent of and stronger than the claim that theories are evidentially underdetermined. Quine's criticism of "the idea idea" thereby renders untenable any doctrine of meaning realism. Yet I want to consider carefully what consequences follow from thus restoring the difference between indeterminacy and underdetermination. Many readers of Quine may take this argument to vindicate Quine's empiricism, and perhaps to restore the supposed differences in kind between the natural and the human sciences, which Rorty had challenged by trying to reduce

4. As I discuss below, Roth (1987, 22) goes astray when he interprets the argument for the existence of observation sentences as an argument to the best explanation of how languages are learned.

indeterminacy to the underdetermination of interpretive theories. Roth, for example, concluded that the need for intersubjective stimuli for language learning warrants a belief in objective evidence (at least within the domain of natural science) that is more than just a matter of social agreement. He also claimed that the difference between theories about the publicly available world and theories about meanings accords with a fundamental epistemological difference between the natural and social sciences: "Natural science is about this intersubjective domain [that provides the basis for language acquisition]. Hence theories in natural science posit a fact of the matter" (Roth 1987, 19). Both of these inferences are mistaken, however, and the reasons for their failure are philosophically instructive.

Empiricism and Language Acquisition

The place to begin is with the notion of "objective empirical evidence," whose existence is supposed to be warranted by an argument from the conditions under which a natural language would be learnable. To assess this inference, I will examine more closely the difficulties language acquisition seems to pose for Quinean holism. Holism says that evidence does not bear upon sentences individually, but only collectively. But if that is true, then it may seem impossible to distinguish (as indeterminacy requires) theories about the public world, for which there is supposedly a fact of the matter, and theories about meanings, for which no facts obtain. In choosing one translation over another, one certainly appeals to evidence. Indeed, one appeals to the same body of evidence that could support natural scientific theories, namely, the physical events (picturesquely described by Quine himself as "surface irritations") through which human neural receptors encounter their surroundings. The applicability of evidence therefore cannot be restricted to physical theory alone without contravening holism. So "the dilemma is how there can be no facts for a linguistic theory for which we nonetheless have evidence" (Roth 1987, 46).

Quine's response was that at least some sentences must be initially learned, before their interanimation with other sentences can provide the evidential base provided by a "total theory." To learn sentences is to learn how and when to use them. If one cannot presuppose that some other sentences have already been learned and made available to guide further language learning, then the only resource available for interpreting the initially learned sentences must be a publicly available world. And so Quine concluded that some sentences function in language learning as "observation sentences," that is, sentences that one can learn to use more or less correctly based only upon the circumstances that occasion their use.

Consider first what this argument does *not* purport to establish. Which sentences are observation sentences (i.e., are cued to the surrounding environment with sufficient directness to function as observation sentences) cannot be picked out independently of a theory. Nor can one identify independently of a particular theory what the evidence is that cues the production of observation sentences, or even that it is the same evidence, that is, the same features or aspects of the shared environment, to which teachers and learners attend.[5] Being an observation sentence is not a property of the sentence, but of the use of the sentence in various environmental circumstances within a particular discursive context, and that use itself cannot be explained or described independently of a theory that would characterize the circumstances (Quine 1965, 159).

Furthermore, this argument cannot establish that there is objective evidence for the truth of any particular sentences. For Quine, what remained of empiricism was a theory of evidence, but not a theory of truth. Empirical evidence does not provide a foundational justification for knowledge, but only a basis for explaining from within scientific theory as a whole how the theory itself could have been acquired. The surrounding environment thus cannot determine which sentences are true apart from the interanimation of sentences, even though it can be used to connect sentences (evidentially) to particular (sorts of) stimulations. In particular, it provides the common reference point for teachers and learners of a language.

Now consider what Quine meant by calling this nonconstructive argument for the existence of observation sentences an account of "evidence." Quine's project of naturalizing epistemology involved the hope that evi-

5. Thus, Quine suggested that "different persons growing up in the same language are like different bushes trimmed and trained to take the shape of identical elephants. The anatomical details of twigs and branches will fill the elephantine form differently from bush to bush, but the overall outward results are the same" (1960, 8). Strictly speaking, however, it is not clear why the elephants need be *identical* in order for speakers to share a language; they need only be *recognizable* as elephants. Thomas Kuhn (1989, 1993) made a similar point in his later reconception of incommensurability, insisting that people who share a "lexicon" use terms in ways that have the same structure, without in any way requiring that users acquire that lexicon by appealing to the same manifest features of the items it structures. Davidson (1986) extends this criticism further, arguing against the claim that linguistic communication requires an underlying convergence between "languages" understood by speaker and interpreter.

dence would be a physical concept that could eventually explicate language acquisition in neurophysiological terms. The "stimulus meaning" of sentences was supposed to be expressed by similarity groupings in the perceiver's physical interaction with the world. But Quine himself recognized the problem with the claim (without seeming to grasp fully its implications for his epistemology):

> Unaided by language, we might treat a great lot of sensory events as recurrences of one and the same sensation, simply because of a similarity between each event and the next; and yet there can have been a serious cumulative slippage of similarity between the latest of these events and the earliest of them. (Quine 1977, 176)

An account of stimulation at the neurophysiological level thus cannot do the requisite epistemological work of accounting for how a common environment mediates among speakers' perspectival differences to enable effective communication.[6] Quine's physicalistic notion of evidence is inescapably individualistic, just at the juncture when his epistemology needs to transcend the individual subject; it amounts to a kind of "linguistic solipsism" (Roth 1987, 71). This problem for Quine parallels Neurath's earlier criticisms of Carnap's asocial empiricism, suggesting that the underlying problem for Carnap was his epistemological individualism rather than his autopsychological appeal to first-person experience. Only publicly accessible correlations between the surrounding world and what others say and do will suffice to account for the possibility of language learning.

Hence the evidence necessary for language learning is irreducibly situated amid the interpersonal uses of linguistic expressions. There is no way for Quine to account for the public dimension of language (and hence, knowledge) in terms of the physical interaction between an individual subject and her surrounding environment without reference to what other speakers say under those circumstances. Indeed, discursive practice may be crucial in establishing a perceptually shared world with respect to which language may be learned. The criteria for what is similar and what is different in speakers' perceptions are provided in part by correlations with what

6. Roth (1987, 62–72) extensively developed this point, concluding that the behavioral aspects of language-world relations cannot be eliminated from Quine's account of observation sentences. In chapters 6 and especially 7, I devote considerable attention to reconceptualizing this understanding of speakers having different "perspectives" on their shared environment.

others do in those circumstances, including what they say.[7] Even a public environment with respect to which one can first learn a language thus may have to incorporate a fairly extensive discursive practice already underway.[8]

Merely recognizing the irreducible socially interactive correlations between utterances and world that are established in language acquisition does not go far enough, however. Language learning is a form of social interaction that does not occur between social equals. The occupants of one social role (teacher) are, within the context of language acquisition, authoritative. In that context, there is no publicly available check on the correctness overall of the teachers' utterances (it is important here that the role of the teacher is filled collectively by those whose speech the learner seeks to master for herself, that is, by anyone who is generally recognized to be a competent speaker).[9] Teachers speak in the presence of the learner and correct the learner's utterances on various occasions. The learner gradually adapts her own utterances to conform to the open-ended collection of available models of correct use and corrections of her own incorrect use. But how does the learner ever get beyond this unequal role, so that she can generally succeed in making utterances that are taken first as correct, then eventually as being as authoritative as those of her teachers? The explanation that is offered is that "a public and shared sense of the environment" (Roth 1987, 57) or, in Quine's preferred terms, a body of sufficiently similar stimulations, is available as an initial entry point into language, after which the interanimation of sentences can take over. This argument is supposed to establish indeterminacy in its strong Quinean form, because semantics has no such publicly available resource to appeal

7. For a parallel insistence that linguistic practices partly *constitute* the shared circumstances for their appropriate application, see Wittgenstein's discussion of "expecting" (1953, 1: par. 442–45)

8. The implications of this point for how to understand the origins of language will be addressed in chapter 9.

9. Among those appropriately recognized as competent speakers, the circumstances in which utterances are taken to be appropriately used can provide a check upon the correctness of those utterances, such that the entire community of competent speakers can be mistaken in their beliefs. In the context of language learning, however, teachers' utterances *must* be taken to be mostly true in their circumstances of appropriate use, so that the learner can grasp what the utterances are supposed to say. They need only be *mostly* taken to be true, however, not only to leave room for explicable errors, but also for the discursive practices of joking, testing the learner's ability to recognize absurdities, and so on. The underlying philosophical significance of this distinction between interpreting meaning and assessing truth will be discussed further in chapter 8.

to. Teachers and learners alike can appeal to a surrounding world for cues and clues for the use of sentences, in a way they cannot appeal to meanings.

Recognition of the social and epistemic inequality of the language learning situation shows the first difficulty with any attempt to vindicate Quine's notion of objective evidence. Roth construed the argument that holism and language learning presuppose a shared opening onto the world as an inference to the best explanation of the fact that languages are regularly acquired successfully. But an argument to the best explanation begins with a fact in need of explanation and posits some other matter of fact that one is warranted in believing to account for the original explanandum. Now the "fact" that is supposed to be explained by the argument is the success of language learning on the basis of finite circumstantial and behavioral cues. From the retrospective view of the learner, this may indeed seem to be what needs explaining, because it may seem simply apparent that what was once incomprehensible is now evidently understood.

From the socially authoritative view of the teachers, however, the question is significantly altered. For the teacher, it cannot be simply given that the language is actually being learned. A language speaker is one who no longer merely imitates the sounds or marks produced by others, but speaks in her own right. In an older idiom, one would say she produces those sounds or marks "meaningfully" rather than by simple repetition. But it is precisely this issue about which Quine insists there is no fact of the matter, and hence nothing to be explained in the usual scientific manner. The point is not that no distinction can be made between speakers and imitators, for this is done quite readily. Rather, it is that this distinction is not an independent explanandum, for which the explanans is to be the shared environment that cues observation sentences. The ability to correlate another's utterances to a shared environment is constitutive of her counting as a speaker.[10] Speaking a language is a normative matter, not simply an objectively determinable fact. That is why the language learner's perspective cannot be authoritative, for absent the responses of teachers, the learner has no basis for distinguishing having learned the language from merely seeming to have learned it. Thus, the argument from the conditions for language learning is not an inference to the best explanation, which

10. Wittgenstein (1953, 1: par. 156–71) makes a similar argument about the conditions under which someone appropriately counts as being able to read. Consider also Brandom's discussion (1979) of how translation is "socially normative" rather than objectively determinate.

would warrant belief in "objective evidence" as a matter of fact, but an argument about the appropriate conditions under which marks and sounds should be taken as utterances within a language.[11]

The force of this difference can also be seen by considering the intended target of Roth's argument for the "objectivity" of the evidence that enables language acquisition. Roth wanted to vindicate Quine's insistence upon the role of objective evidence in justification. The target of his criticism was the claim he attributed to Rorty that "evidence, in the guise of observation sentences, is what the community of scientists currently accept on this score, [so that] observation sentences are a purely sociological phenomenon [and] internal coherence [is] the only constraint on the warranting of beliefs" (Roth 1987, 55). Roth was right to reject this claim (whether or not it was correctly attributable to Rorty), but that does not vindicate his presumption that rejecting it requires objective evidence in some strong sense of that phrase.

One should have learned by now to be suspicious about the disjunction "either based upon objective evidence or else just a matter of coherence or consensus." There is indeed no evidence for any beliefs apart from practices of justification and the norms established and sustained within them; but there are no such practices except within language use sustained by continuing reference to a shared environment. Are languages then learnable because one can appeal to the objective evidence of a shared world, or does the notion of evidence simply drop out in favor of a consensus on which sentences to accept? The question is confused and cannot be directly answered in this form. Discursive practice and public world are inextricable, and that is what Quinean reflection upon the conditions for language learning once again shows. Quine has shown not that there really is evidence that provides objective warrant to social consensus on matters of justification, any more than Davidson (1984) showed that speakers really do share most of their beliefs about the world. In both cases, the ability to correlate utterances and beliefs with worldly surroundings is constitutive of the normative status of utterances or beliefs.

11. The reader may note that this interpretation brings Roth's argument about the necessary conditions for language learning very close to Davidson's argument (1984, essay 13) that the ability to interpret a speaker within a language antecedently understood is a necessary condition for the speaker's utterances to be linguistic. Nor should this be surprising; the criteria for determining whether a child has successfully learned a language are whether one can with reasonable coherence interpret her utterances (usually homophonically) by sentences in the language being learned.

Meaning, Evidence, and the Natural Sciences

From these reflections on the argument from language acquisition and the significance of empirical evidence, I now return to the question of the epistemic and methodological implications of semantic indeterminacy for the natural and the human sciences. Roth thought that his argument that objective evidence was needed for language learning provided a basis for restoring a distinction in kind between the natural and the human sciences. He took the domain of the natural sciences to be coextensive with the body of evidence whose existence was shown by the argument from language learning:

> Natural science is about this intersubjective, public domain. Hence theories in natural science posit a fact of the matter (the ontology of a theory) by way of explaining the intersubjectively available evidence that provides the basis for language acquisition. (Roth 1987, 19)

By contrast, the evidential basis for the social sciences is supposedly up for grabs; no comparable argument restricts the human sciences to accounting for a specific body of publicly available evidence.

Interpreting the epistemic significance of Quinean indeterminacy of translation in this way depends upon a specific conception of what it is to understand utterances, actions, or experiences as meaningful. One would have to contrast theories in the natural sciences to semantic or other "interpretive" theories as concerning two different sorts of object domains, one of which contains matters of fact while the other does not. Meaning would thereby remain an "unfilled" object domain, instead of ceasing to be an object domain at all. "Physical object" and "semantic object" would remain contrast classes, and indeterminacy would specifically concern theories about meanings as opposed to theories about objects. Theories about objects, in turn, might then be identified specifically with the epistemic jurisdiction of the natural sciences.

A better alternative to this interpretation of indeterminacy is to see it as a thesis about the meaningfulness of any and all language use, including the semantic articulation of facts. All utterances are holistically connected to the public world through patterns of discursive practice, without any intermediary "meanings" that fix how the connection is to be made. Fact and meaning are not separate epistemic domains, but aspects of every significant utterance (or implicit commitment) in any epistemic domain.

The difficulties with distinguishing physical and semantic theories according to their supposed object domains arise on both sides of the contrast. First, I will argue that the need for some fairly direct connection between language and world in order to overcome the paradox of language learning does not provide a body of evidence or a domain of content that is in any way fixed by reference to the surrounding world. Hence, it provides no distinct subject matter for the natural sciences as opposed to other disciplines. Then I will show that meaning, the focus of Quinean indeterminacy, cannot be localized. Meaning is an ineliminable aspect of all discursive practices, including those of the natural sciences. Indeterminacy thus pervades all discourse, including scientific theorizing and everyday talk about shared surroundings. The indeterminacy thesis is thus the claim that there is nothing prelinguistic "behind" discursive practices that can fix meaning or reference.[12] Once one has learned what speakers are inclined to say in what circumstances, there is no further ground for interpretation. Of course, learning what to say when is simultaneously learning the language and learning about the circumstances; or better, the relevant discursive practices are always part of the circumstances. Moreover, those circumstances also incorporate what speakers might say in those circumstances, so their specification is an inexhaustible enterprise. The lack of further grounds for interpretation thus does not imply the possibility of interpretive finality. Even who counts as a competent speaker is at issue, as became clear in my discussion of language learning. There is no already determinate fact of the matter about who is competent, although judgments of competence are accountable to norms implicit in the discursive practices through which speakers recognize one another as competent speakers within a common environment.[13]

Consider first the question of whether the publicly accessible evidence necessary for language acquisition is a domain with a specifiable content. The argument from language learning showed that the appropriate use of some sentences must be learnable more or less directly from shared envi-

12. In chapter 6, I will develop a more precise characterization of this denial of a "prelinguistic domain": the inferential articulation of discursive practices cannot terminate in perceptual experiences or practical performances that are themselves without further inferential significance.

13. As I shall show, the surrounding environment should not be regarded as an independent criterion for the correctness of discursive practices, but as itself part of those discursive practices. Discursive practices are not just ways of talking, but ways of articulating publicly accessible circumstances. The distinction between objective regularities (matters of fact) and normative accountability will be developed more extensively in chapter 5. Its application to discursive practices specifically will be addressed in chapters 6 and 7.

ronmental cues. But as we have seen, this argument provides no basis for identifying either the observation sentences of a language or the features of the environment that cue them. Nor must teachers and learners attend to the same environmental features in cuing observation sentences; it may turn out that the best translations of a learner's observation sentences into the teachers' idiolect are not homophonic with the teachers' utterances that provided her with behavioral clues for learning how to use those sentences.

Quine thought the meaning of observation sentences was fixed (at least relative to physical theory) because the evidential basis of language learning could be construed physiologically. Stimulus meaning was a way of correlating at least some fragments of language with events determinate independently of language. These correlations must have at least a discursive component, however; only if sentences are situated amid speakers' and learners' responses to environmental cues will there be the requisite shared basis for linguistic understanding. Such correlations must then be interpreted, and indeterminacy thus turns out to be ineliminable from such interpretation.

Any doubts that the behavioral context of language learning invokes Quinean indeterminacy can be removed by considering a correlate to what Steve Fuller (1988, chap. 6) has called "the inscrutability of silence." Utterances and other activity in the context of language acquisition are highly contextual, and Gricean "conversational" considerations (Grice 1988) must play a significant role in the interpretation of both the teacher's and the learner's performance. After all, one cannot presume a prior grasp of a common stock of meanings, since that is precisely what the learner lacks and seeks access to. Thus, teacher and learner rely upon interaction with salient features of their environment, past patterns of interaction, and the practical exigencies of the situation to orient the learner into language. But the language learner attempting to correlate her teachers' utterances and other performances with a common environment must not only take account of what the teachers overtly say and do, but also of what they do *not* say or do. The learner must decide which omissions mark an expectation that the salience and relevance of an environmental feature or practical implicature is so obvious as to need no further attention drawn to it, and which omissions display an insensitivity to or a discounting of its significance. There are thus performative and environmental parallels to the Gricean maxim of quantity: all that is done to help the learner understand the contextual cues for the use of a word or sentence needs to be done.

The moral is that such correlations do not provide a specifiable body of evidence that can fix linguistic interpretation; so the evidence grounding language acquisition has no objectively fixed content. Even in the case of observation sentences, teachers and learners do not have to agree about the correlations to be made between sentences and stimuli; they only need an environment in which they can successfully track one another's responses to their shared surroundings.

We can now see why Quine's use of 'evidence,' with its empiricist overtones, was misleading as a description of the availability of a shared environment as the necessary context of language acquisition. Evidence in the empiricist sense must be a determinate content that can be made present to a knower nondiscursively. Something manifest prior to language (whether surface irritations, sense-data, or perceptually manifest physical objects) would serve as determinate evidence for a linguistically articulated judgment. The argument from language acquisition establishes no such thing, for two reasons. First, whatever is manifest to us perceptually must be already connected to patterns and norms of language use.[14] Second, as the inscrutability of silence clearly indicates, whatever meaningful content can be wrested from the context of language acquisition is indeterminate even relative to any physical theory (whose own indeterminacy is for rhetorical purposes ignored).

If a notion of evidence is appropriate here at all, it would be what occurs most clearly in Heidegger (1996; 1977, essays 3, 5, 10): language acquisition would not be possible without a shared "world" ("clearing" or "opening") making itself evident. That is not evidence *for* anything, but only the evident manifestation *of* a common world. The world in this sense is not some object or objective regularity present prior to or independent of meaningful discursive practices. The world is a field or context within which particular things can show up, but it cannot itself be a determinate item appearing within that context, nor the totality of things thus manifest.[15]

Quine's residual empiricism would be dubious in another way if one were to think of the evidence necessary for language acquisition as coexten-

14. The classic argument to this effect within twentieth-century philosophy is Sellars 1997.

15. Mark Okrent (1988, chap. 4) gives an unusually clear account of Heidegger's notion of evidence, and its connection to more widely known empiricist and phenomenological conceptions.

sive with the evidence for natural scientific theories.[16] Indeed, this identification would be doubly problematic, for it unduly restricts the interpretation of the evidence underlying language acquisition and implausibly restricts the aims of the natural sciences. Consider first the evidence underlying language acquisition. The unargued presupposition underlying the notion of objective evidence is that the manifestation of a shared environment that enables language learning can be appropriately described (or at least explained) in the terms of the natural sciences. I have already shown that this evidence may itself be partially constituted through discursive performances (e.g., the willingness of competent language speakers to use the same word or sentence in various circumstances may be a constitutive feature of the perceptual similarity of those circumstances). But the "non-natural" aspects of that evidence may go much deeper, such that the shared phenomena that cue language acquisition would seem quite remote from the domain of the natural sciences.

One telling example is that the observation sentences through which some speakers enter into language may concern the mood-inflectedness of the surroundings they share with others. Once evidence is freed from the empiricist shackles that confine it to physical surface irritations or their sense-data correlates, cheerfulness, sadness, or menace seem to be as readily available to the observant language learner as colors, shapes, or movement, and perhaps more so. After all, the affective significance of the world, the way it *matters* to us in our ongoing activities, may be more readily manifest than abstracted sensory features. In any case, neither the sentences that actually function as observation sentences nor the observed characteristics that cue them can be identified without reference to a theory. So a dilemma confronts anyone claiming that the publicly shared environment needed to resolve the paradox of language learning is coextensive with the evidence base for natural science. Either the natural sciences include such phenomena as mood-inflection within their domain of explanation, or else a theory of language acquisition that posited moods as the proximate referents of first language would be ruled out a priori, in conflict with Quine's rejection of epistemology as first philosophy. Neither alternative seems acceptable.

There are other ways to construe the evidence for language acquisition

16. Recall that Roth 1987 construed the relation between natural science and the shared environment for language acquisition in just this way.

that would undercut their identification as grist for the mill of physical theory. The salient features of a shared environment might be neither objects nor sensory qualities, but rather purposive complexes of practices and equipment. A child's first utterance of 'car' could well denote not a physical object, but rather its connectedness with traveling and various significant possible destinations (Granny's house, the doctor, the market). The best translation of the child's plaintive utterance 'bottle' might connect it to the emotional/physical ambiance of feeding, such that the child's implied request would not be appropriately met by providing her an empty bottle, a bottle filled with cold milk, or by failing to hold her while she eats.[17] Yet I trust no one would say that the purposive links between equipment, social practices, and the satisfactions they can produce provide the objective evidence for which natural science is to account. So even if there were a determinate content to empirical evidence in Quine's sense, there seems little reason to confine it to what might be plausibly accounted for by natural science, let alone claim that it belongs to "a notion of evidence satisfying the tenets of empiricism" (Roth 1987, 41). In the end, Quine's pragmatism is indeed contravened by his residual empiricism, and the force of these arguments is that the empiricism is what must be abandoned.

Comparable problems also arise from the other side of any correlation of linguistic evidence with natural scientific evidence. There is a serious ambiguity in the claim that "natural science is about this intersubjective, public domain [of evidence for language acquisition]" (Roth 1987, 19). This claim could mean that the historically evolving practices collectively identified as the natural sciences have de facto aimed to account for precisely those intersubjective stimuli that have provided the initial focus of language acquisition. Alternatively, the claim could be that in order to count as a science of nature, a discipline must at least putatively account for those same stimuli. Neither version is attractive. The latter seems both arbitrary and contrary to Quine's pragmatic naturalism. How could and

17. Heidegger (1996) is the obvious source for both of these examples of nonempiricist manifestation. Okrent (1988, chaps. 4–5) has explicitly extended Heidegger's discussion of equipment, consistent with Quinean holism and behaviorism, to argue that the evidence needed to make possible the understanding (and hence the acquisition) of language is evidence of purposiveness in a context of practical activity, notably what various tools are to be used for and how they are to be used. This second example, therefore, goes well beyond merely imagining possible ways in which a shared environment could successfully cue language acquisition in the face of the supposed paradox of language learning.

why should the meaning and reference of "natural science" be fixed prior to the actual unfolding of scientific inquiry? What first philosophy could possibly accomplish this confinement of scientific inquiry, given holism and indeterminacy?

But the former option seems little better. The evolving history of the natural sciences has seen substantial change in their evidence base. The phenomena that scientists investigate empirically, and those that scientific theories most directly explain, are increasingly remote from the sorts of observations that could cue language acquisition, not only owing to the increasing importance and complexity of theory (for which everyday observational evidence might still offer Quinean boundary conditions), but also because the phenomena that are of scientific interest are often complex, skillful constructions (Hacking 1983, part 2). The natural sciences undoubtedly offer rich resources for creating empirically plausible stories about human beings' physical encounters with their everyday environment, but scientists generally do not provide more rigorous treatment of such matters, and one should certainly not take it to be the ground or measure of the sciences' epistemic enterprise. I thus conclude: even if, contrary to my earlier arguments, determinate content could be assigned to the publicly shared circumstances needed as a context for language acquisition, there are no plausible grounds for identifying it with the evidential domain of the natural sciences.

Meaning

I have been considering the suggestion that indeterminacy of translation has quite different significance for the natural and the human sciences because they differ essentially in their object domains: on this hypothesis, the objects of the human sciences are meaningful thoughts, actions, or institutions, whereas the natural sciences are directed toward the shared physical environment presupposed by any linguistic understanding. So far I have taken up only one side of this supposed contrast. I claim to have shown that arguments for the indeterminacy of translation based on indispensable conditions for language acquisition neither establish a determinate domain of objective evidence nor provide special evidential warrant for or constraint upon the natural sciences. I now consider the other side of the question. Is semantics (in the broad sense that also encompasses the meaning-content of thoughts, actions, and practices as well as sentences or utterances) an object domain distinct from that of physical theory, albeit a domain shown by the indeterminacy of translation to be empty of any matters of fact?

A preliminary question is why one would ever think that "meaning" demarcates an epistemic domain distinct from that of the surrounding (natural) environment. This suggestion has been prominent within the philosophy of the social sciences, as an account of why the study of human beings must be methodologically different from the study of nature. Thus, Wittgensteinians like Peter Winch and hermeneuticists such as Charles Taylor or Hubert Dreyfus have urged that understanding human agents involves grasping the meanings informing their actions and institutions. Meanings are allegedly embodied in semantic rules or practical skills, which must be interpreted, and which are not susceptible to natural scientific explanation. So the human sciences would be concerned to account for the domain of meaningful activity, while the natural sciences account for the brute, meaningless processes of the physical world. Similar issues have been addressed from a very different direction in Anglo-American philosophy of mind, where the dominant views take intentional entities (thoughts or meanings) to be supervenient upon, reducible to, or eliminable by their causal-functional role in the physical world. In the Anglo-American context, Quine's naturalism and his anti-intensionalism have been very influential.

Paul Roth thought that Quinean indeterminacy fatally undermines any attempt to prescribe or constrain methodology in the human sciences on the basis of the need to account for meanings or accommodate rationality. In the exemplary cases of Winch and Martin Hollis (and, mutatis mutandis, Charles Taylor), he claimed that their apparently contrasting views of rationality each presuppose an unwarranted "meaning realism," which "asserts that the meaning of words is fixed by virtue of some special relation, whether this special relation be to rules or concepts, or to objects in the world" (Roth 1987, 151). Meaning realism in this sense is made untenable by Quinean indeterminacy. Roth concluded from this line of argument that "any purported problem that makes realist assumptions with regard to meaning or a priori standards of rationality constitutes a pseudoproblem in social science" (151). Whatever the merits of such criticism of Winch and Hollis, however, Roth's argument is misleading in its suggestion that meaning realism is an assumption characteristic of the *social* sciences. Meaning realism turns out to play a crucial role in some interpretations of *natural* scientific knowledge, which therefore run afoul of Quinean indeterminacy of translation in a similar way.

Appeals to meaning realism in the natural sciences can most easily be seen in the case of the family of views known as "scientific realism." Contemporary realists like Richard Boyd (1984, 1996) recognize the epistemological significance of Quinean holism for the natural sciences in undercutting any straightforward empirical foundation for knowledge. Boyd, however, tried to turn holism into a premise of an indirect argument for realism about the referents of successful scientific theories. According to Boyd, the theory-dependence of scientific methods, including methods of experimentation and measurement, makes it highly unlikely that those methods could be systematically successful in coping with the world around us (which is also the shared environment presupposed by language learning), unless the theories that sustain those methods were at least approximately true. Boyd's explanatory project demands a strong correspondence sense of truth. There must be a fact of the matter about how the world is, independent of anyone's beliefs, norms, and practices, to account for the sciences' epistemic and technical success. There must also be a fact of the matter about whether the relation of correspondence holds (at least approximately) between scientific theories and the facts about the world; otherwise, successful correspondence would not be explanatory. But correspondence could not itself be a fact unless there were also a fact of the matter about what the relevant theories say the world is like. Scientific realism is therefore committed to meaning realism about the claims of natural science.

The connection between scientific realism and meaning realism is often obscured by scientific realists' objections to Fregean accounts of reference as determined by meanings. But the realist alternative to a Fregean conception is a causal account of reference. One can retranslate, say, J. J. Thomson's claims about (supposedly hard, determinate, particulate) electrons into claims (some true, some false) about quantum mechanical wave packets, because there is a determinate cause to the phenomena Thomson was investigating. The realist charitably interprets Thomson to refer to whatever really caused (most of) the phenomena that prompted his claims. But this project cannot intelligibly go through unless at a minimum the meaning of 'cause' is already determinate, fixed by the interactions of the objects really present in Thomson's experimental setup. The realist cannot, that is, treat 'cause' in the same way as 'electron,' as having been fixed by what (really) caused our interactions with things we first labeled as causes. The meaning of 'cause' as the scientific realist uses it must be fixed by what it *really* means to be causally efficacious.[18]

Those who believe that meaning realism and realism about the objects of natural scientific theories are logically independent theses do so because they believe that intentionally intermediary entities and physical entities compose two distinct domains of possibly extant objects, toward which different epistemic and existential commitments might well be possible. But my reflections upon the implications of realism about physical objects shows why this independence thesis is unfounded: one cannot intelligibly be a realist about physical objects, such that realism is an explanatory hypothesis, without being a realist about at least some meanings. The scientific realist employing a causal account of reference shows this point especially clearly precisely by attempting to avoid commitment to a Fregean realism about meanings, but failing in the crucial case of 'cause' itself. The meanings of the terms or sentences of physical theories must be fixed in some way for the realist's assertions about approximate correspondence between theory and world to be factually determinate. Thus, even if realists were to abandon causal theories of reference in favor of some other way of establishing an explanatory correspondence between theories and things, they would invoke meaning realism about whatever determined the actual representational content of theories. The crucial conclusion to draw about the constitutive role meaning realism plays in articulating the claims of scientific realism is this: meaning and intentionality are not a localizable epistemic domain, but instead condition any possible interaction with things that could amount to knowledge of them.

Another example of a surreptitious commitment to meaning realism informing an account of natural scientific knowledge is even more telling. Reflection upon my earlier discussion of Quine's physicalist version of naturalized epistemology shows how that arch-extensionalist nevertheless held to an objectionable meaning realism in theorizing about natural scientific knowledge. Recall quickly what aspect of Quine's view was at issue in my earlier critical reflections upon Quine's physicalism. As a naturalized epistemologist, Quine was committed to the project of providing a scien-

18. A more extensive account of how scientific realism is inextricably bound up with meaning realism is developed in Rouse 1987, chapter 5, and developed further in Rouse 1996b.

tific explanation of how scientific knowledge has been acquired. Roth has elaborated upon what it would take for this project to succeed:

> Natural science, in order to fulfill its promise as epistemology, must enlighten us with regard to how people make shared sense of the world and the language used to discuss it. . . . It is the stimulus meaning [of observation sentences] that provides the causal connection between language and the world. Yet, if what science seeks is to understand this causal connection as a basis for warranted belief, then what needs explaining is how people are stimulated the same. (Roth 1987, 60)

Quine insisted that this explanation must ultimately be provided at the level of neurophysiology, or "receptual similarity," since that is the most fundamental relevant level of physical science.

I showed earlier, however, that the necessary socially interactive basis for knowledge can never be provided at the neurophysiological level, because the surface irritations through which the environment physically affects us are different for different individuals (who occupy different locations within the environment and possess physically different receptors). Knowledge (even at the most basic level of language acquisition) requires shared circumstances with which to correlate the utterance of observation sentences. In order to mediate these differences in individuals' physical reception of stimuli, the epistemologist must appeal to correlations of what people say and do with what can be noticed going on around them. Only thus can one account for the differences in perceivers' similarity groupings that Quine called differences in "salience."

Roth considered such a modification of Quine's naturalized epistemology as a relatively friendly amendment that would "surrender none of [Quine's] critical theses and none of his empiricist tenets" (Roth 1987, 72). My discussion of language acquisition shows that the criticism is hardly so innocuous. What the role of practical-discursive correlations in language acquisition instead demonstrates is that Quine's naturalized epistemology is another unsuccessful version of meaning realism that "asserts that the meaning of words is fixed by virtue of some special relation to objects in the world" (Roth 1987, 151), in this case by a relation of observation sentences to physical similarities in reception. An empiricist reading of Quine would require that speakers' differential perceptual and practical responses to the surrounding environment have the same factual determinacy of content that Quine ascribed to neurophysiology. There must be an objectively determinate fact of the matter about what we are doing when we respond to what goes on around us.

But speakers' responsiveness does not have objectively determinate content in this way. The relevant activities include not just what speakers overtly do, but also what they do not do, and do not respond to. My suggested extension of Fuller's argument for the inscrutability of silence showed that indeterminacy of translation pervades the interpretation of all of the public interactions underlying language acquisition. The conclusion to be drawn can be stated two ways. One reiterates my earlier conclusion from the argument that a holistic account of language acquisition shows that indeterminacy of translation is not just the empirical underdetermination of semantic theories. This argument does not justify the claim that knowledge claims are answerable to a determinate body of evidence about the world, only that knowing presupposes a shared opening onto a world whose configuration is indeterminate except relative to interpretation in an already understood language. The other way to state the conclusion is that what is objectionable in Quine's epistemology is really his attempt to fix the stimulus meaning of observation sentences by a relation to events whose interpretation is not subject to indeterminacy. Quine's proposed physical science of science is objectionably meaning realist, and meaning realism is thus not just a view about intensionality as a distinct epistemic domain.

This recognition that Quine was a meaning realist shows a convergence between my discussion of indeterminacy and Davidson's criticism of Quine's commitment to the "third dogma" of empiricism, the distinction between a linguistic or conceptual scheme and a preconceptual experiential or worldly content (Davidson 1984, chap. 13). Davidson also rejected the idea that some determinate body of objective evidence makes utterances true or false. Truth or falsity can only be assessed from within a disclosure of the world that is already conceptually articulated. Such disclosure cannot be construed as evidence *for* some determinate truth-claim, but only as what *becomes* evident through discursive practices. As Davidson put this point:

> Nothing, no *thing,* makes sentences and theories true: not experience, not surface irritations, not the world, can make a sentence true. *That* experience takes a certain course, that our skin is warmed or punctured, that the universe is finite, these facts, if we like to talk that way, make sentences and theories true. But this point is put better without mention

> of facts. The sentence 'My skin is warm' is true if and only if my skin is warm. (1984, 194)

Understanding the world and understanding its linguistic articulation are not separable, correlatable achievements.

Indeterminacy, Naturalism, and "First Philosophy"

I now return to Rorty's post-Quinean strategy for discounting any essential differences between the natural and the human sciences. Rorty's criticism of hermeneuticism in the human sciences began by assimilating the interpretation of meaningful actions and utterances to the empirical underdetermination of theories. Interpretation is just theorizing from Dennett's intentional stance. Instead of positing electrons occupying distinct energy levels to explain the ways in which chemical substances interact, interpreters posit contentful beliefs and desires or meaningful utterances to explain what people say and do. But these explanatory strategies are supposedly not different in kind, only in subject matter; in both cases, the complex webs of theoretical sentences speakers construct are justified by holistic interconnections to their beliefs and desires. Rorty therefore concluded:

> The author of "Two Dogmas of Empiricism" *should* have said that concepts and meanings are harmless if posited to give explanations of our behavior, and become harmful only when treated as the source of a special kind of truth and of a special sort of authority for certain assertions. In particular, . . . the reasons normally given for translating languages one way rather than another . . . are justified simply by their internal coherence, and such practices as translation and ascription of intentional states are justified by their social utility. (1979, 194)

Rorty's diagnosis of the temptation to differentiate such interpretation from scientific theorizing does not challenge hermeneuticist accounts of the human sciences, but instead rejects the traditional portrait of the natural sciences, which they depend upon for contrast. On this view, the kind of certainty typically ascribed to mathematics or the natural sciences "will be a matter of conversation between persons rather than a matter of interaction with a nonhuman reality" (157). Even though rarely achieved in the human sciences, the prediction and control that is a characteristic achievement of the best natural scientific research can be explained in part by differences in the kinds of things people are interested in predicting and controlling (Rorty 1991, 40) and in part by historical coincidence:

> For all we know, it may be that human creativity has dried up, and that in the future it will be the *non*human world which squirms out of our conceptual net. It might be the case that all future human societies will be humdrum variations on our own. But contemporary science may soon come to seem as badly off as Aristotle's hylomorphism. (Rorty 1979, 351)

In any case, both natural science and human science are, Rorty thinks, best understood as attempts to articulate vocabularies within which to continue a coherent and interesting conversation. The successes of the sciences should be marked up as signs of social solidarity rather than objectivity:

> Notions like "unforced agreement" and "free and open encounter"—descriptions of social situations—can take the place in our moral lives of notions like "the world," "the will of God," "the moral law," "what our beliefs are trying to represent accurately," and "what makes our beliefs true." (1991, 42)

Rorty is not wrong to reject any notion of rationally or objectively compelled consensus about the world, but I think he is fundamentally wrong to see the only alternative in terms of conversation, social solidarity, and the achievement of unforced agreement. To talk about the sciences solely in terms of their theoretical coherence, and the social practices and institutions that enable its consensual achievement, is to overlook the significance of the specific ways in which the sciences are material practices. Rorty talks in these terms to emphasize his rejection of the notion criticized above, that the material surroundings of our discursive practices provide a body of linguistically unarticulated "objective evidence" to which those practices are accountable. Instead, he argues, "we have to see sentences as connected with other sentences *rather than* with the world" (1979, 372; my emphasis). But the contrast is misplaced. The practices through which utterances are connected to their publicly accessible surroundings are indeed not an encounter between already "interanimated" sentences and something alien to language and social norms, but that is because the practices that connect utterances to their circumstances are already part of the constitution of those utterances *as* meaningful sentences (and simultaneously of those surroundings as intelligible situations).

The development of a science involves new ways of talking and new ways of encountering and dealing with its objects. Neither meaningful sentences or theories nor articulated objects can be manifest except through their mutual interrelations. This point shows, pace Rorty, why the capacities for prediction and control that emerged with many of the natural sciences are not merely contingent, and are not just the result of a prior interest in controlling aspects of our physical surroundings. To understand the predictive capacities of a science, it is not enough to characterize either some general characteristic of its objects (for example, as physical, nonintentional, causally interconnected) or the purposes for which the science was developed; there is no substitute for a historical accounting of the detailed ways in which patterns of talk and other interactions were worked out together. Before one has any conceptual handle on scientific predictions, one needs to talk seriously about laboratories, the creation of phenomena (Hacking 1983, chap. 13), and the transformation of one's surroundings that results from the extension and adaptation of laboratory phenomena (Rouse 1987, chaps. 4, 7), but these developments are not independent of discursive and conceptual practices. Perhaps the deepest insight of Kuhn's work was the recognition that paradigms, "accepted examples of actual scientific practice which include law, theory, application and instrumentation together" (Kuhn 1970, 10), are more basic to scientific research than either the theories or empirical evidence extractable from them, and that scientific work normally consists of "articulating" these examples theoretically and experimentally.[19]

This priority of a practical field of concepts and conceptually articulated objects reinforces a common criticism of Quine on the underdetermination of theories by evidence: far from there being many theories equally compatible with the evidence in a domain, it is often all too difficult to work out even one minimally adequate theory. Theories are not groups of meaningfully interanimated sentences that then collectively confront a body of evidence as a whole. If theories were intelligible apart from the manifest phenomena around which they are articulated, one could well imagine artificially reorganizing their structure, with complementary changes to one concept to compensate for arbitrary revisions in another.[20] But articulating a theory is a matter of simultaneously developing experimental or other disclosive practices, uncovering a coherent domain of phenomena, and articulating concepts that can be resiliently used in sustaining and extending the relationships between practices and phenomena. Coher-

19. Rouse 1998 and 2002a provide a more extensive discussion of this reading of Kuhn.

20. The relations between scientific theories and experimental phenomena are discussed more extensively in chapter 8 below.

ently keeping these heterogeneous elements in line is hard, such that most aspects of the world, natural or social, are not objects of a science at all.[21] Where such coherence is achieved is indeed contingent in one sense, namely, that for all one knew at the outset, it might have been unattainable, and for all we know now it may eventually break down. Yet in another sense it is not at all contingent that ways of talking and ways of engaging and manipulating things function together—they only emerged at all through their mutual interaction.

A shared mistake thus connects Rorty's conversational model of science with the physicalist/behavioralist underpinnings of Quine's naturalized epistemology. Both Quine and Rorty take for granted the notion that causal interactions among things and normative relations between intensions are distinct domains of possible objects of knowledge. Quine then argued that the intensional domain is empty: there are no semantic facts, only natural facts. Talking about the world is just more behavior, which counts as the same or different solely on the basis of its pragmatic correlations with other facts. The relations between theory and evidence then turn out to be just that part of the natural world that neurophysiology considers: surface irritations and the neural activities they prompt. Rorty argued instead that interpretations were theories just like any other, and meanings were just more theoretically useful entities. Explicative and justificatory relations between sentences are involved when one is dealing with people's utterances and actions rather than other kinds of things. Of course, the natural sciences themselves then turn out to be entirely within the domain of actions and utterances: understanding a science is understanding the relations among its constitutive concepts and beliefs, for what can justify or explicate a belief can only be another belief. If Quine eliminates the intensional domain, Rorty renders the causal domain unintelligible. Interactions with the world may provide occasion for what scientists say and do as scientists, but only further talk can explicate or justify it.

This confusion is correctable in light of Quine's claim that there are no semantic facts. That claim is true, if semantic facts are supposed to be a distinct domain of possible knowledge, separate from natural facts. But

21. Haugeland begins to work out a philosophical consideration of why these mutual articulations are hard to achieve at all, and thus why theoretical underdetermination is almost never a practical difficulty in research. Haugeland also usefully characterizes the sense in which the articulation of scientific concepts and related skills must be "resilient" if they are to be accountable to phenomena in the world (1998, 322, 334).

in that sense, there are no natural facts either. Facts, all facts, are true claims, from which semantic or normative considerations are ineliminable.[22] Meaningfulness is not a distinct feature of some kinds of objects (sentences, thoughts, actions, institutions), but is the situated disclosure of anything as what it is (or, mistakenly, as what it is not). There can be such disclosure only within ongoing patterns of practical interaction among agents in partially shared circumstances. But such disclosure is a configuration of the world as intelligible, not a property of those agents or their interactions. The apparent opposition between what is objectively or causally determined and what is the socially determined outcome of conversation is an artifact. What should be learned from the indeterminacy of translation is that one cannot have one of these alternatives without the other. My reconstruction of the Quinean argument shows that conversation and consensus are only contentful through people's mutual interaction with a shared environment, and thus are never "merely social"; yet that shared environment is not a domain of "objective evidence" in any sense that would escape involved, interpretive disclosure through ongoing practices.

Quine argued that naturalizing epistemology abolished any sense of "first philosophy" that would take priority over the ongoing course of inquiry. What I have shown instead is that Quine offered an alternative candidate for first philosophy: an implicit metaphysics of nature as the already determined object of what natural science *can* be about, and the already unified subject of knowledge in the physical bodies of individual human organisms. What physical bodies, the physical world, and their causal interactions are like is to be determined by ongoing scientific inquiry. That these are physical bodies, interacting causally, and that the "subject" of knowledge is the individual organism to which the rest of the physical world is "external" have nevertheless been specified in advance as the prior conditions for epistemology having been naturalized.

One initially plausible response to the limitations of Quine's naturalism, starting from a recognition of the ineliminable socially interactive dimension of knowledge, would be to substitute sociology for psychology as the core discipline of a naturalized (or "socialized") epistemology. The aspiration to incorporate a science of science within an overall scientific or empirical outlook would be retained, but that aspiration would be real-

22. Brandom (1994, chap. 5) usefully proposes that facts are true claims (in the sense of what is claimed rather than the claiming of it).

ized on an apparently more plausible and accessible level, namely, the ways in which knowledge becomes authoritative and transmissible through the interactions of knowers. The sociology of scientific knowledge thus reflects a reconstruction of Rorty's "conversational" conception of science as solidarity, in the form of an empirical research program. Indeed, many of the sociologists who have undertaken such a reconstruction have explicitly recognized the limitations of a thinly social or conversational model, and focused extensive attention upon the sociology of experimental science.

The problem that results from sociologizing knowledge remains parallel to Quine's, however: the domain of the "social" and its role in constituting "knowledge" as a distinct domain of inquiry are specified in advance by the now-sociological remnant of the aspiration to first philosophy. Indeed, the problem with both Quine's naturalism and recent sociologists' "sociologism" is less in their accounts of nature or society as such, than in the residual role those accounts play in explaining knowledge, and hence their vestigial commitment to epistemology as first philosophy (it is in this sense that I earlier suggested that "naturalized epistemology" is an oxymoron). In chapter 4, I highlight the significance of this assumption by developing a contrast between constructivist sociology of science and feminist science studies. The interesting difference between them, in the light of my reflections upon Quinean indeterminacy, concerns whether recognizing the socially interactive character of knowing requires a sociological theory of knowledge. Feminist science studies will thus point the way toward an understanding of scientific inquiry that would dispense with these last vestiges of first philosophy. In doing so, feminist science studies also point toward the intertwining of the semantic/epistemic and practical/political significance of scientific practices. In the chapters that follow, I shall explore further the resulting conception of the normativity of practices, culminating in a return to the question of how to understand the project of a philosophical naturalism.

FOUR

Feminist Challenges to the Reification of Knowledge

ALTHOUGH TODAY MANY philosophers would regard themselves as "social epistemologists," the most influential and provocative attempts to account for scientific knowledge as a fundamentally social accomplishment have emerged from the Edinburgh "Strong Programme" for the sociology of scientific knowledge and its sociological successors and competitors.[1] Whereas earlier programs in the sociology of knowledge had exempted the natural sciences and mathematics from their purview, and the dominant Mertonian approaches to the sociology of science had confined their studies to scientific institutions, motivations, and organizational norms, the advocates of the Strong Programme proposed to explain the content of scientific knowledge in the same way that they would explain any other system of beliefs and practices. To do otherwise, they often argued, would invoke a scientifically unjustifiable a priori decision to exempt the sciences from empirical sociological investigation.

This emphasis upon the empirical study of knowledge, as well as the rejection of a priori normative principles, has strong affinities to Quine's naturalized epistemology. Moreover, the philosophical significance of constructivist sociologies of scientific knowledge has often been construed by their proponents as an appropriate response to Quinean underdetermination of scientific theories, and especially to Wittgenstein's reflections upon following a rule.[2] Many sociologists have argued that since empirical evidence cannot unequivocally determine the correct choice of scientific theo-

1. I interpret the scope of the Strong Programme broadly; my discussions of the sociology of scientific knowledge will presume that its central adherents are the Edinburgh school (Barnes, Bloor, Shapin, Edge, MacKenzie, Pickering), the Bath and York groups (Collins, Pinch, Mulkay, Gilbert, Woolgar, Yearley), the early sociological ethnographers (Latour and Woolgar, Knorr-Cetina), and the social history of Shapin and Schaffer.

2. See Barnes and Bloor 1982; Bloor 1983, 1991, 1992; Collins 1992; and Lynch 1991.

ries, and past practice cannot determine how one ought to go on, only a sociological explanation can adequately account for the beliefs that scientists adopt, their assessments of experimental competence, the methods they employ, and the practices they engage in. The shared interests or form of life of a community, or the detailed social negotiations through which agreement in practice is achieved, have been proposed as the only way to explain how scientific controversies reach closure, and thus how scientific knowledge acquires a definite content.

These inferences from Quinean and Wittgensteinian themes to a constructivist sociology of knowledge depend upon two premises that have not often been articulated in the debates sparked by the Strong Programme and its sociological successors. First and foremost, social constructivism presupposes that knowledge, the content and justification of scientific beliefs or practices, is itself a coherent domain of determinable facts susceptible to and in need of explanation. Otherwise, the proffered sociological explanations would have no purchase; there would be no determinate phenomenon to be explained. Second, the sociologists' appeal to a distinctively social explanans (interests, negotiations, shared form of life) as defining what can count as the natural world presumes that the social "world" and the natural "world" constitute relatively autonomous domains whose articulated descriptions then need to be brought into an appropriate relation to one another. Collins and Yearley draw a rather stark version of this distinction between the social and the natural:

> We provide a prescription: stand on social things—be social realists—in order to explain natural things. . . . Others will be standing on natural things to explain social things. That is all there is to it. (1992b, 382)

Sociologists (and their philosophical critics) have differed among themselves as to which are the most powerfully explanatory social things and whether they could successfully explain what natural things are or what counts as knowledge of them. These very disputes, however, typically take for granted the reification of natural scientific knowledge as a determinate explanandum and the reification of some aspect of the social world as potential explanans.[3]

Recent feminist studies of science suggest a rather different understanding of what it should mean to understand the natural sciences as social

3. For an extended critical discussion of the claim that knowledge or scientific knowledge is a coherent object of theoretical reflection, see Rouse 1996b, chap. 7.

practices. Feminist science studies initially examined and criticized the ways in which biological, psychological, and social scientific studies of women and men, and of sex and gender more generally, have been androcentric. Their aim was often to develop an explicitly feminist science alongside feminist transformations of scholarship in the humanities and social sciences. Scientific and philosophical resistance to these initial criticisms of sexism in the guise of science has encouraged more general reconceptions of scientific inquiry that recognize feminist criticism as a constructive contribution to science.[4] In the resulting reconceptions of the sciences as politically engaged practices, sex and gender have become aligned with race, colonialism or imperialism, sexual orientation, and other politically significant categories. The concern to challenge sexism specifically has remained a powerful motivation for feminist science studies, but sex and gender are no longer privileged or isolable analytic categories.

Feminist scholars, like constructivist sociologists, are committed to recognizing the socially interactive and politically significant articulation of the world in natural scientific work. Nevertheless, I shall argue, the best feminist work develops an understanding of scientific knowing that accords with implications appropriately drawn from the Quinean indeterminacy of translation, and it thereby challenges the reifying presuppositions that make a social explanation of scientific knowledge intelligible.

In what follows, my concern is less to argue for or against a sociological constructivist project than to articulate what I take the deep and interesting differences to be between sociological constructivism and feminist interpretations of scientific practices. Feminist science studies have often been thought to seek an intermediate position between traditional philosophy of science and the sociology of scientific knowledge: whereas constructivist sociologists reject any normative account of the objectivity, rationality, or truth of scientific claims or methods, feminist theorists would revise norms of rationality or objectivity rather than abandon them.[5] I shall challenge

4. Among the more prominent critical assessments of gender in science are Bleier 1984, Hubbard 1990, Birke 1986, and Fausto-Sterling 1985. My discussion, however, will focus upon the broader feminist reconceptions of science found in Haraway 1989, 1991, 1992a, 1992b; Barad 1996, 1998, 1999, forthcoming; Longino 1990, 1992, 1993; Nelson 1990, 1993; Keller 1985, 1992; Wylie 1991, 1992, 2002; Harding 1986, 1991, 1992; Alcoff and Potter 1993; Potter 2001; Addelson and Potter 1991; and Addelson 1993.

5. Feminist scholars in epistemology and science studies often do reject the possibility of "sovereign" epistemic norms that stand outside of ongoing practices as the predetermined standard for the legitimacy of those practices, but they then still insist upon an accountable

this interpretation, despite its familiarity and initial plausibility. The supposedly sharp differences between normative philosophy and descriptive sociology presuppose a shared conception of knowledge, which I characterize as "epistemological." Feminist scholars, I argue, have been developing a different ontology of knowing, whose articulation displays and challenges the continuity between epistemological philosophies and sociologies of science.

Before initiating this comparison, I offer two significant caveats. First, my discussion deliberately overlooks contested issues within the feminist and sociological traditions in order to accentuate some shared conceptions that frame those internal disputes. My aim is not to minimize or suppress these disagreements, which in other contexts remain important, but only to recognize some interwoven themes connecting otherwise disparate approaches. Some feminist theorists of science may well dissent from my presentation of these themes, but my hope is that dissenters will still find it useful in clarifying such differences. Even for my purposes, some recent work in the sociological tradition, notably by Bruno Latour, Leigh Star, Andrew Pickering, Karin Knorr-Cetina, Michael Lynch, and others, challenges orthodox sociology of scientific knowledge in ways that may encourage rapprochement with feminist studies of science. Yet the significance of any such rapprochement, and perhaps its limits, should also become clearer in light of my contrast of feminist and sociological approaches to scientific knowledge.

Second, my comparison brackets the most obvious difference between feminist and sociological studies of science, namely, that gender and the sex/gender distinction play almost no role in most recent sociologists' empirical case studies and methodological reflections. The relative absence of sociological attention to gender in science should be surprising, given the importance of sex and gender as socially salient categories, and especially given both the historical predominance of men within the sciences and the widespread use of gendered imagery to characterize both science and scientific interpretations of nature. Even if the relative unimportance of sex and gender for understanding science were an outcome of empirical research in sociology, one might expect that this conclusion would be highlighted and supported by extensive argument. No such argument is appar-

responsibility of scientific work both to the world and to those affected by scientific dealings with it. For a discussion of epistemic sovereignty and its significance, see Rouse 1996a.

ent in the sociological literature, and its absence ought to be regarded prima facie as a serious problem for the sociological tradition. Nevertheless, there are good reasons to bracket it for my purposes here. Sociologists' inattention to gender might reflect only a faulty application of their theoretical frameworks; perhaps the frameworks themselves leave ample room for a full appreciation of the significance of gender relations as a social explanans for the content of scientific knowledge. Moreover, bracketing questions of sex and gender can usefully emphasize that feminist science studies are not limited to studies of gender in science. Feminist scholars have called for a much more far-reaching reconception of the sciences as socially and culturally significant practices, only one consequence of which would be to enable critical studies of gender to occupy a more prominent place in the cultures of science.

The differences I shall identify between the sociological and feminist traditions emerge from a background of three important shared themes. First, although one may still reasonably speak of individuals as knowers, both traditions understand scientific knowledge as a socially interactive achievement. Second, both traditions argue that important aspects of traditional epistemologies and philosophies of science are not merely false, but ideological. Sociologists of scientific knowledge argue that philosophical explications of the rationality or verisimilitude of scientific methods and practices misconstrue the actual development of science as revealed by empirical sociological studies, and thereby unjustifiably legitimate the cultural and political authority of the sciences. Moreover, they argue, the exceptional or asymmetrical treatment accorded to the sciences by such philosophical accounts obscures important similarities between the sciences and other less authoritative practices and belief systems. Many feminists, meanwhile, have argued that familiar conceptions of scientific knowledge as achieved or possessed by individuals, as disinterested and apolitical, or as rational and cognitive in ways that exclude the affective and embodied aspects of human experience have served to distort understanding of science, to rationalize male dominance in science and elsewhere, and to reinforce the alienation or exclusion of many women from effective participation in science.

Third, as their insistence upon the ideological character of individualist and rationalizing interpretations of scientific knowledge suggests, feminist and sociological studies of science are both politically engaged projects. Political commitment has been more readily apparent in feminism, since feminist science studies have always been projected as part of a larger political and cultural movement to criticize sexism and empower women. Yet beneath the sociologists' frequent insistence upon the need for methodological detachment and a symmetrical treatment of all belief systems has also been a broad political commitment, to a humanism clearly articulated by Collins and Yearley:

> The effect of SSK [the sociology of scientific knowledge] has been to show that the apparent independent power of the natural world is granted by human beings in social negotiation. Because the special power and authority of natural scientists comes from their privileged access to an independent realm, putting humans at the center removes the special authority. . . . Symmetry between the true and the false requires a human-centered universe. (1992a, 310–11)

Thus, both feminists and sociologists of science argue that important aspects of science are contingent and alterable, even though traditional epistemologies present them as natural and immutable, and that recognizing their contingency makes a political difference.

The Sociology of Scientific Knowledge as Epistemological

From this shared background, however, the two traditions diverge significantly. Many of their differences result from sociologists' commitment to an epistemological conception of their project, whereas many feminist science studies begin to develop a philosophy of science that rejects an epistemological conception of knowledge. But what do I mean by an epistemological understanding of science? The intelligibility of epistemology as a field of study presumes that 'knowledge' demarcates a coherent, surveyable domain of inquiry. Michael Williams has recently challenged the commonplace philosophical acceptance of that presumption:

> It is tempting to use "human knowledge" and "our knowledge of the external world" as though it were obvious that such phrases pick out reasonably definite objects of study. But it isn't obvious, or shouldn't be. We can talk of "our knowledge of the world," but do we have any reason to believe that there is a genuine totality here and not just a loose aggregate of more or less unrelated cases? (1991, 102)

The corresponding temptation in philosophy and sociology of science would be to presume that scientific knowledge is a "reasonably definite object of study," either on its own or as a result of its subsumption within knowledge in general.[6] To see why the program for a sociology of scientific knowledge remains committed to epistemology in this sense (and why most feminist science studies are not), consider the epistemological project more closely. Williams succinctly summarized the characteristic forms in which this commitment has been deployed within philosophy:

> The traditional philosophical examination of knowledge . . . points to four central ideas: an *assessment* of the *totality* of our knowledge of the world, issuing in a judgment delivered from a distinctively *detached* standpoint, and amounting to a verdict on our claim to have knowledge of an *objective* world. (1991, 22)

Sociologists of scientific knowledge have given a distinctive spin to each of these ideas, but all four remain important to their project.

The ideal of detachment has clearly played a central role in the recent sociological tradition; indeed, sociological criticisms of mainstream philosophy of science often focus upon philosophers' insufficient detachment from the norms and cultural familiarity of the natural sciences. The classic statement of sociologists' conception of an appropriately detached standpoint to study science is David Bloor's programmatic demand that the sociology of scientific knowledge be "impartial with respect to truth or falsity, rationality or irrationality, success or failure, . . . and symmetrical in its style of explanation [of] true and false beliefs" (Bloor 1991, 7). But relativism has not been the only methodological commitment resulting from sociologists' desire for a detached standpoint from which to understand the sciences. Latour and Woolgar deliberately presented their account of Latour's participation in a neuroendocrinology laboratory from the perspective of a relative stranger to the culture of science:

> We regard it as instructive to apprehend as strange those aspects of scientific activity which are readily taken for granted. . . . The uncritical acceptance of the concepts and terminology used by some scientists has had the effect of enhancing rather than reducing the mystery which surrounds the doing of science. . . . For us, the dangers of "going native" outweigh the possible advantages of ease of access and rapid establishment of rapport with participants. (Latour and Woolgar 1986, 29)

6. Rouse 1996b argues that the principal research traditions in philosophy of science and sociology of scientific knowledge are committed to epistemology in this sense and develops in much more detail what a nonepistemological philosophy of science might look like.

More recently, Collins and Yearley (1992a) have suggested an alternative image of detachment from epistemic norms and commitments. The sociologist of science must be able to alternate between competing beliefs, methods, and epistemic ideals, understanding each from the "inside" but without the insider's epistemic commitment. Moreover, the sociologist must combine this "promiscuous" epistemic alternation with a "metaalternation" that also permits the sociologist to accept naively the categories and norms of everyday life and of sociology itself. From Collins and Yearley's perspective, Latour and Woolgar's methodological estrangement is thus not detached enough. Ultimately, what is called for is a disciplinary antagonism between natural and sociological realisms, hence, Collins and Yearley's insistence above that one must choose between standing on social things in order to explain natural things and standing on natural things to explain social things.

Whichever form of sociological detachment is preferred, the result is to situate the sciences within a larger epistemic totality. Claims to scientific knowledge are regarded as examples of a more general kind, which can be surveyed and assessed as a whole. Sociologists characterize this more general object of study variously, as beliefs (Barnes, Bloor), order (Latour and Woolgar, Collins, Shapin and Schaffer), discourse (Gilbert and Mulkay), inscriptions (Latour and Woolgar), or representation (Woolgar), but underlying these differences is a shared commitment to subsuming the sciences under a more general analytical category. Of course, the new sociologists of science rightly pride themselves on the wealth of detailed case studies that exemplify their methodological and theoretical commitments. By 1982, Steven Shapin's prominent review already had listed a wide range of such sociological studies, which has since greatly expanded. Yet the supposed significance of these studies is oddly disconnected from their empirically rich particularity. MacKenzie and Barnes recognized this disconnection in reflecting upon the implications of their own study of early-twentieth-century controversies between Mendelians and biometricians:

> The general point is not that the goal-directed character of scientific judgment implies its relationship to any particular contingency, or to external factors, or political interests; what is implied is that any such contingency *may* have a bearing on judgment and that contingent sociological factors of some kind *must* have. (1979, 205)

What matters to constructivist sociologists is that all scientific practices and achievements exemplify a more general kind of entity (knowledge) that can be accounted for by appeal to distinctively social contingencies.

These general categories that sociologists of science have used to define their object of inquiry have been remarkably continuous with the "semantic ascent" that so often characterizes philosophical reflection upon science. Sociologists, too, have shifted from accounting for scientists' interaction with the world to explaining scientists' beliefs, accounts, inscriptions, or representations, or at still further remove, the content of their beliefs or representations. Undoubtedly the sociologists have multiple reasons for semantic ascent. The Strong Programme was explicitly defined in opposition to Mannheim's and Merton's sociological approaches, which had exempted the content of knowledge from sociological study. The sociologists also intended to challenge normative philosophical accounts on their own turf, and that challenge might be facilitated if the contested turf were described commensurably. Finally, a sociology of science needs a sociologically accessible domain. Collins and Yearley emphasize this concern in challenging Michel Callon's (and Bruno Latour's) attempts to recognize natural objects as "actants":

> There is only one way we know of measuring the complicity of scallops [or other natural objects that scientists work with] and that is by appropriate scientific research. If we are really to enter scallop behavior into our explanatory equations, then Callon must demonstrate his [natural] scientific credentials. (Collins and Yearley 1992a, 316)

If scientific knowledge can instead be characterized in terms of collectively accepted systems of belief, consensus, discourse, order, representation, or other forms of mediation, then sociological credentials may regain their relevance. Whatever the reason, however, the new sociology of science defined its domain of inquiry in ways that significantly abstract from and generalize over the diverse forms of scientific practice.

The sociology of scientific knowledge may nevertheless initially seem to diverge from epistemological philosophy of science by rejecting the normative perspective from which philosophers typically have assessed the rationality or verisimilitude of scientific knowledge. Sociologists have hoped to explain why scientists accept some beliefs rather than others in ways that are carefully severed from whatever justification they might have for accepting those beliefs themselves. As Barnes and Bloor put the point:

> The incidence of all beliefs without exception calls for empirical investigation and must be accounted for by finding the specific, local causes of this credibility. Whether the sociologist evaluates a belief as true or rational, or as false and irrational, he must search for the causes of its credibility. (1982, 23)

But this shift from a classically normative epistemological stance to an empirical and explanatory account is a move within epistemology rather than against it; the sociological accounts thereby make common cause with the increasingly widespread philosophical commitment to naturalized epistemology.[7] Moreover, such naturalized (or socialized) accounts of knowledge do not thereby lose all normative force. Explanations of the acceptance or credibility of beliefs may then reinforce or undermine one's own inclination to accept them. Knowing the causes of certain beliefs might well increase confidence in them, by displaying reliable causal connections to aspects of their intended objects. Adherents of the Strong Programme's commitment to explanatory symmetry typically go in the opposite direction: they suggest that the causes for the acceptance of scientific beliefs are disconnected from any reason for the analyst to believe them, and hence that "all beliefs [including scientific beliefs] are on a par with one another with respect to the causes of their credibility" (Barnes and Bloor 1982, 23). To that extent, the most prominent adherents of the sociology of scientific knowledge do engage in the epistemological project as Williams described it: they assess the totality of scientific beliefs as claims to knowledge of an objective world, and find them to have no greater (but also no less) justification than any other collectively accepted systems of belief. The new sociology has not abandoned justification; rather, it has engaged in the quite general epistemological project of showing why standard philosophical defenses of the rationality or truth of scientific claims should instead be regarded as ex post facto rationalizations. In Latour and Woolgar's words, "'reality' cannot be used to explain why a statement becomes a fact, since

7. Both naturalized epistemologists and sociologists of scientific knowledge emphasize their break from classical normative epistemology, because they seek to explain rather than to justify knowledge. I am emphasizing their underlying continuity in taking 'knowledge' to be a theoretically coherent domain that needs to be accounted for as such, whether the account is an explanation or a justification.

it is only after it has become a fact that the effect of reality is obtained" (1986, 180).

The distinctively epistemological character of the new sociology of science can be seen clearly in another way. Despite their rejection of realist philosophy of science, few of the new sociologists of science espouse a metaphysical antirealism. Barnes and Bloor are especially clear about this point, for they are metaphysical realists. They only insist that truths about the natural world, whatever those truths are, are useless to explain the diversity of conflicting beliefs about that world. Latour and Woolgar, whose views differ in so many important ways from Barnes's or Bloor's, nevertheless concur in distinguishing sociological from metaphysical antirealism: "We do not wish to say that facts do not exist or that there is no such thing as reality. . . . Our point is that 'out-there-ness' is the *consequence* of scientific work rather than its *cause*" (1986, 181–82).[8] These sociologists separate commitment to an epistemological relativism from metaphysical antirealism by arguing that the world's effects upon our beliefs or representations are always socially mediated to an extent that makes beliefs referentially opaque. Social practices, interests, or interactions, and the holistic interconnections among beliefs (and among the concepts in which they are framed and the experimental and observational practices that provide them evidential support) always intervene between knowers and the world, in much the way that strict empiricist antirealists take experience to mediate the world opaquely. The sociologists' commitment to relativism is a natural consequence of conjoining the social mediation of belief or representation with acknowledgment of diverse social practices among humans. Sociological relativism is thus epistemologically parallel to the relativism strict empiricists would presumably accept if there were compelling evidence that human beings differed significantly in their sensory modalities and capacities.

One last feature of the new sociology of science reveals its continuity with the epistemological tradition. Among David Bloor's four tenets of the Strong Programme is the call for a reflexive account of one's own claims to sociological knowledge (Bloor 1991, 7). Subsequent theorists in the sociological tradition, most notably Woolgar and Ashmore, argue that neither Bloor nor the ensuing sociological tradition have taken reflexivity seri-

8. Note Latour and Woolgar's use of quotation marks, indicating that the supposed consequence of scientific work is not reality itself, but the *effect* of reality, what could *count* as reality for socially situated agents.

ously enough, while others (Collins and Yearley 1992a,b) respond that there must be limits upon reflexive criticism if the sociological project is to proceed at all. Yet these disagreements conceal an unexamined presumption that the important question to ask reflexively is epistemological. Bloor makes very clear that internal consistency and completeness are at stake in reflexivity: the sociology of scientific knowledge should not be self-refuting, nor should it accept arbitrary limits upon its scope. Woolgar (1988) and Ashmore (1989), by contrast, sometimes characterize their reflexive concern as rhetorical. Yet the central rhetorical problem they pose is that the authoritative voice of sociological narrators and the apparent transparency of their referential prose are in pragmatic contradiction to the explicit content of their "findings" about the social construction of all accounts and representations. Woolgar and Ashmore are far too sophisticated to ask that sociological rhetoric accurately represent the findings it expresses, but they do ask that sociologists adopt literary forms that disrupt and interrogate readers' too easy acceptance of their claims as transparent representations of the social construction of scientific knowledge. The "other voices" that Woolgar hopes will be included in sociologists' texts are there not because of their own need to be heard, but to satisfy the sociologists' need not to be believed too readily. Collins and Yearley object to Woolgar's project primarily because they do, after all, want their own accounts to be believed straightforwardly.

Feminist Reconceptions of Knowing

Feminist insistence that scientific knowledge is socially constructed and politically engaged appears in a new light when juxtaposed to my subsumption of the sociology of scientific knowledge within the epistemological tradition. Feminist science studies scholars differ from the new sociologists most evidently in their opposition to relativism, their normative stance toward particular scientific claims, and their willingness to retain and employ suitably revised conceptions of evidence, objectivity, and a distinction between belief and knowledge. Yet in many cases, these familiar differences are a consequence of feminist scholars' working toward "postepistemological" conceptions of knowledge, evidence, justification, and objectivity, and thereby opposing a framework shared by traditional philosophies of science and the new sociologies of scientific knowledge.

I outline five ways in which this feminist scholarship transcends epistemology. First, feminist science studies shift their primary object of study from the semantic content of knowledge or belief to a concern with relationships, or "intra-actions," between knowers and known.[9] Second, feminist studies take up a participatory stance toward scientific practices and scientific knowledge, rather than trying to explain or assess scientific knowledge as a totality. Third, the primary concern of feminist science studies is less with the present state of knowledge than its future possibilities. Fourth, many feminist reconstructions of the concept of objectivity dissolve any sharp conceptual distinction between epistemic and political criticism, a transformation that also prohibits reducing one category to the other. Finally, some feminists develop a more adequate conception of reflexivity and its epistemic, rhetorical, and political significance for science studies.

My first point concerns the ways feminist science studies have typically worked with a conception of knowledge that is less austere and abstract than the various forms of semantic ascent that characterize much recent philosophy and sociology of science. This is not to say that feminist science studies are an epistemological analogue to an ethics of care,[10] but rather that feminist scholars conceive of 'knowing' as concretely situated and as more interactive than representational. Knowledge is not merely a propositional attitude (belief or acceptance) toward some ideal or abstracted propositional content, but a relationship between knower and known, a situation that guides what knowers do and how the known responds and can be understood. Evelyn Fox Keller points clearly in this direction:

> Although scientific theories cannot be understood as faithful reflections of either culture or nature, perhaps they can be understood as good enough reflections of the forms of interaction that speaking and desiring social actors seek to implement with that mute but nonetheless responsive world of actors we call nature. (1992, 95)

Yet even Keller's formulation preserves an unnecessary vestige of the representational idiom; it would be consistent with her overall argument to omit

9. Karen Barad (1996) introduces the term 'intra-action,' which avoids the connotation implicit in 'interaction' that the relevant actants are definitely bounded systems independent of their interaction, for reasons that will be discussed more extensively in chapter 8.

10. Some feminist studies of science or other ways of knowing may indeed fit under this heading, such as Belenky et al. 1986, or the conception of knowledge that Evelyn Fox Keller (1983) attributes to Nobel laureate Barbara McClintock. My point, however, is perhaps best exemplified by theorists like Haraway and Barad, for whom feminist reconceptions of how we might engage the world epistemically require suspicion of metaphors of organicism, holism, caring, and so forth, yet do strongly suggest that new forms of knowledge are as much changes in practices as changes in beliefs or other representations.

altogether any talk of "reflections," and to say that theorizing is indispensable to the "forms of interaction" that she takes as central to science.

Undoubtedly this emphasis upon knowledge as a concrete relationship to its intended object gains its centrality and its urgency from initial feminist concern with the human sciences. Feminist theorists saw early on that knowledge claims about women, or about human beings in general, that excluded or diminished the humanity or rationality of women are not simply instruments that can be used to oppress women or justify that oppression, but are themselves integral to patterns of domination. These concerns with scientific knowledge as itself a form of action were reinforced by parallel reflections upon the politics of the human sciences within postcolonialist anthropology, where Western ethnographers' interactions with and writings about the people they study are increasingly seen to embody power and not merely to serve it. Observing, writing, and reading do not merely propose or accept the content of certain beliefs, but are themselves actions with consequences (one must consider to whom one writes, in what language, available to whom, drawing upon what patterns of interaction, using what narrative conventions and authorial stances, and who is permitted or enabled to respond, with what effects).[11] Yet feminist science theorists apply these lessons more generally. As Donna Haraway has noted:

> The agency of people studied itself transforms the entire project of producing social theory, . . . but the same point must apply to the other knowledge projects called sciences. A corollary [for] the sciences as a heterogeneous whole, and not just in the social sciences, is granting the status of agent/actor to the 'objects' of the world. Actors come in many wonderful forms. Accounts of a 'real' world do not, then, depend on a logic of 'discovery,' but on a power-charged social relation of conversation. (1991, 198)

Feminist science studies have as a consequence often been explicitly concerned with different ways in which knowers might interact with the objects of knowledge. Criticism of the sexual politics embedded in some epistemic models and practices is one manifestation of this concern. Feminist scholars have directed attention, for example, to Bacon's vision of scientific mastery over a feminized nature, to Boyle's articulation of a distinctively masculine modesty as an epistemically constitutive virtue, to

11. For this perspective, see work by Clifford and Marcus (1986), Marcus and Fischer (1986), and Rosaldo (1989), among others.

molecular biologists' invocation of their project as "a calculated assault on the secret of life," and to sadism within the scientific vision, narrative structure, and experimental practices of Harry Harlow's Wisconsin Primate Laboratory.[12] Lynn Hankinson Nelson emphasizes that these issues

> are not about language or mysterious metaphysical agendas. The commitments to linear and hierarchical relationships, to executives and controllers, and to laws that phenomena obey are incorporated in our theories, methodologies, and models, and a commitment to "dominating nature" is incorporated in many of our scientific practices. (1990, 213–14)

Feminists' concern for different ways of knowing have also led to the articulation of possibly more constructive relationships between knowers and knowns. Most familiar, perhaps, are Evelyn Fox Keller's biographical evocation of Barbara McClintock's "feeling for the organism," and many feminist scholars' defense of holistic and interactive explanatory models in various scientific fields.[13] But these projects do not exhaust feminist reconstructions of knowing. Haraway has described her own erotic response to some "rigorously analytical and biotechnical" procedures in cell biology as a "knowing love [that] took shape in quite particular, historical-social intercourse, or 'conversation,' among machines, people, other organisms and parts of organisms" (1992a, 71, 72). She has also argued for a feminist reconstrual of vision as an epistemic model:

> The 'eyes' made available in modern technological sciences shatter any idea of passive vision; these prosthetic devices show us that all eyes, including our own organic ones, are active perceptual systems, building in translations and specific *ways* of seeing, that is, ways of life. (1991, 190)

Feminist defenses of holistic models, then, are best regarded not as constitutive of feminist reconstructions of knowing, but as one specific group of contestable moves within a more widely shared concern to understand knowledge as embedded within specific ways of engaging the world. Thus, Haraway crucially captures this feature of feminist science studies as a criti-

12. On Bacon, Keller 1985, chap. 2; on Boyle, Potter 2001; on the "secret of life," Keller 1992, part 2; on Harlow, Haraway 1989, chap. 9.

13. Keller (1983), Longino (1990), Nelson (1990), and Hubbard (1990), among others, defend specific holistic, interactive, and/or dialectical models as epistemically and politically preferable to more simplified causal accounts of the same domains.

cal project: "The point is not new representations, but new *practices,* other forms of life rejoining humans and not-humans" (1992a, 87).

Sociologists of scientific knowledge might reasonably object that I have overstated the contrast between the feminist and sociological traditions. After all, most recent constructivist sociologists of science emphasize that scientific knowledge is situated within forms of life, while Shapin and Schaffer's influential sociological reconstruction of seventeenth-century conflicts over experiment perhaps goes further in claiming that "solutions to the problem of knowledge *are* solutions to the problem of social order" (1985, 332; my emphasis). Differences reemerge, however, upon considering the place of feminist and sociological accounts of scientific knowledge with respect to the forms of life within which knowing is to be situated. The crucial difference, marking the second of my five main points, is located in the sociologists' aspiration both to explain the content of scientific knowledge and to achieve a detached standpoint from which to proffer their explanations. Feminist science studies have generally eschewed detachment and explanatory projects, in favor of a participatory stance. Feminists claim a place within and seek to have an effect upon the cultures of science that they study.Their project in part is to legitimate the changes they hope to effect.

Feminists' opposition to the sociologists' explanatory detachment has multiple motivations. Latour and Woolgar's hesitation to "go native" among the scientists, for fear of being taken in, ironically contrasts to feminists' recognition that for women, being taken into science has too often not been an option. Adopting a stance of estrangement or antagonism toward the culture of science would risk endorsing an exclusion of feminist concerns from that culture in ways that would dangerously reenact its history of excluding women. A thoroughgoing constructivist detachment from science can threaten to be just "one more excuse for not learning any post-Newtonian physics and one more reason to drop the old feminist self-help practices of repairing our own cars. They're just texts anyway, so let the boys have them back." (Haraway 1991, 186). Such detachment can also serve as one more rationalization for those who take feminist criticism as a sign that women do not belong in science.

Perhaps more fundamentally, however, feminists often regard sociological aspirations to a detached and totalizing explanation of the content of scientific knowledge to be objectionably androcentric. Feminist theorists have been suspicious of attempts to escape (metaphorically, methodologically, or theoretically) from the concrete particularity of bodies and social relationships.[14] Sociologists of scientific knowledge may initially seem to share feminists' concern, and to respond by insisting that knowledge claims are always situated within particular forms of social life. Yet their explanatory aspirations require that these particular forms be surveyable as a totality: as determined by interests, as the outcome of negotiations, as the manipulation of inscriptions, and so forth. As Bruno Latour has suggested, such explanations are attempts to act at a distance, and thereby to exercise power (Latour 1987, chap. 6; 1988). The "strategic position" from which explanatory abstractions can account for the totality they survey (in this case, for scientific knowledge in its diverse manifestations) is precisely what Nancy Hartsock (1984, 240–47) once critically characterized as the standpoint of abstract masculinity, and what other feminists have since characterized as suppression of difference. Feminists object not simply to the desire for "explanatory power" in its dual political and epistemic senses, but to the androcentrism of this aspiration.

Feminists have also objected to the specific forms that sociological detachment has taken, for they cannot afford to follow Bloor in his "impartiality to truth and falsity." Whereas the political commitment of the sociology of scientific knowledge has been to challenge the cultural hegemony of natural scientific knowledge in general, feminist science studies were initially concerned with the adequacy of particular scientific projects and proposals damaging to women. Feminists have been concerned to substitute scientific approaches and accounts that are not harmful, but it matters to them that those proposals provide not merely more congenial beliefs, but more adequate understanding. Feminists certainly hope to change important aspects of the way the world is, but they want the political struggles on behalf of their utopian aspirations to be responsive to their actual situa-

14. Feminists' concern not to overlook the concrete particularity of bodies and relations does not preclude all generalization, or any feminist theory. The objection is rather to a specific kind of theoretical detachment that seeks to overlook particularity by appeal to free-floating, ahistorical categories. As Fraser and Nicholson concluded, "[Feminist] theory would be explicitly historical, attuned to the cultural specificity of different societies and periods and to that of different groups within societies and periods. [Theoretical] categories would be inflected by temporality, . . . non-universalistic, . . . pragmatic and fallibilistic. It would tailor its methods and categories to the specific task at hand, using multiple categories when appropriate and forswearing the metaphysical comfort of a single feminist method or feminist epistemology" (1990, 34–35).

tion, and that calls for more reliable theorizing. A feminist understanding of scientific knowing thus requires not detachment or neutrality,[15] but a reflective and self-critical participation in the assessment of particular scientific projects and knowledge claims.

The ambitions of some recent sociologists of scientific knowledge are problematic for feminist science studies for another reason. Sociologists have often aspired to explain scientific knowledge in a very stringent sense that would require the sociological categories describing the explanans not to be dependent upon or interdependent with the categories whose application is to be explained (bear in mind Collins and Yearley's call, cited above, for sociologists to "stand upon social things to explain natural things"). Yet feminist science studies have neither sought nor provided such independence from the terms of their interpretations. Elizabeth Potter's study of Robert Boyle (2001), for example, does not try to explain the content of Boyle's scientific project by reference to a given conception of gender, but instead interprets his work as itself a linked reformulation of received conceptions of both gender and natural philosophy: Boyle was simultaneously "making gender [and] making science." Haraway has insisted upon the importance of this point more generally, in articulating her "nervousness about the . . . appropriationist logic of domination built into the nature/culture binarism and its generative lineage, including the sex/gender distinction" (1991, 198), while Judith Butler has trenchantly argued that feminists need to problematize the discursively conditioned experience through which sex and gender emerge as possible categories:

> Whether gender or sex is fixed or free is a function of a discourse which seeks to set certain limits to analysis. . . . The locus of intractability, whether in "sex" or "gender" or in the very meaning of "construction," provides a clue to what cultural possibilities can or cannot be mobilized through any further analysis. (1990, 9)

I shall return to Butler's concern to understand the effects and limits of feminist analysis itself, when we consider sociological and feminist conceptions of reflexivity.

The differences between feminist and sociological approaches to science are also manifest in their temporal orientation, my third main point. The

15. See Proctor 1991 and Rouse 1991 on criticisms of recent social constructivist views of science for their detachment and their aspiration to value neutrality, and Harding 1992 and Haraway 1991, chap. 9, on feminist alternatives to neutrality and detachment.

sociologists' explanatory project would account for the present state of scientific knowledge as an outcome of a social history and the current social situation. An important aim of such a project is to display the contingency of scientific beliefs and practices to counteract the appearance of their natural or rational necessity. The sociology of scientific knowledge does not by itself point toward specific changes in scientific belief or practice, but instead opens a space of contingency within which human agency can be exercised more freely: "scientific choice is in principle irreducible and open, [even though] historically, options are foreclosed according to the opportunities perceived for future practice" (Pickering 1984, 405). As I noted earlier, the underlying cultural politics of the classical sociology of scientific knowledge is thus humanistic, allowing broader scope to human freedom in constructing views of the world and practices within it.

The temporal orientation of feminist science studies is more specifically futural and transformative. The aim of feminist science studies is not to expose scientific knowledge as in general contingent and alterable if "we" choose, but rather to show it as in need of alteration in specific respects, and as potentially responsive to that need. Feminist science studies are thus specifically oriented toward a "successor science" in a way that the sociology of scientific knowledge has not been.[16] This concern has a specifically utopian dimension that would encourage new ways of organizing specific scientific fields and the cultural politics in which they are situated.[17]

This difference in orientation emerges especially clearly in contrasting uses of a superficially similar rhetorical and argumentative strategy. Both feminists and sociological constructivists frequently display alternatives to well-established historical or contemporary scientific programs, and argue that there was or is no compelling epistemological justification for a choice many scientists actually made. Yet when Pickering, for example, argues for the contingency and underdetermination of high-energy physicists' commitment to the Weinberg-Salam model of electroweak interactions in

16. I use the phrase "successor science" here more expansively than some feminist theorists do, to denote any transformation of scientific practice or belief motivated by feminist criticism. In chapter 9, I introduce a modal concept, 'real possibility,' that would aptly express what I here describe as feminist orientation toward a successor science.

17. To some extent, this utopian dimension is present in any feminist imagination of alternative ways of organizing scientific culture and practice that would escape or confront the androcentrism or sexism revealed by feminist criticism. It is perhaps most strikingly manifest, however, in Haraway's explicit use of SF literature (1989, chap. 16; 1991, chap. 10; 1992b) as a way of posing the utopian imaginative task set by feminist science studies.

the face of apparently contrary experimental results (1984, esp. chap. 10), the point is emphatically not to advocate the resurrection of the contingently defeated points of view.

Feminists use comparable juxtapositions to quite different ends: when Longino (1990, chap. 7) compares the evidence for and against linear hormonal and neural selectionist explanations of higher brain functions, when Keller (1985, chap. 8) argues for the underdetermination by evidence of biologists' preference for pacemaker cell models of aggregation in cellular slime molds, or when Haraway (1989, chaps. 8, 14, 15) recontextualizes debates over the human evolutionary models of "man-the-hunter" and "woman-the-gatherer," their aims are specifically to encourage reconsideration of the merits of the less dominant view, and to show how "one story is not as good as another" (Haraway 1989, 331). Feminist science studies have been specifically concerned to criticize androcentrism and sexism in the development and acceptance of scientific work,[18] to envision less constraining and differently revealing scientific practices, and to enhance recognition of the contribution feminist inquiry can make to science and the cultures of science.

Feminists resurrect ideals and norms of objectivity in the context of this concern to reconstruct science, and to secure a place for feminist inquiry in that ongoing reconstruction. This continuing endorsement of objectivity marks the fourth main point of feminist transcendence of epistemology. Yet this endorsement is more often read as an attempt to find an intermediate or compromise position between a strongly internalist claim that scientific knowledge and practice are (or ought to be) fully determined by reason and evidence alone and a thoroughgoing sociological constructivism whose explanations of knowledge as the outcome of interests, ideologies, or the contingencies of social negotiation would leave no rationalist residue.[19] I think feminist science studies are better understood as attempting to rescue a conception of objectivity from the clutches of *both* epistemology and sociology of knowledge. Epistemological accounts of objectivity or rationality have traditionally been surrogates for realism: if objects of knowledge cannot directly regulate practices of inquiry from "outside," perhaps

18. As noted above, scholarship has been identified as specifically "feminist" based on its critical focus upon sexism and androcentrism, but feminist scholars have been as much or more concerned to examine racism, heterosexism, colonialism, and other forms of domination and exclusion *within* feminist politics as well as elsewhere.

19. McMullin (1992, 22), exemplifies this familiar reading of feminist science studies as "less epistemically radical" than the sociology of scientific knowledge.

the concern for objective representation can regulate them from within. The new sociology of science would deny even this last vestige of transcendence, but that denial is only significant if one accepts the epistemological framing of the question (Rouse 1996b, introduction, chap. 7). Feminist science studies would reclaim objectivity not by finding a new route to transcendence of our all too human epistemic limitations, but by carefully distinguishing the desire for objectivity from desires for transcendence, and by questioning the familiar epistemological metaphor of the world as "external" to what is "inside" knowers' thoughts or culture.

A plausible reading of these feminist reconstructions of objectivity might initially characterize them as sociopolitical rather than epistemic. The crucial virtue that objectivity would better serve would then be justice rather than truth.[20] The exclusion or marginalization of groups of knowers, whose lives, concerns, or needs are thereby prevented from contributing to the critical assessment of knowledge claims, is after all a characteristic failure of objectivity cited by many feminist theorists, and the objectionable consequence seems to be domination mediated by misrepresentation rather than vice versa. Yet thus contrasting epistemic to sociopolitical virtues and ends risks seriously misreading and understating many feminists' interest in better understanding. For most feminist theorists of science, knowing is neither external to nor merely instrumental for justice, but is itself a valued end for which justice is integral.

Understanding this point requires taking very seriously my earlier claim that feminist theorists are construing knowledge as multidimensional relationships between knowers and knowns, rather than as a simple relation of representation and correspondence—or of intertextuality, for those who like Woolgar (1988) would deny any transcendence of representation. Moreover, such relationships are intertwined, so that scientific knowing always involves complex relations among knowers as well as to the proximate object of knowledge.[21] The question is how all parties involved can

20. In recent political philosophy, 'justice' is often used to refer only to fair distribution of rights, goods, and obligations; feminists sometimes criticize not merely particular conceptions of justice, but the focus upon justice as the most central political virtue (Baier 1987). Here, I use 'justice' in a more expansive sense, as a placemarker for whatever criteria or concerns should be used in the moral and political assessment of actions, and the institutions and relationships that facilitate or sustain them.

21. I have underscored this point elsewhere (Rouse 1987, chap. 7) in discussing the kinds of disciplines that must be imposed to enable laboratory knowledge, and the extension of those disciplines that must accompany the extension of knowledge outside of the local setting of the laboratory.

be accorded what is due them. Representation is a power-charged relation that involves not merely speaking for other people and things, but also the ability to shape their circumstances. Feminist reconstructions of objectivity are attempts to hold knowers accountable for what they do, and to determine to whom and to what they need to be held accountable. These attempts still recognize that inquiry and representation are inevitably partial, perspectival, and interested. The demand for justice cannot be an impossible demand for completeness or equal significance, but must instead call for recognition of partiality, openness to criticism and to alternative practices of inquiry and the concerns that motivate them, responsibility for one's actions and position as inquirer and authoritative knower, and accountability for their effects.

The concern to make knowledge more accountable is manifest at multiple levels. Feminist science studies are centrally interested in questions of evidence not as a vestige of a theory of confirmation or rational belief, but as an aspect of understanding accountability. Feminist theorists such as Longino (1990, 1992) and Wylie (2002) contextualize questions of evidence, and focus upon how phenomena come to count as evidence and what they can be evidence for, what other assumptions, concerns, and practices play a role in constituting evidential relations, and how one can come to more critical assessment of evidential relations in their many dimensions. Similarly, Haraway's concern to reconceive vision should be understood as a more richly articulated account of the partiality, activity, malleability, and relationality of seeing, and with it a recognition of more complex and far-reaching possibilities for criticism and transformation of what technologies of vision can make evident. In these feminist recontextualizations of seeing and making evident, "the goal of an epistemology and politics of engaged, accountable positioning remains eminently potent; the goal is better accounts of the world, that is, 'science'" (Haraway 1991, 196). This goal cannot be achieved by subordinating or reducing the epistemic to the political, but instead requires recognition of how familiar conceptions of both "domains" are transformed when their boundaries dissolve. In that light, Barad advocates a *"realism [that] is not about representation of an independent reality, but about the real consequences, interventions, creative possibilities, and responsibilities of intra-acting within the world"* (1996, 188).

Feminists' moral-epistemology/epistemic-politics also raises questions about how knowledge claims and practices of inquiry become significant and authoritative. For example, not all or even most recognized truths about the world count as "scientific" truths. Science as an ongoing practice of inquiry discounts truths that are trivial, marginal, anomalous, arcane, or otherwise "uninteresting," in order to focus resources and attention upon others that are taken to be significantly revealing.[22] Feminist objectivity would incorporate self-critical assessment of judgments about what is interesting or important as well as what is well confirmed. Not surprisingly, for example, feminists see the coupling of an obsessive interest in researching possible sex-linked differences in cognition with an abiding lack of interest in research on diseases afflicting women as both an epistemic and political failing. But the critical assessment of scientific significance extends much further than just the exposure of blatant sexism. The feminist quest to hold judgments of scientific significance accountable extends to critical narrative reconstructions of the cultural significance of what is at issue in whole fields of inquiry, as in Keller's account (1992) of the reorientation of biology around a molecular biological quest for "the secret of life," or Haraway's examination (1989) of how Western primatology came to be focused upon questions of origins.

These questions about how knowledge becomes significant are closely related to feminists' critical examination of how knowers are positioned, a further reminder of why knowing is a multidimensional network of relationships from which epistemic and political judgments cannot be readily disentangled. Feminist assessment of knowing and understanding is directed not merely at what is said on what grounds, but also at who gets to speak, who is heard as authoritative, whose concerns and possible responses must be taken into account in constructing knowledge claims, who has access to the material and social resources needed for research, what sustains or compromises these various forms of credibility, and how the resulting authorization of knowers and knowledge constrains or enables people's lives. The normative aspirations of feminist science studies are addressed not only to the content of knowledge and justification, but also and inseparably to questions of who knows, with what effects. Yet these aspirations are intertwined without being subordinated to one another. The aim is better understanding and a better world, together.

This broadening of the normative questions at stake in feminist science studies points to the fifth and final contrast I draw between feminist science

22. For further discussion, see Elgin 1991 and Rouse 1996b, chap. 6.

studies and the sociology of scientific knowledge. I maintained earlier that sociological discussions of reflexivity have been guided by epistemological concerns about consistency and completeness. Even Woolgar's and Ashmore's questioning of sociological rhetoric turned out to be focused upon a pragmatic inconsistency between what sociologists of science have said and how they have said it. Reflexivity has also incorporated a political dimension for feminist science studies. This contrast does not exempt the rhetoric of science studies from reflexive criticism. Sharon Traweek, for example, criticizes the same rhetorical strategies that concern Woolgar, but to different ends. Traweek notes that, like scientists,

> almost all those writing the newer social studies of science and technology also account for everything and reject all other stories. Almost all these stories, whether about nature, scientists, or science, are narrative leviathans, producing and reproducing all-encompassing stories of cause and effect through the same rhetorical strategies. (1992, 430)

In criticizing such rhetoric, Traweek's concern is not to "interrogate" and defamiliarize representational practices generally. From her feminist perspective, Woolgar's stories constitute yet another narrative leviathan, about how all representations (including his own) are projections of "the Self."[23] Moreover, this conception of "the Self" as encountering only its own constructions is widely recognized in feminist theory as characteristically masculine. For Traweek and other feminist theorists, by contrast, reflexivity discloses partiality and situatedness, not self-enclosure. It exposes the illusion that representation is autonomous and self-projecting; feminists respond that we can never encounter or understand ourselves (and especially not "the Self") except through our interactions with others in partially shared surroundings. If rhetoric is always situated, then reflexive concern for one's own authorship cannot remain internal to texts. The textual self-presentation of the author is subject to reflexive criticism only as part of a larger concern for writing and speaking as forms of action. What do these writings and sayings do? To whom and about whom are they expressed? In what ways do they allow for and acknowledge, or foreclose and not hear, the responses of those they speak to, about, or past? Above all, to whom are they accountable? Critical reflection upon knowledge claims is thus always both moral/political and epistemic, and feminist

23. This is Woolgar's own capitalization (1988, 109).

reflexivity would reconfigure the politics of science and science studies (including the consequences of the knowledges they produce) in reconstructing their rhetoric. Reflexive attention to one's own practices of speaking and writing would encourage a science, and a political engagement with science, that would be appropriately modest and self-critical.[24] Such a reflexive science would be attentive to the effects of its own investigations, including the foreclosing of some questions and concerns by its own theoretical categories and experimental practices.[25]

Beyond Epistemology

I have been arguing that the contrast between feminist and sociological science studies points toward a postepistemological conception of science and scientific knowledge. Williams's characterization of the epistemological project offers a useful way to summarize the contrast. Feminist science studies as I understand them would abandon the epistemological aspiration to a detached assessment of the totality of scientific knowledge and its relation to an objective world understood as something apart. The alternative is engaged and self-critical participation in the making and remaking of scientific understanding of the world we live in. Such participation requires an abiding interest in questions of justification, consistency, clarity, and so forth. These questions now arise more locally, however, while drawing upon more wide-ranging considerations. Thus, what is at issue in feminist accounts is not scientific knowledge as a totality, but particular scientific practices, projects, and claims, understood as ongoing interactions between knowers and the world known. What is at stake in feminist participation in science and science studies is at once better knowledge and a better world.

How should we then understand the relationship between scientific practices and the activities of feminist scientists, philosophers, historians, or social scientists who interpret those practices? Feminist science studies offer a distinctive answer to this question, one that assimilates the epis-

24. Haraway's discussion (1989, chap. 10) of the scientific career of primatologist Alison Jolly provides an illuminating example of how such reflexive modesty might be realized in one very particular setting. Jolly's career and its scientific and cultural setting are unusual in ways that would strongly discourage taking her work as a model for politically engaged scientific practice, but it nevertheless illustrates Haraway's conception of a reflexive rhetoric and politics. Haraway 1997 articulates more extensively the theme of knowing as a form of "modest witness."

25. Butler 1995 focuses on this concern for the contingent limits of one's own theoretical foundations, but for a natural science such concern must certainly be extended to the material practices at laboratories and field sites.

temopolitical norms of science and science studies. Feminism, after all, is not a distinct discipline, nor an epistemological or metaphysical doctrine, but must instead be a practical orientation that can be taken up in any meaningful and fulfilling activity. If there were a significant and worthwhile human practice whose appropriate pursuit was at odds with a feminist stance, that would constitute a significant limitation to the aspirations of feminism. It is no accident that feminist science studies began with feminist scientists' critical reflection upon their own disciplines, and only gradually evolved into more general interpretations of the sciences that would accommodate and even vindicate the aspiration to a feminist science. Hence, feminist interpretations of science and feminist interpretations of science studies have emerged in intimate conversation.

From this conversation a new model for understanding the sciences as practical engagement with the world has begun to take shape. Scientists are involved with their objects of study rather than aspiring to detachment; scientific understanding is situated and intra-active rather than totalizing and explanatory; the telos of scientific work is transformative and futural rather than representational and retrospectively reconstructionist. The commitment to objectivity requires justice to both the objects of inquiry and the diverse agents who participate in or otherwise encounter science; and this commitment can only be fulfilled in a reflexively critical reflection upon one's entire practice, not merely the results abstractable from it. This account is not a utopian aspiration to a different kind of science (although it may be informed by a utopian imagination that seeks to expand the apparent limits to what is possible), but a feminist reading of the normative commitments implicit in scientific practice and scientific knowing.

Such a postepistemological conception of scientific knowing takes scientific *practices* as the primary focus of interpretation, and recognizes that the sciences are ongoing ways of dealing with the world practically and discursively. In an earlier book (Rouse 1996b), I offered an initial account of how to understand scientific practices philosophically. In the chapters to follow, I shall show how that conception of scientific practices connects to the reconception of knowing that has emerged from feminist science studies, explore the significance of understanding the sciences both as discursive practices and as bodily intra-actions with material surroundings, and consider how such a conception reconfigures conceptions of nature and philosophical naturalism.

FIVE

Two Concepts of Scientific Practices

MY DISCUSSION OF feminist science studies in the previous chapter highlighted the need to direct philosophical discussion of science toward scientific practices rather than scientific knowledge. This shift in focus requires some important clarification, however. Practice talk has been rampant within late-twentieth-century philosophy, social theory, and science studies, but there has been too little attention to divergent uses of the term. A shift in philosophical focus from scientific knowledge to scientific practices will lead in a better direction only if practices are understood in the right way.

My principal aim in this chapter is to introduce and clarify a distinction between two fundamentally different conceptions of practices. Practices are often identified with regularities or commonalities in the performances or presuppositions of some community of human agents. Yet one can also characterize a practice in terms of the normative accountability of various performances. Understood in the latter sense, a practice need not involve any underlying regularity or commonality.[1] Indeed, I shall argue, the normative accountability that binds a practice together need not even presuppose an uncontested formulation of norms to which all phenomena belonging to the practice are accountable. There can be something at issue and at stake in these phenomena that accounts for the integrity of the practice as a whole, even though the appropriate formulation of those issues and their stakes is itself part of what is at issue in the practice.

1. The central insight of Davidson 1986 was that natural languages are practices that do not involve regularities or commonalities among their participants. Davidson concluded that understanding and using a language involved "no learnable common core of consistent behavior, no shared grammar or rules, [and] no portable interpreting machine set to grind out the meaning of an arbitrary utterance" (445). The result is that "there is no such thing as a language, not if a language . . . is supposed [to operate on the basis of shared conventions, rules, or regularities]" (446).

This normative conception of practices will turn out to have very important implications for a philosophy of science. In particular, it changes how one should think about the semantic, epistemic, and political dimensions of scientific practices, and about the ways they are interconnected. It also changes how one should think about the relation between scientific practices and the philosophy of science, in ways that matter to my overall discussion of philosophical naturalism. Philosophical reflection upon practices conceived in this way must critically engage with the practices themselves, in ways that are also accountable to what is at issue and at stake in those practices.[2] To this extent, an adequate philosophy of science must be conceived as belonging within the field of scientific practice.

Before addressing this central ambiguity in the more general use of the term 'practice,' however, I note several possible misunderstandings that should be avoided when thinking about the extension of the concept to encompass scientific practices specifically. One mistake crucial to avoid is reading the concept of scientific practices in terms of a distinction between theory and practice. This mistake is especially tempting because a prominent understanding of 'practices' in science studies emphasizes the autonomy of experimental and instrumental practices from theoretical determination.[3] The contrasting claim to the widespread insistence upon the theory-ladenness of observation and methodology that emerged from the postempiricism of the 1960s and 1970s has been that "experiment has a life of its own," and similarly that the development and use of scientific instruments cannot be fully accounted for in terms of their supposed theoretical presuppositions.[4] The appropriate target of these criticisms, however, was any conception of theory as systematic representation. Thus, an

2. My use of 'engage' here emphasizes that any adequate philosophical reflection upon the sciences must itself be *responsive* to scientific practices and practitioners, and *accountable* to what is at stake in scientific work. Just as genuinely engaging in conversation requires the participants to listen and respond to one another rather than simply imposing a single will, so a reflective engagement with the sciences must take seriously what is at issue and at stake in involved participation in scientific research.

3. Among the most influential contributors to this experimentalist philosophy and historiography of scientific practices are Hacking 1983; Galison 1987, 1997; Rheinberger 1997; Knorr-Cetina 1981; Latour 1983, 1987; Shapin and Schaffer 1985; Pickering 1995; Ackermann 1985; Franklin 1986; Gooding, Pinch, and Schaffer 1989; Gooding 1990; Kohler 1994; Buchwald 1995; Radder 1988, 1996.

4. Hacking 1983; Rheinberger 1997; Galison 1997; Shapin and Schaffer 1985; Van Helden and Hankins 1994.

important complement to the rediscovery of experimental practices has been to recognize that theorizing is also misconstrued as the development of static representational structures. Scientific theory is better understood in terms of theoretical practices of modeling particular situations or domains; articulating, extending, and reconciling those models and their constituent concepts and techniques; and connecting theoretical models to experimental systems, rather than in the classical sense of *theoria* or through more recent analyses of theories as axiomatic or model-theoretic systems.[5] I shall return to these themes in chapter 8 in considering how to reconceive the interrelations of theory and experimentation if sciences are understood as practices rather than as bodies of knowledge.

A second mistake would be to understand the turn to practices as an account of the individual or collective doings of scientific researchers, as distinct from the material setting of what they do. Such an interpretation would take 'practices' as an alternative account of scientific subjectivity. Knowing subjects, instead of being understood as disembodied minds representing the world, would be taken as embodied agents acting upon or within it. This idea is attractive, and up to a point, not mistaken. But it founders as soon as one asks whether the doings of such agents are intelligible and coherent when considered apart from their material surroundings. When understood as mental representation, subjectivity might plausibly have a kind of intelligibility apart from the world it represents; there is a reasonably well-articulated sense of what it would be to (purport to) represent something, in the absence of any object or situation corresponding to the representation. Similarly, it might initially seem plausible to speak of social practices (socially instituted practical norms) as swinging relatively free from their material instantiation.[6] This plausibility would be reinforced by thinking of practices as regularities or commonalities of perfor-

5. Useful treatments of theorizing as practices include Cartwright 1983, 1999; Hacking 1983, chap. 12; Morgan and Morrison 1999; Rouse 1987, 30–40, 80–95, 111–19; Rouse 1996b, 127–33, 166–78, 224–33; Galison 1997, 1998a; Keller 2000; Warwick 1992, 1995; Hughes 1998; and Krieger 1992.

6. Thus, Brandom 1994 and Haugeland 1998 take greeting practices as paradigm cases of socially instituted practices whose correctness or incorrectness depends solely upon what the relevant community accepts as a greeting correctly performed. In the end, I think that the difference between such "socially instituted" practices, and "constitutive" practices that are normatively accountable to the world (as practically configured), is at most a difference of degree in the *directness* of their normative entanglement with their worldly surroundings. This issue will be discussed at greater length in chapters 6 and 7.

mance or belief, rather than as patterns of normative accountability. But the accountability of practical activity, including scientific practices, cannot be adequately conceived in this way.

Consider some simple cases of practical performance. To open a door or pour a liquid in the absence of the door or the liquid can only be construed as pretending to act, performing some other action (reaching for a door or tipping an empty bottle, each of which engages a different material setting rather than none), or utterly failing to act at all in the requisite way.[7] Representational opacity might plausibly seem to make sense for thoughts or utterances, but it surely cannot do so for practical performances. More needs to be said about what it means to begin an analysis of practices with the relational complex of materially situated activity rather than with the agent's activities or the material setting considered apart from one another, but for now think of scientific practices as practical configurations of the world rather than as agents' performances abstracted from that setting.

A final danger to avoid right from the outset is the construal of scientific practices as clearly bounded and distinct from an extrascientific "context." This construal is implicit even in accounts that acknowledge extensive interrelation between the sciences and their social or cultural context. The metaphor of interrelation suggests relatively autonomous components of a larger system, with interrelation or interaction confined to a well-defined, narrow-bandwidth interface. As John Haugeland has recently pointed out:

> Interrelationist accounts retain a principled distinction between [science and context]. The contrary of *this* separation—or battery of separations—is not interrelationist holism but something I would like to call *intimacy*. . . . The term 'intimacy' is meant to suggest more than just necessary interrelation or interdependence but a kind of *commingling* or *integralness* [that] undermines their very distinctness. (1998, 208)[8]

7. I speak of "utter" failure, because there are many ways in which action can fail to be successfully completed (the door might be locked, missing a handle, too heavy, or too symbolically portentous for me to open it), but in the absence of an appropriate material setting, the action cannot even get started in a way that would allow for a failure of completion. By the end of the book, it should be clear that speech acts are also practical performances in this sense. If my vocalizations or other performances are not appropriately connected (perhaps by complicated inferential patterns) to a suitable, publicly accessible material setting, they cannot be contentful or referentially indicative.

8. I have substituted "science and context" for Haugeland's "mental and corporeal." Haugeland was arguing specifically for the intimate entanglement of mind, body. and world rather than of science and its social and material context. I can readily adapt his argument,

The point is not that scientific practices are indistinguishable from, say, nature, religion, government, or business. Rather, such distinctions (made along different lines for different purposes) do not demarcate a separable component of the world whose interface with other components is readily localizable. The traffic in all directions across whatever boundaries can be usefully demarcated will be too extensive and wide-ranging (traveling through "high-bandwidth" channels, in Haugeland's evocative analogy) to allow for any significant autonomy of a domain of scientific practices.[9] The sciences are contextually integrated in multiple ways: in acquiring funding and material resources; in recruiting and training scientists and auxiliary personnel; embedded in institutions such as universities, academies, corporations, bureaucracies, or foundations; in the articulation and application of social norms, cultural forms, and bodily disciplines; but also in acquiring and answering to much of what is ultimately at issue and at stake in scientific practices.[10]

Moreover, the extension of scientific practices and achievements into new settings cannot always be dismissed as merely the subsequent extrascientific application of antecedently intelligible scientific achievements. The

---

both because the distinction between interrelation and intimacy can be applied to other supposed interrelata, but more pointedly because the most familiar ways of distinguishing science from its social and material realization are simply variations upon traditional distinctions of mind from world: what is typically "internal" to science, for both internalists and their interrelationist critics, is scientific understanding and reasoning, and what is "external" (both to science, and to one another) are human society and nature. Laboratories, as both material settings and institutions, thus play an ambiguous role in such accounts comparable to the role of the body in conceiving interrelations between mind and world. The relevant interface is sometimes located between mind and body, and sometimes between embodied mind and surrounding world; laboratories are likewise variously placed on the near or far side of scientific thought. This parallel role for bodies and laboratories is more than coincidental. In chapters 6–8, I show how scientific laboratories, and other experimental, observational, or even theoretical equipment should be recognized as aspects of a prosthetically embodied agency.

9. The reverse is also true, that is, what is left of the social or cultural "world" apart from scientific practices is not autonomously intelligible. It is a truism to remark on the importance of scientific practices and achievements within the "modern" world, but this truism is often neglected in practice. If as a thought experiment, for example, one were to consider the image of the social and cultural world constructible by an imaginary overlay of the social science and humanities curricula of typical American universities, one of the more striking divergences between this image and the world it aspires to make intelligible would likely be that the image provides a seriously diminished role for the natural sciences.

10. Mario Biagioli's discussion (1993) of Galileo's articulation of a new role as a courtly natural philosopher shows this point especially clearly by the constraints that emerged within absolutist court culture upon what *could* be at stake in natural philosophical controversies.

intentional directedness of scientific practices is often only fulfilled by such extensions beyond the laboratory and outside the field. How, after all, does the manipulation of successive generations of *Drosophila* come to be about genetics, or the carefully monitored interaction of artificially purified or even synthesized substances come to be about geochemistry? The spatiotemporal extension of scientific practices and materials beyond the laboratory is often the vindication of the *semantic* extension of scientific concepts.[11] Scientific practices are often construed as apart from any surrounding culture, and even free from culture,[12] but such construals are not adequate to the richness and complexity of scientific work. Recognizing the intimate entanglement of the sciences with other practices does not diminish or blur their significance, but instead acknowledges their pervasiveness throughout the world.

Against this background, I now turn to the central ambiguity in the concept of practices, which I will explicate by reflecting upon some recent critical discussions of practice theories by Stephen Turner (1994) and Steve Fuller (1993, 1992b). Turner's and Fuller's criticisms turn on a failure to recognize the possibility of a conception of practices that does not invoke an underlying regularity or commonality. Seeing how that very conception is implicit within Turner's and Fuller's criticisms will provide a clearer sense of why this alternative understanding of practices matters philosophically.

Practices as Regularities versus Practices as Normative

Stephen Turner's criticisms of "the social theory of practices" begins by recognizing an important philosophical rationale for practice talk in Wittgenstein's discussion of rule-following and Heidegger's account of understanding and interpretation.[13] The ability to follow a rule or to make an assertion seems to presuppose an unarticulated (but interpretable) understanding-in-practice in order to halt regresses of explicit interpretation and justification (if all understanding is by recourse to rules, how does one correctly understand a rule?). The practice theorists that Turner criticizes argue that one makes sense of the explicit commitments of a community,

11. For a more detailed discussion of this point, see Rouse 1987, chaps. 4, 7, and Rouse 1996b, part 2.

12. Traweek (1988) interprets the self-understanding of the physicists she studied as their belonging to "the culture of no culture."

13. Wittgenstein 1953; Heidegger [1953] 1996.

and of the integrity of their development over time, by interpreting them against a background of practices. Within science studies, Philip Kitcher's discussion of the practice of classical genetics exemplifies this strategy:

> The corpus of statements about the inheritance of characteristics accepted at a given time is only one component of a much more complicated entity that I shall call the *practice* of classical genetics at that time. There is a common language used to talk about hereditary phenomena, a set of accepted statements in that language (the corpus of beliefs about inheritance mentioned above), a set of questions taken to be the appropriate questions to ask about hereditary phenomena, and a set of patterns of reasoning which are instantiated in answering some of the accepted questions (also: sets of experimental procedures and methodological rules, both designed for use in evaluating proposed answers . . .). The practice of classical genetics at a time is completely specified by identifying each of the components just listed. (1984, 352)

Historians of science have often been skeptical of claims that scientific communities share any such corpus of beliefs, procedures, or patterns of reasoning.[14] Turner's criticism goes further, however, arguing that appeals to shared practices to account for regularities, continuities, and commonalities in social (or scientific) life are pseudoexplanatory, and thus fail to avoid the worrisome regresses. According to Turner, the inference from common behavior to its supposedly underlying source in shared presuppositions or practices cannot be justified, the causal powers of practices are inevitably mysterious, and the transmission or reproduction of practices over time and from one practitioner to another cannot be accounted for. Fuller's criticism both echoes Turner's objections to practice talk as pseudoexplanatory and adds a political dimension: Fuller argues that any recourse to the *geisteswissenschaftlich* interpretation of shared, tacit understandings implicit in practices is deeply conservative and antidemo-

14. Thomas Kuhn, often mistakenly cited as a proponent of such wide-ranging community consensus as a defining characteristic of "normal science," actually denied that such consensus could be found, or that community consensus was necessary to understand the coherence of normal scientific research. As Kuhn noted in retrospect (1977, xviii–xix), his concept of a paradigm (exemplary problem solutions that scientists could use as models for subsequent work without any underlying agreement about how to interpret or use these models) was introduced to understand how scientists could build on one another's work in the absence of any such underlying consensus. Kitcher, following the passage just quoted, acknowledged that "the community of geneticists was [not] always distinguished by unusual harmony of opinion," while reassuring the reader that "it is not hard to relax this simplifying assumption [to the contrary]" (1984, 352).

cratic, an argument that is buttressed at least ad hominem by reflection upon the political commitments of Heidegger and Wittgenstein, philosophical ancestors of the practice industry.[15]

I take Turner's and Fuller's criticisms very seriously, but they have misunderstood the significance of their own arguments. Turner in particular concludes that practice talk is altogether bankrupt in philosophy and social theory, a conclusion that leads him to resurrect in its place the long abandoned explanatory appeal to 'habit.' Turner's criticism thus turns on a failure to recognize the ambiguity between practices conceived as *regularities* and a *normative* conception of practices. The two uses have a rough correspondence to the respective uses of the term in the tradition of social theory (Durkheim, Weber, Oakeshott, or Winch) and in philosophy (Heidegger and Wittgenstein), a distinction that is nevertheless complicated by widespread misappropriation of Heidegger and Wittgenstein within social theory. Perhaps because his focus is upon social theory, Turner only identifies one side of the ambiguous usage. His arguments are indeed telling objections to the conception of practices (as regularities) that he actually addresses, but the upshot of the argument is instead to highlight the importance of conceiving practices normatively.

So what are practices? The question arises with some force, because of the diversity of things sometimes incorporated by the term. Turner notes that 'practice' has been used interchangeably with 'tradition,' 'tacit knowledge,' 'paradigm,' 'presupposition,' and much more. Practices are sometimes taken as tacit propositional attitudes and sometimes as inarticulable competences or performances. In either case, he claims, the concept of practices is typically invoked to explain continuities or commonalities among the activities of social groups. Turner argues that to do the explanatory work thus attributed to them, practices must be objectively identifiable regularities. If they are presupposed propositional commitments, they must have some psychological reality; if they are practical competences, they must have some causal efficacy.[16] In either case, moreover, the content or pattern embodied in a practice must be transmissible so as to preserve its

15. Turner himself (1989) has argued that the ad hominem consideration of Heidegger's politics must be taken as a quite serious prima facie problem for any conception of practices that draws extensively upon Heidegger's work.

16. Note that Turner's demand for the psychological reality of propositional commitments is at odds with my endorsement of Quinean indeterminacy. A more modest analogue to his demand could still be made, however, by asking whether these ascribed commitments play an integral and economical role in the interpretation of what practitioners do.

identity across practitioners. Turner ultimately rejects any explanatory appeal to social practices just because of the allegedly intractable difficulties in justifying the psychological reality, causal efficacy, or transmissible identity of any regularities underlying more readily manifest human activities.

Turner, however, fails to recognize the possibility of an alternative conception of a practice whose constituent performances are appropriately regarded as answerable to norms of correctness or incorrectness. Not all practitioners perform the same actions or have the same presuppositions, but practitioners and other constituents of a practice are accountable for performances or presuppositions that are inappropriate or otherwise incorrect.[17] Of course, not all improprieties are actually held to account. So the sanctions and other differential responses that would mark the incorrectness of some performances are themselves normative practices. It is always possible that such chains of proprieties come to an end in some kind of objectively recognizable regularity. But, as Robert Brandom has noted, "we can envisage a situation in which *every* social practice of [a] community has as its generating response a performance which must be in accord with another social practice" (1979, 189–90). Such a network of practices need not be identifiable as regularities of action or belief, even as a whole. Brandom therefore argues that the difference between regularities and normative responsiveness should itself be regarded as a normative matter, that is, as a distinction between those patterns appropriately explained in causal terms and those appropriately understood as subject to interpretation and normative response.[18]

Turner's own arguments against the integrity of practices conceived as

17. It would be a mistake to assume that such normative assessment applies only to "practitioners"; as Haugeland rightly insists, the constitutive standards governing practices also apply to the behavior of all of their relevant constituents. Take a game like baseball, for example: "The bases must lie on the ground (in a diamond, 90 feet apart), remain visible, and not move. The ball, when pitched, must sail through the air toward the strike zone, deviating at most slightly from a ballistic trajectory; when hit by the bat, it must rebound (not stick to the bat or explode) and sail off in another direction; and so forth. These requirements are not regulations governing the actions of any players; yet they are constitutive of what it is to be bases and balls, hence indirectly, also of other baseball phenomena" (1998, 321). Haugeland goes on to consider differences between games and sciences as practices subject to constitutive standards, but I shall take up this consideration in subsequent chapters.

18. I show in chapters 8–9 that the patterns (phenomena) appropriately explicable in causal or nomological terms themselves constitutively belong to practices subject to norms, such that one cannot demarcate a domain of causal relations or causal structure independent of normative assessment. I am therefore only making heuristic use here of Brandom's distinction between objective regularities and normative proprieties.

regularities ironically often point toward such a normative conception of practices. Turner argues, for example, that Marcel Mauss's identification of culturally distinctive ways of walking could not easily be captured in terms of "culture-free causal categories," in part because "one might acquire the 'same' [external] walk by mimicking or by a kind of training which corrected various untutored walks—and corrected them in different respects—to produce a walk which is externally the same" (1994, 22). Moreover, Turner claims, the description of practices typically depends upon classification schemes that presuppose acquaintance with other practices: such descriptions are only identifiable by contrast to other local, cultural expectations with which they conflict (24). Practices are only identifiable as practices at all against a background of other practices, and thus can never be reduced to objective regularities. Turner still fails to grasp a normative conception of practices as a genuine alternative, however; what normative-practice theorists would regard as normative responsiveness, he dismisses as instead an instrumentalist appeal to regularities (37).

Turner objects to such an instrumentalist conception of the appeal to regularities of social practice because it "fails to connect the stuff of thought to the world of cause and substance, [leaving] no basis for using our past understandings or interpretations to warrant future interpretations" (1994, 37). But this objection only makes sense if the domain of practices is conceived narrowly, in two respects. First, it presumes that the "world of cause and substance" is somehow distinct from the "social world" of meaningful practices, such that the two are in need of reconnection (in my terms, Turner explicitly endorses the dualism of nature and normativity). Practices should be understood as intimately causal and substantial rather than as a distinct, ethereal realm of thought and meaning that only interacts with material things at localizable interfaces. Second, Turner's objection presumes, more subtly, that practices are distinct from linguistic representation: he takes practices to be ontologically suspect, whereas linguistic meaning and reference are not. Here we must consider the relations between practices and 'language,' which has often been a source of confusion in discussions of practices.

Practice theorists have been ambivalent about the significance of language for practices, and vice versa. On the one hand, unarticulated or even inarticulable practices are frequently contrasted to explicitly articulated assertions or rules. On the other hand, the domain of practices is often extended to incorporate conceptual or linguistic practices, perhaps even as the paradigm case of practices. Shared conceptual schemes or presuppositions are often the focus of practice talk. Recognizing the difference between regularist and normative conceptions of practices, however, highlights a correlative distinction between two conceptions of linguistic (or discursive) practices. Those who identify practices with regularities (including shared beliefs or conceptual schemes) typically situate language outside the domain of practices; shared practices may account for particular intentional contents (beliefs or conceptual schemes) expressible *within* language, but linguistic intentionality itself is then modeled by a representational semantics instead of a normative pragmatics of discursive practices. By contrast, a normative conception of practices is best understood as incorporating a general conception of intentionality. Brandom (1994) has most explicitly worked out such a conception of intentionality as altogether pragmatic, but once the distinction is clear, Heidegger and Wittgenstein, and more recently Donald Davidson, are best understood as giving normatively pragmatic rather than representational accounts of intentionality.

A pragmatic account of language, and of intentionality more generally, treats language dynamically and normatively, without reifying meanings, reference, or shared languages. Shared meanings or beliefs are not preexisting facts that explain the possibility of communication, but norms presumptively invoked in the course of interpreting someone as communicative.[19] Only by interpreting a speaker as mostly acting appropriately so as

19. Interpretation (or as Turner too narrowly construes it, "attributing assumptions") is here understood in a thoroughly pragmatic way, as adopting a practical attitude rather than offering an explicit account: I "interpret" someone as communicative simply by listening and responding in appropriate ways. A useful parallel is Heidegger's discussion of *Auslegung* in *Sein und Zeit,* in which one interprets something as a hammer (and oneself as a hammerer/carpenter) by hammering with it; no explicit attribution of properties, meanings, or functional roles is required. In chapter 6, I shall say more about Brandom's modeling (1994) of interpretive practices as discursive scorekeeping. An interpreter can, of course, try to make explicit the practical attitudes adopted in making sense of someone or something, but such explications always come to an end in further unexplicated proprieties: "this is what we do." This Wittgensteinian phrase is often misunderstood by taking "what we do" as a behavioral regularity, but that cannot be right in the light of his discussion of rule-following. Appeals to "what we do" are supposed to halt the regress of explications of a rule (whether the rule is supposed to govern the interpreted performances or is only a regularity the interpreter takes them to exhibit). Such a regress *cannot* end in another regularity, but only in a propriety (to adapt an example from Samuel Wheeler, we should understand "this is what we do" in the sense of "we don't hit other children, do we?"). "What we do" always includes further practices of *correcting* (but sometimes ignoring) deviant performances.

to make sense (within a field of linguistic and other practical proprieties that enable me to make sense of myself as making sense) can I acknowledge her activity (or my own) as discursive, and thus as linguistic. Note that this characteristic feature of Davidson's and Brandom's interpretive semantics (namely, that truth, meaning, language, and other semantic categories can only be explicated via interpretation in an *unanalyzed* home language)[20] is precisely the feature of practices that most vexes Turner as unacceptably instrumentalist:

> The assumptions [one] attributes to [another] are identifiable as assumptions only because [one] is in a position to make a specific comparison [to one's own understanding of the "same" situation]. Starting from a different comparison . . . would produce different misunderstandings, and different assumptions would need to be attributed. . . . Such 'assumptions,' then, are *not natural facts,* but hypotheses that solve specific comparative problems. In the case of practices conceived causally, there was a gap between the 'practices' that could be constructed on the basis of specific comparisons (between the French and American walk, for example) and the causal facts that the practice (the walk, in this case) consists of. (Turner 1994, 33–34; my emphasis)

But for Davidson, Brandom (and Wittgenstein and Heidegger, I would argue) linguistic practices are *non-natural* in precisely this sense, and cannot be explicated on the basis of underlying (non-normative) causal facts.[21] The issue closely parallels the point I made concerning Quine in chapter 3. The alternative sets of "analytical hypotheses" that Quine postulates as an indispensable background to translation do not constitute a distinctive domain of semantic facts over and above the natural facts.[22] The appropriate conclusion to draw, however, is not Quine's withdrawal to a more

20. The phrase "relative to an unanalyzed home (or background) language" is the canonical characterization of a point common to Quine, Davidson, and Rorty. Strictly speaking, however, such views ought to hold no place for any determinate concept of a language, as Davidson (1986) came to recognize, for discursive practice always outruns any previously stable linguistic structure. The background to interpretation must therefore instead be unexplicated discursive practices, which Brandom (1994) models as scorekeeping practices (I discuss this point more extensively in chapter 6).

21. Whether causal facts are themselves determinate apart from the practical normative accountability that constitutes their proper bounds, and their appropriate comparisons to relevantly similar causal patterns, will be taken up in chapter 8.

22. A thoroughly pragmatic account of discursive practices would not, of course, construe the background to interpretation (Quinean translation) as composed of analytical hypotheses.

austere ontology that eschews intensions, but a recognition that discursive normativity is an ineliminable dimension of all practices, including those scientific practices that disclose natural facts.

With this background, I now turn to Turner's worries about the connection between practices and "the world of cause and substance." Turner clearly takes practice theories to be objectionably antirealist (either instrumentalist or social constructivist) about the causal powers and susceptibilities of practices. But, as I have argued elsewhere, a normative conception of linguistic and other practices challenges the shared commitments of realists and antirealists alike to a representationalist semantics.[23] The attitudes and responses that identify a practice (including a discursive practice) are only contentful amid ongoing intra-actions with the world.[24] Utterances are interpreted by making sense of what is said when (that is, on which occasions) and where (in what worldly circumstances). (Thus, for Davidson, interpretation proceeds via prior acquaintance with the utterances' "truth conditions"; for Heidegger, interpretation is an aspect of being-in-the-world). To ask how our representations can ever get a foothold in the world is to presume, erroneously, that one could ever make or understand representations without already having a foothold in the world.[25]

Thus, taking seriously a normative conception of practices suggests that there is no such thing as the "social world" (or the "natural world") except as reified abstractions from *the* world.[26] The meanings, agency, institutions, or forms of life with which social constructivists would explain how nature becomes manifest are themselves senseless apart from those manifestations; they cannot be an independent explanans. But as subsequent chapters will show, the natural kinds and causal capacities that supposedly constitute the preexisting "natural world" likewise only acquire their

23. Rouse 1987, chap. 5; 1991; 1996b, introduction and chaps. 7–8.

24. I have discussed elsewhere (Rouse 1996b, chap. 5) the need to incorporate such material intra-actions within any adequate conception of practices.

25. The classic philosophical source for this criticism of representationalism is the introduction to Hegel's *Phenomenology of Spirit* (1972): the representationalism underlying both realism and antirealism is what Hegel called "the fear of error that reveals itself as fear of the truth" (par. 74).

26. Brandom (1994) still identifies practical proprieties as instituted within social practices, but in a sense he describes as an "I-Thou" model of social interaction rather than "I-We." It makes more sense to drop the term 'social,' which almost inevitably has connotations of supraindividual entities. But dropping that term should not suggest that Brandom's theory instead accounts for intentionality as individual. The agency and identity of individual speakers is instead only established within ongoing practices. See chapters 6–7 below for further discussion.

constitutive counterfactual import from normative applicability within scientific practices of experimental manipulation and theoretical modeling.

The Consequences for Science Studies

I have begun to articulate how a normative conception of practices is distinct from mere interpersonal regularities of behavior or belief. I now want to consider, in a similarly programmatic way, some possible consequences of attending to scientific practices as normative, instead of as social regularities.[27]

One topic that becomes central here is the temporality of science.[28] Post-empiricist philosophy of science has emphasized the importance of historical changes in the methods, concepts, and standards of the sciences, whether construed as sharp, "revolutionary" breaks or as more gradual, reticulated developments. Understanding scientific change in ways that do not render it unintelligible as rational was a central preoccupation of the philosophical responses to Kuhn and Feyerabend. Such interpretations of scientific change do not yet attain to an adequate conception of the temporality of scientific practices, however. If one takes scientific communities to share specific beliefs, values, conceptual contents, or activities, then it would make sense to ask how those communities and their commitments change over time. Kitcher's discussion (1984) of the "practice" of classical genetics nicely exemplifies such a project. But such conceptions presume that there is an already determinate character to the community itself, and to the shared commitments that define its boundaries. Such determinacy cannot be presumed once scientific practices are conceived normatively, for such practices are constitutively temporally extended. What the practices are *now* depends in part upon how their normative authority and force are encountered and sustained in ongoing practice, through responses to continuities and variations in what is done and what happens. But those future responses are not yet determined.[29] The present con-

27. A more detailed discussion of some of these issues can be found in part 2 of Rouse 1996b.

28. The temporality of scientific practices has also been a central concern in recent work by Andrew Pickering (1995) and Hans-Jörg Rheinberger (1991, 1994, 1997). Rheinberger's work figures prominently in the more extensive discussion of the temporality of practices in chapter 9 below.

29. This context shows more clearly why there might be an explanatory tradeoff between a normative conception of scientific practices and an alethic-modal interpretation of what there is in terms of *possible worlds*. Empiricist attempts to express the counterfactual force of scientific laws foundered on their inability to distinguish nomological necessity

tent of scientific practices is thus subject to reinterpretation and semantic drift.

Hans-Jörg Rheinberger points to just that conclusion in his historical study of the intertwining of research on oncogenesis, virology, and the functional structure of cells, insisting that "the virus of 1950 must be seen as the condition of possibility for looking at [Peyton] Rous's [1910] agent as that which it had *not* been: the *future virus*" (1994, 77). Samuel Wheeler (2000) argues that such conclusions are ontological rather than epistemic: present semantic content is comparable to whether a runner crossing the plate in an early inning has made the game-winning run, in being not yet settled by any facts. Understanding practices normatively helps us see why this is so: what a practice is, including what counts as an instance of the practice, is bound up with its significance, in terms of what is at issue and at stake in the practice, to whom or what it matters, and thus with how the practice is appropriately or perspicuously described.[30] The temporality of scientific practices as "a machine for making the future" (Rheinberger

---

from contingent regularity on the basis of any finite body of *actual* evidence (Hempel 1965; Salmon 1989; Miller 1987). Possible worlds semantics seems to offer a way to construe actual regularities as the manifestation of the more fundamental structure of what is (actually) possible (one classic treatment of natural or causal necessity in these terms is Lewis 1973). Possible worlds semantics, when applied to this problem of how to fix the content of present scientific practice (where 'practice' includes the objects of scientific inquiry as semantically articulated) is thus metaphysically naturalistic in one crucial sense, since modally robust natural laws would be the only determining constraint upon present practice. (Natural laws determine which possible worlds are nomologically "accessible" from the actual world.) As I shall show in chapters 8–9, however, there is a more fundamental sense in which a metaphysics that allows other possible worlds to occupy the normative role of the actual world as background for the interpretation of meaning and truth violates naturalistic scruples. Moreover, such a multidimensional metaphysics of possible worlds desperately needs, but cannot easily provide, an account of how the modal structures invoked are *normatively binding* upon scientific practice. I cannot adequately explore these issues, however, without intervening discussions of normativity and discursive practices in chapters 6–7.

30. Wheeler's example highlights an underremarked feature of Brandom's discursive scorekeeping model of semantic contentfulness (1994). "Scorekeeping" suggests an underlying determinacy to semantic content; there is always a (normatively regulated) fact of the matter about what the score is. But this determinacy is an artifact of the analogy to games. A score of three to two is a significant fact only via the proprieties that connect it to an eventual final score at the end of the game. Just as there is (normally) minimal significance to which team scores the first points in a basketball game, because the initial score is only tangentially connected to what is at stake (victory) that the scorekeeping is supposed to track, so the discursive score only matters in relation to what is at stake in the normative practices that it regulates. Unlike games, discursive scorekeeping has no conventionally mandated conclusion. So for discursive scorekeeping to confer significance as well as merely formally specified "semantic content," there has to be something at stake *within* ongoing practices with respect to which the discursive score matters.

1994, quoting François Jacob), and hence their constitutive openness to genuine future disclosure, therefore cannot be appropriately conceived in terms of scientific change.

More adequate assessment of the significance of scientific practices requires dynamic, normative accounts of language, knowledge, and power. I have already made similar claims in *Engaging Science* (Rouse 1996b), although there I placed less emphasis upon the constitutive normativity of each of these aspects of scientific practices and talked instead about their deflationary uses, to emphasize that appropriate use cannot be grounded in underlying regularities of social practice or causal connection. As a further development of these claims, I focus on why such normative, nonreified conceptions of language, knowledge, and power are *mutually* implicated. In the opening section of this chapter, I suggested a conception of language as discursive practices that involve dynamic interactions among speakers and their surroundings. On such an account, various performances acquire semantic content not through the mediation of conventional meanings, structured languages, or psychological representations (beliefs), but by their responsiveness to partially shared situations that have a trackable history. One consequence of recognizing the dynamics of practices highlights the importance of tropes, whose contrast class is not "literal" meanings, but familiar or uncontested uses.[31] Within the sciences, models (including mathematical, verbal, physical, pictorial or schematic, and experimental models) are especially important examples of tropes. Models, I have argued (1996b, chap. 8), should be thought of as simulacra rather than representations. The crucial difference is that representation too often denotes a semantic content that intervenes between knowers and the world,[32] whereas simulacra are just more things in the world, with a multiplicity of relations to other things. What makes them models, with an *intentional* relation to what they model, is their being taken up in practices, ongoing patterns of normatively accountable use.

The constitutive role of proprieties and sanctions in discursive intra-actions thus already shows the indispensability of concepts of power and

31. The loci classici for such a conception of metaphor are Davidson 1984 and Wheeler 1991. For its explicit application to scientific discourse, see Bono 1990 and Rouse 1996b, chap. 8.

32. This sense of representation is indifferent to whether representations are conceived as thoughts accessible to individual minds, or as concepts, conceptual frameworks, languages, or forms of life shared by social groups.

resistance for understanding language. Whether an unfamiliar way of speaking about or dealing with a situation is taken as an innovation, a mistake, a curiosity, an irony, or a humdrum variation on the familiar depends crucially upon asymmetries of power among those who encounter it. Yet the recognition of models as simulacra extends the interconnection of meaning and power beyond the immediate relation between speakers and their interpreters. To see why this is so, consider a question sometimes asked rhetorically about meaning: how could merely representing things differently possibly have a causal influence on them? A similar question about simulacra cannot have the same rhetorical effect: simulacra *are* transformations of the world, and more significantly, they transform the available possibilities for human action. They do so both by materially enabling some activities and obstructing others, and also by changing the situation such that some possible actions or roles lose their point, while others acquire new significance.[33]

So far, I have sketched how considerations of power might become relevant to conceptions of language and meaning as emergent from the normativity of discursive practices. If such an account were construed in terms of a *reified* conception of power, the result would be a reduction of meaning to something like rhetorical force, and language to a technology of persuasion, a move that is not unknown within science studies.[34] Such a reduction of meaning or significance to rhetorical or material effects is nevertheless a fundamental mistake. The mistake can be avoided by also conceiving of power dynamically, not as a regularity of social life, a thing possessed or exercised by dominant agents, but as a situated and temporally extended relationship among agents and their surroundings.

Thomas Wartenburg (1990) has developed a partial model for such a dynamic conception of power. He began by noting how power is mediated

33. The sense in which scientific practices transform the available possibilities for human action is more extensively discussed in Rouse 1987, chaps. 6–7. That discussion should be connected to my more detailed subsequent account (Rouse 1996b, chap. 8) of both theoretical modeling and experimental systems as simulacra rather than representations, and to the further discussion of these points in chapters 6–9 below.

34. Shapin and Schaffer 1985, Latour 1987, and Pickering 1984 are among the works in science studies that prominently interpret meaning as rhetorical force, or reasoning as a "technology of persuasion." It is less clear to what extent Pickering 1995 departs from his earlier account, for although the latter book does encourage a more dynamic conception of power, it strikingly omits discursive dimensions of practices from among the many elements supposedly "mangled" through the ongoing realignment of resistance and accommodation.

by "social alignments": one agent's actions effectively exercise power over another only to the extent that other agents' actions appropriately align with the actions of the dominant agent. For example, judges exercise power over prisoners only if the actions of bailiffs, guards, appeals courts, and others align in the right way with what the judge does. Power relations are dynamic, because the presence and effectiveness of an alignment depend upon how the alignment is sustained or transformed over time, in response to subordinate agents' efforts to resist or bypass them as well as dominant agents' attempts to utilize, strengthen, or extend them. Power is thus not something possessed or exercised by an agent, or even a relation between two agents, but is instead dispersed and deferred across a field of possibilities. Wartenburg's model is nevertheless only partial, because he mistakenly restricts the mediation of power to social alignments of human agents. A more adequate conception would recognize the materially circumstantial configuration of power, such that tools, processes, and physical surroundings more generally all belong to dynamic alignments of dominance, subordination, and resistance. Thus, just as practices should not be reduced to social practices, power should not be reduced to social power.

Trying to understand knowing without reifying knowledge may initially seem more difficult. Dynamic accounts of language have helpful precedents in Davidson, Derrida, and Brandom; Foucault likewise makes familiar a dynamics of power (even if the significance of a dynamic conception has not always been adequately appreciated in either case). Yet Wartenburg's discussion of dynamic power alignments offers a useful analogue to a dynamic conception of knowing.[35] Power is only *effected,* such that it enables or constrains action, through dynamic alignments that bring one action to bear upon another. I have argued (1996a,b) that knowing is likewise only *informed* through dynamic alignments that enable one thing (a statement, a model, an image, a skillful performance, and so forth) to be correctly taken to be *about* another. Philosophers of science nowadays emphasize the role of "background knowledge" in establishing inferential relations between hypotheses and evidence, but that concept is too homogeneous and static. Knowing is mediated not just by a background of

35. The connection is more than just an analogy. Wartenburg's approach to power is deeply influenced by Foucault, whose reflections on the dynamics of power were introduced as an analytics of power/knowledge. Foucault's work of the 1970s suggests a dynamics of knowing, even though it is less extensively articulated than his correlative conception of power. See Rouse 1993, 1994.

assertional commitments, but also by models, skills, instruments, standardized materials and phenomena, and situated interactions among knowers, in short, by practices. Moreover, a dynamic account of language as discursive practices obliterates any clear distinction between the representational content of knowledge and its material or social construction or implementation.

The result is a dynamic and deflationary conception of knowledge, modeled on deflationary, prosententialist conceptions of truth.[36] In the latter case, the truth predicate and the capacity to use it are recognized as indispensable to linguistic and epistemic practices, even though no underlying nature of truth unifies or reifies the instances of its appropriate application.[37] A deflationary account of knowledge likewise denies that 'knowledge' (or, mutatis mutandis, 'scientific knowledge') constitutes a theoretically coherent kind. There are many appropriate ascriptions of 'knowing' within the multifarious practices of assessing, attributing, relying upon, or contesting understanding and justification, but there is no *nature* of knowledge underlying these ascriptions. This claim has far-reaching consequences. As I have noted elsewhere, participation in the wholesale legitimation or critique of scientific claims to knowledge that has characterized much of late-twentieth-century philosophy and sociology of science almost invariably proceeds from a conception of scientific knowledge as theoretically coherent and surveyable as a whole. It only makes sense to claim that scientific knowledge as a whole is approximately true, rationally arrived at, socially constructed, or interest-relative if there is such a (kind of) thing.[38] Scientific realists, historical metamethodologists,

36. It is crucial to distinguish two oppositely directed uses of Tarski's schema for a deflationary (or semantic) conception of truth. For Tarski or Hartry Field, the semantic conception shows how 'truth' can be eliminated (at least for formal languages), and thereby rendered innocuous within a thoroughgoing metaphysical naturalism. Reduction, however, is inevitably a two-way street. For Davidson, Brandom, Fine, Williams, or me, the Tarski schema is deployed in the opposite direction, making 'truth' the fundamental semantic category, and dispensing with reifications of meaning (or convention), reference, or shared languages.

37. The standard technical objections to deflationary conceptions of truth arise from their proponents' inability to articulate what to say about the uses of the truth predicate in conditionally embedded, nominalized, or quantified contexts. Brandom's development (1994) of Grover, Camp, and Belnap's prosentential theory(1975), and its extension to 'refers' as a "proform-forming operator," has resolved the principal technical problems, however, which ought to place a burden of argument upon those who would still call for a more robust and substantive conception of truth.

38. Elsewhere I have articulated and criticized the commitments to the "legitimation project" that are necessary to make sense of the disagreements among scientific realists,

empiricists, and social constructivists regularly argue for their preferred accounts of the nature or goal of scientific knowledge, but typically and mistakenly take the unity and theoretical integrity of scientific knowledge for granted. The alternative deflationary conception of knowledge thus expresses constructively the rejection of an epistemological understanding of science that I have attributed to many feminist theorists of science.

Relocating Science Studies

Conceiving of practices normatively rather than as composed of underlying social or natural regularities and refusing to reify language, power, and knowledge thus challenge many of the most familiar philosophical and sociological approaches to science studies. An important aspect of this challenge concerns the theoretical and political location of science studies themselves, often discussed under the heading of reflexivity. Philosophers and sociologists alike have typically attempted to achieve a standpoint that I have elsewhere described as epistemic sovereignty,[39] a theoretical position "outside" or "above" scientific practices from which to establish or undermine their legitimacy and authority once and for all. The conceptual link between aspiring to epistemic sovereignty and conceiving scientific practices as regularities should now be clear. The various participants in the legitimation project hope to characterize the underlying regularities that most fundamentally govern scientific practices, whether socially or psychologically predetermined goals or causally efficacious features of the world (the relevant causal features might in turn be taken as modally robust natural kinds that scientists actually encounter, social interactions that determine the interpreted outcome of those encounters, or perceptual capacities of human knowers). By thus ascertaining from the "outside" what science "really is," its indispensable nature or goal, adherents of the legitimation project would provide an incontrovertible ground for assessing its successes and failures, as well as its appropriate place among other human practices and achievements. Such an aspiration to epistemic sovereignty should, however, be anathema to philosophical naturalists, for whom scientific understanding must always be situated within the world understood.

historical metamethodologists, empiricists, and social constructivists (Rouse 1996b, introduction, part 1).

39. Rouse 1994; 'epistemic sovereignty' was chosen as a deliberate parallel to Foucault's discussion of how the tradition in political theory has been shaped by conceptions of sovereignty as both an analytic focus and a metatheoretical standpoint.

Understanding scientific practices normatively allows no such determining standpoint above or outside of ongoing practice. Davidson's and Brandom's semantics offer useful parallels; they take natural language to be its own metalanguage, so that discursive practice can only be explicated from within. Science must likewise be understood to include its own interpretive and critical metapractices, which only permit epistemic and political explication and criticism from within. 'Science' should thus be construed broadly as referring to the cultures of science, the practical, discursive nexus that incorporates everyone and everything in relation to which 'science' is an intelligible and significant category, however contested. Understanding science and philosophy in this way has obvious affinities with what is at stake in philosophical naturalism, in its refusal to tolerate either arbitrarily "external" restrictions upon science or explanatory appeals to concepts that would only make sense from a postulated standpoint outside the natural world.

This naturalistic insistence that science studies are inescapably internal to the culture of science may then raise Fuller's principal worry, that an interpretive engagement with scientific practices is necessarily conservative.[40] Fuller objects that interpretive engagement with scientific practices abandons any aspiration to hold science accountable to norms not of its own choosing.[41] But Fuller's criticism instead shows only his commitment to a conception of practices as (social) regularities: he limits science to the set of practices already conventionally recognized as science, and its norms to the goals and standards to which its practitioners already subscribe. He then concludes that critical perspective can only come from elsewhere. By contrast, if one recognizes scientific practices as constitutively normative, what science or knowledge *is* is not already determined, but is instead at issue in scientists' and others' practical responsiveness to their circumstances. Various epistemically significant practices are normatively accountable,[42] but such practices also offer competing interpretations of the

40. Strictly speaking, as this criticism goes, there is an alternative to a conservative continuity with current practice, namely, an arbitrary imposition upon current practice, but this alternative would then be inherently unjustifiable, because justification would be conceptually linked to continuity with current practices and norms.

41. Fuller 1992b provides a clear and provocative account of this objection, while Fuller 2000 argues that the most prominent developments in philosophy and sociology of science of the past half century have been conservative in this way.

42. A crucial point I have been developing throughout this book is the need to understand how scientific (and other) practices are simultaneously and "intra-actively" accountable both to other knowers/practitioners, and to the surrounding world.

epistemic and cultural/political norms within which they are situated. The interpretive resources for science studies thus include a multiplicity of sciences and metascientific discourses, together with various marginal and oppositional epistemic practices. Science studies do not come in from outside to settle the differences among these coexisting and competing practices, but are already situated among them and engaged with them.

Such a conception of scientific practices as normative may nevertheless seem to pose troubling concerns that should be familiar from my discussion of Taylor and Kuhn in chapter 2. Emphasizing the normativity of scientific practices, and the ineliminability of interpretive conflict, may seem to push the sciences toward one pole of the dualism of nature and normativity. The question then arises how such practices could ever get beyond the contested issues and norms endemic to human cultures, to become accountable to a nature indifferent to human concerns. After all, the more intimately claims to knowledge seem to connect to relations of power and social criticism, the more remote they may seem to be from the world disclosed by the sciences, and from the material environment of human action. A long tradition in philosophy, whose grip on theoretical imagination has barely loosened, proclaims that one only encounters nature in its brute causal unfolding by turning away from human cultural significance and political stakes. The one lesson preeminently taken from the history of modern science is that the sciences disclose a natural world bereft of normative significance. Faithfulness to this insight would require allowing things to manifest themselves as they are, were, and always will be, except to the limited extent that the similarly brute causal powers of human agents can effectively rearrange them. Moreover, such an encounter with independent causal capacities may seem to be the only nonarbitrary constraint upon the inventive and constructive powers of imagination and human cultures.

These intuitions undoubtedly help account for the persistence and even dominance of representationalist construals of intentionality and knowledge, which begin with the separate constitution of representational content and the metaphysical natures or capacities of things represented, and ask how they can ever be appropriately reconnected. Unfortunately, the most familiar answers to this question are that representations and what they represent cannot effectively bear on one another. These answers thus reveal an implicit awareness within the representationalist tradition of the dualism of nature and normativity, the difficulty, that is, of understanding how norms and causes could ever make contact. Typically, representationalist philosophies of science undertake either an antirealist strategy (representations or practices of inquiry do not respond to things as they really are apart from us, but only to their manifestations in experience or social practice) or else a naturalist, realist strategy (representational content is reduced to its causal/functional role within the natural world). Antirealists then have trouble understanding how experiences, actions, and social practices or institutions are accountable to the world, except through the brute, unintelligible resistance of empirical anomalies and practical failures. Realist naturalists have similar difficulties understanding how the modally robust causal powers and rigidly designated natural kinds that supposedly determine the content of thoughts, utterances, and actions are normatively binding upon human agents except through the brute, unintelligible failure of some of their projects. The turn to representationalism has so far not addressed the problem: it has reinscribed the dualism of nature and normativity precisely by rendering unintelligible any encounters between the causal efficacy of nature and the normativity of scientific understanding.

These considerations reinforce what I take to be the most fundamental lesson to be learned from Heidegger and from Sellars: once the normative and the causal dimensions of the world have been conceived incommensurably as self-enclosed domains, they can never effectively be brought back together. Causality must be understood as always already normative, and normativity as always already causally efficacious. Causality must enter into the space of reasons (and vice versa) for Sellars, while Heidegger insists that there can only be practical norms or causal powers within the clearing of being-in-the-world (or, as he once somewhat cryptically described it, in the indispensable belonging together of the intelligibility of world and the opacity of earth).[43] The disturbing thought remains, however, that these claims only name the problem, and do not solve it. The *intra*twining of the normativity of understanding and action with the causal powers of things must actually be intelligibly articulated.[44] "Being-in-the-world" or the conjoining of the scientific and the manifest images are no more than words without a display of sufficiently rich connections to various situated sayings and doings so as to confer content and significance upon these philosophical invocations. The chapters to come are my initial attempt to make headway in this task, and thereby also to satisfy the dual criteria for an adequate philosophical naturalism.

43. Heidegger 1960.
44. 'Intratwining'is a variation on Barad's 1996 use of the term 'intra-action.'

SIX

Perception, Action, and Discursive Practices

THE FUNDAMENTAL PROBLEM for recent naturalistic philosophy has been to understand the relation between the normativity of human understanding and action and the seemingly anormative inexorability of causal processes in nature. In the preceding chapters, I have proposed approaching this problem by developing a normative pragmatics of scientific practices. Scientific practices are part of how human beings interact with their surroundings and with one another, in ways that can be and are held accountable as correct or incorrect, appropriate or inappropriate, successful or unsuccessful, just or unjust, insightful or not. The aspiration of a normative pragmatics is to understand how such interaction with the world is normatively accountable at all, and how it is rightly accountable in one way rather than another.

Philosophers of science have almost invariably approached this problem more narrowly by substituting a representational semantics of scientific knowledge for a more general pragmatics of scientific practices. The principal aim of scientific practices, this substitution presumes, is to represent the world (or experience of it) accurately and insightfully. Philosophers can then focus attention upon how scientific claims represent the world in one way rather than another, and how those representations are accountable to what they purport to represent.[1] The rich complexity of scientific practices can thus be largely set aside for philosophical purposes, because what matters scientifically is simply the representations the sciences produce, and their correctness and explanatory power. The distinc-

1. The normative accountability of such representations has had two distinct dimensions. Science must represent both accurately and insightfully. If its representations are not accurate, they cannot appropriately connect to *what* is to be understood. Without insight (usually construed as an explanation of *why* experience or the world experienced is as it is in various respects), these connections can produce no understanding.

tive concerns of metaphysical naturalists in turn focus upon representation, and its associated norms of correctness and insight, as natural phenomena to be understood (accurately and insightfully represented) scientifically.

A principal motivation for my return to the more general problem of understanding the pragmatics of scientific practices has been the suspicion that this representationalist simplification of the problem fails. Scientific practices are not merely regularities of performance, but more complicated, temporally extended patterns of mutual accountability among practitioners' performances and their circumstances. What is at issue and at stake in such practices is thus also more extensive and complex than just representational accuracy and explanatory insight. Moreover, the various attempts to develop a philosophical conception of representation adequate for these purposes have been singularly unsuccessful, although defending that claim is beyond the scope of this book.[2]

There is a very important reason why representationalist accounts of scientific knowledge have had few if any serious competitors. A distinctive feature of scientific work, and of human activities more generally, has been the structural complexity and fine-grained articulation characteristic of discursive practices.[3] The goals and other norms that can plausibly be attributed to nondiscursive beings are hopelessly vague and unarticulated by comparison to the normative accountability made possible discursively.[4] What the sciences allow one to *say* about the world must therefore be central to any account of scientific understanding. Any adequate account of scientific practices must attend to the semantic content of scientific understanding. Heretofore, however, the logical and semantic apparatus developed to explicate the notion of linguistic representation has seemed to provide the only plausible way to express clearly and systematically the fine-grained semantic articulation of discursive normativity.[5] A representa-

2. Except to the extent that my articulation and defense of Quinean indeterminacy of translation amounts to a criticism of any attempt to understand discursive practice as representational.

3. The term 'articulation' is especially revealing here. It is most commonly used to apply specifically to discursive expression. More generally, however, it characterizes any distinctly segmented system, and the normative accountability expressible discursively is appropriately paradigmatic of fine-grained segmentation.

4. Okrent (forthcoming, chap. 7) provides a useful discussion of this issue.

5. I have argued (Rouse 1996b, chap. 8) that Donald Davidson's interpretivist semantics can also be appropriated for the purposes of an antirepresentationalist conception of scientific practices. It is telling, however, that Davidson still placed the familiar apparatus of Tarskian representational semantics at the core of his account, even though his use of that apparatus runs directly counter to its traditional uses.

tionalist conception of scientific understanding may thus seem unavoidable.

This situation has been fundamentally changed, however, by Robert Brandom's monumental book *Making It Explicit* (1994). Brandom's important achievement is to show how to express the full logical and semantic complexity of human language in terms of the normativity of discursive practices, without placing representationalist conceptions of truth or reference at the center of the account. In the first part of this chapter, I briefly summarize the most important ways in which Brandom's analysis contributes to my project of understanding scientific practices. Brandom does not merely level the playing field by accommodating in normative-pragmatic terms the finely articulated discursive expression that has usually been explicated representationally; his approach goes further by successfully rectifying some fundamental difficulties that vex representationalist conceptions of language.

Although I endorse and appropriate the core of Brandom's account (his explication of the semantics of language and thought in terms of pragmatic discursive norms), I cannot accept his understanding of normative authority and force. These are, alas, crucial components of his program. Brandom himself recognizes that his account falls short unless he can show how discursive performances become accountable to something outside language and how the norms implicit in discursive practice are binding upon practitioners. I think his account does fail in these two crucial respects; indeed, in both respects, Brandom's analysis fails precisely through its mistaken adherence to the now familiar dualism of nature and normativity. Hence, to make my own account of scientific practices philosophically viable, I need to show where Brandom's account of the authority and force of discursive norms goes awry and suggest a remedy.

My responses to these two difficulties, developed here and in chapter 7, crucially set the stage for my discussion of scientific practices and a naturalistic understanding of nature in chapters 8 and 9. In this chapter, I address the first difficulty, a pernicious vestige of representationalism that persists in Brandom's account of perception and action. His effort to incorporate perception and action within discursive practices fails to show how causal interactions with objects acquire normative authority over what people say and do. This failure is instructive. The difficulty can be overcome, I argue, by turning Brandom's conception inside out. Brandom tried to expand a rationalist conception of discursive norms to incorporate their accountability to the world in perception and action. I show instead how to understand discursive practices as extensions of practical-perceptual bodily interaction with our surroundings.

This chapter proceeds in three stages. In the first section, I provide an overview of Brandom's account of discursive practices and why it is important for my project. Against this background, the second section introduces the pivotal issue of how discursive practices are accountable to the causal interactivity of perception and action. Brandom's treatment of perception and action implicitly responds to McDowell's criticisms of Davidson (discussed in chapter 2) on precisely this point. Davidson failed because he treated norms of meaning and justification as having no intelligible connection to perception and action as *causal* interactions. Brandom responds to such criticisms by trying to incorporate the causal dimension of perception and action within his account of discursive normativity, so as to vindicate the objective accountability of discursive practices. On his account, however, perception and action play importantly different roles in linking discursive practices to their surroundings. Ultimately, he argues, these different roles account for differences between semantic/epistemic and practical normativity: they explain the unity of truth in contrast to the diversity of goods.

I begin the third, critical section of this chapter by challenging these supposed differences between the discursive roles of perception and action. Perception is itself a skillful activity evaluable in practical terms, while action has an ineliminable perceptual dimension; there is no sharp or clear distinction in kind between perceiving and acting. I then show why my argument has striking consequences for the overall significance of Brandom's own project, even though it preserves the core of his analysis. One consequence of my revisions is to give a more prominent and integral role for ethical and political normativity within a general account of understanding. The most significant revision, however, removes the relative autonomy Brandom confers upon the intralinguistic aspects of discursive practices—discourse is more tightly integrated in practical dealings with the world than Brandom acknowledges. Instead of incorporating perception and action within a rationalist conception of human understanding, my revised conception incorporates the fine-grained normativity of discursive practices within a more naturalistically acceptable account of practical/perceptual interaction with the world.

Brandom's Discursive Pragmatics

Brandom undertakes three crucial strategic transformations of more familiar philosophical accounts of the semantic articulation of thought and speech. He first shifts the locus of semantic content from the thoughts or performances of individual speakers/agents to their discursive and practical interactions. The shift to discursive practices requires a second, corollary shift in what bears semantic content. The bearers of semantic content are not mental states, but quasi-institutional normative statuses (which Brandom construes as the deontic statuses of commitment or entitlement). Finally, as a result, the familiar problem of how mental states can represent objects is replaced by the problem of understanding how discursive performances acquire the normative status of semantic articulation. This last problem turns out to be analogous to the familiar questions about representation, however. Instead of asking how beliefs or sentences represent objects or states of affairs, Brandom develops a more elaborate account of how discursive practices become normatively accountable to causal interaction with surrounding circumstances. It is here that Brandom's account of perception and action turns out merely to recapitulate rather than resolve the familiar dualism of nature and normativity, albeit in ways that I shall remedy constructively.

These three shifts can be understood as responses to internal criticism of a more familiar strategy for understanding intentionality. A common philosophical starting point nowadays is that intentional concepts like 'belief' and 'desire' arise in the context of interpreting or explaining other agents. Interpreters attribute beliefs, desires, and in some cases meaningful utterances to something, in the course of making sense of what it does. To take such an "intentional stance" (Dennett 1987) toward something is to understand it as acting for reasons, and thus as accountable to rational norms. The beliefs and desires it "has" are mostly the ones it rationally *ought* to have.[6] One *ought* to attribute such rational beliefs and desires to something in turn, if this strategy can succeed in making sense of it as rationally normative. Intentionality is thus an inherently normative concept, at two levels: its use by an interpreter is governed by norms, and in using it, the interpreter ascribes rational normativity to what she interprets.

6. One can be a rational agent without being a perfectly rational agent. That is why the beliefs and desires attributed need only *mostly* be the ones the agent rationally ought to have.

What are the norms for applying intentional concepts like 'belief,' 'desire,' or 'meaning'? There are two aspects to intentional normativity, on virtually everyone's account, which are necessary to prevent intentional interpretation from being entirely arbitrary and unconstrained. First, any one belief is only attributable if one can consistently attribute many. Interpretation of an agent must be constrained by holistic norms of internal, rational consistency. If an agent had only one belief, or if one could attribute massive internal inconsistencies in her beliefs and desires, any course of action might follow from them, and hence no course of action would be explained. Avoiding incompatibilities among the attributed beliefs and desires is an important constraint upon any such interpretation. Internal consistency is not enough, however, for there might be multiple ways of consistently attributing commitments and entitlements to a speaker/agent, and thus no good reason for taking any of them to be correct or incorrect. Hence, intentional interpretation must also be subject to "external" constraints from the circumstances in which it occurs. If one could attribute to an agent internally consistent networks of belief and desire that were wildly inappropriate to her circumstances, once again any course of action and utterance might follow from these attributions, and hence none would be explained.

Dennett (1987) then argues that taking the intentional stance is appropriate wherever it would yield successful prediction of the actual behavior of an interpreted system, on the hypothesis that the system mostly acts rationally. Both the intentionality of the system and the specific network of beliefs and desires attributed are vindicated by empirical verification that the system is appropriately constrained by its circumstances.[7] Dennett's approach cannot provide a complete account of intentionality, however, for one cannot readily explain in the same way the intentionality of the system that employs the intentional stance. I do not explain my own behavior in interpreting someone else intentionally by its contribution to better

7. Dennett's view has the consequence, for relatively simple systems such as thermostats, that they are both physically determined systems, completely explainable by natural laws, and intentional systems, explainable by the internal consistency and external appropriateness of their attributed beliefs and desires. For Dennett, the choice of which strategy to use in such cases is entirely pragmatic. Indeed, the same is true for more complicated systems such as ourselves. The only relevant difference between us and thermostats is that in our case, the physical stance is made pragmatically inappropriate by our physical complexity, leaving the intentional stance (and the biological-functional design stance) as the only useful strategies for understanding most human activity.

prediction of my behavior; if I tried to do so, assessing the confirmation of these predictions would still remain outside the scope of the explanation. On Dennett's construal, therefore, the explanatory scope of the intentional stance is limited by its ineliminable asymmetry.

Donald Davidson tries to remedy this limitation by offering a more symmetrical account of intentional interpretation that modifies its external constraints. For Davidson, what vindicates the intentionality of the interpreted agent, as well as the beliefs and desires attributed to her, is not that she becomes mostly predictable, but that one can attribute to her systematically interconnected beliefs that are mostly true. To be sure, they are recognized as mostly true from the standpoint of the interpreter, but there is now nothing privileged about that standpoint. The interpreter's own beliefs and desires can in turn be interpreted as mostly true beliefs in the idiom of someone else (perhaps that of the person interpreted). Indeed, these two (or more) standpoints could be reflexively integrated. I do, after all, interpret others as interpreting me (and as interpreting me interpreting them, and so forth, until one's head spins).[8] Hence, the interpretive character of the intentional stance itself becomes part of what is being interpreted, and the intentional domain as a whole is thereby expressed within the overall conception of intentional interpretation.

The problem, however, is that the entire interpretive enterprise then loses its indispensable external constraints, just as McDowell (1994) argued by a different route.[9] On Davidson's construal, what is being interpreted is not just someone else, but the entire interactive discursive practice that incorporates both interpreter and interpreted.[10] Davidsonian radical interpretation treats another's utterances not as an internally coherent system in its own right (a conceptual scheme), but as sufficiently coherent with the interpreter's own beliefs and desires to be intelligible to her. But how is coherence with the interpreter's beliefs supposed to help? The inter-

8. That such reflexive interpretation cannot be fully carried out is not a problem, for Davidsonian interpretations are modeled as recursive theories, which could generate these endless extensions from a finite (and hence in principle comprehensible) set of axioms.

9. See my earlier discussion of McDowell's criticisms at the end of chapter 2. As I noted there, the point discussed here is essential to my reconstruction of McDowell's argument against Davidson.

10. Brandom's implicit point, that Davidsonian radical interpretation concerns a two-person (or more) interaction rather than just an interpretation of one speaker, has a parallel in Fuller's argument (1992a) that almost all experimental work in cognitive psychology makes a similar mistake. Cognitivist experiments study interactions between experimenters and subjects rather than just the cognitive processes of individual subjects.

preter's beliefs are not intrinsically meaningful, but are instead likewise only meaningful because they are interpretable as a coherent system of mostly true beliefs in the idiom of an interpreter. The hope is to vindicate together the rational normativity and intentional contentfulness of the interpreted, the interpreter, and the whole interconnected community of agents whose activities can be construed sensibly as part of the same complex discursive enterprise.[11] Davidson gains the symmetry and comprehensiveness that eluded Dennett, but at the cost of losing the external normative constraint needed to make the entire interpretive enterprise determinate. The appearance of external constraint is maintained, because the beliefs attributed to one speaker seem to be constrained by objective facts (as expressed in the idiom of the interpreter). This appearance is fundamentally misleading, however, because the one speaker is only being interpreted as a believer at all through incorporation of her utterances and other commitments within a larger holistic network of beliefs.[12] Their character as beliefs (as contentful claims) in turn is only vindicated by interpretive incorporation within the *same* holistic interpretive network. For Davidson, what justifies a belief is always other beliefs.

Brandom thus makes explicit a mostly implicit feature of Davidson's interpretivist account of intentionality. Both accounts are interpreting not speakers but interactive discursive practices, within which speakers' various sayings and doings acquire the status of meaningful performances. This emphasis upon discursive practices distinguishes Brandom's and Davidson's work from virtually every other familiar account of intentionality. Alternative accounts not only ascribe intentionality primarily to states or performances of individuals; they also foreground the intentionality of one particular kind of state or performance (either mental states, instrumen-

11. The intended inclusiveness of Davidson's interpretive enterprise and the underlying difficulty are thus shown especially clearly by his arguments against the very idea of a conceptual scheme. There cannot be a radically different conceptual scheme, because its inclusion within the overall system of discursive interpretation is criterial for its character as conceptual. That is just another way of saying, however, that this system cannot be accountably connected to anything outside the system as a whole, and hence is not subject to any content-conferring, external normative constraint.

12. Brandom (1979) once thought that the shared commitments of a discursive community were sufficient to establish norms for the correctness or incorrectness of the intentional states and performances of its individual participants. He has eventually (1994) recognized that the community's shared commitments can themselves count as normative attitudes (and hence confer normative status) only if they are themselves responsive to norms. He accounts for this collective responsiveness by understanding them as adopting a stance toward *objects* to which we take ourselves to be answerable in our judgments.

tally rational actions, or linguistic representations), and treat the other two as derivative. Brandom and Davidson argue, however, that these three elements are integral aspects of discursive practice, and that none of them can be understood independently. To attribute intentionality requires the attribution "all at once" of purposive action, meaningful linguistic performances, and additional intentional commitments not manifest in such performances.[13] Language still takes a certain priority in Brandom's view, in the sense that he thinks intentionality in the fullest sense can only be attributed by and to language-using beings.[14] Expressions can only be understood as linguistic, however, by considering how they are used in discursive practices.

This shift from interpreting individuals to interpreting discursive practices leads fairly directly to Brandom's second strategic move. What are usually construed as mental states (beliefs or desires) turn out to be, not properties of individuals, but quasi-institutional statuses publicly instituted within ongoing discursive practices. This practical-discursive character of mental states was already implicit in a pervasive ambiguity within ordinary talk about beliefs, desires, and other supposedly psychological states. An example highlights two different senses in which one can be said to have a belief (and likewise for desires, intentions, hopes, and other propositional attitudes): Before reading this sentence, did you believe that zebras do not wear polka-dotted underwear? In one sense of 'belief,' obviously not, because the thought had likely never crossed your mind. Presumably no such claim had ever been explicitly uttered or entertained by you, nor could it plausibly be thought to have been somehow unconsciously represented in your mind. Yet in another equally important sense, of course you believed that zebras do not wear underwear of any design whatsoever, and have believed that ever since you were capable of forming beliefs of any kind about zebras. The ambiguity is thus between claims actually endorsed by the believer and claims that the believer *ought* to endorse whether or not she ever actually does so (Brandom 1994, 236–37).

13. Recall, however, that whether or not an entity is a language user in the relevant sense cannot be assessed as a straightforward matter of fact, but is a normative question. As I argued in chapter 3, correct ascription of linguistic competence depends not only upon what putative speakers say and do, but upon the significance of what they do *not* say or do.

14. His view thus contrasts to those views that foreground either rational agency or mental states, which often emphasize continuities between the intentionality of language users and that of nonlinguistic animals.

Brandom proposes to remove this ambiguity by no longer treating beliefs (or other propositional attitudes) as de facto psychological states. He highlights this point by replacing 'belief' with the explicitly normative 'commitment.'[15] Commitments are not private mental states, but publicly accessible normative statuses. Whether I am committed to a certain claim depends upon what has been said and done when, not upon something accessible introspectively. Such commitments then take two importantly distinct forms: commitments avowed and commitments attributed by others. These two forms correspond to the ambiguous senses of belief, but the notion of commitment is not similarly ambiguous. Only avowed beliefs could even plausibly be construed as internal, de facto mental states. Both forms of commitment, by contrast, are appropriately construed as public norms. Even avowals (or implicit dispositions to them) count as having a semantic content only if they are appropriately subject to normative assessment in public discursive practice.[16]

Brandom thus foregrounds the normativity of discursive practices, including the normative status of implicit practical and assertional commitments (what are usually thought of as beliefs, intentions, or desires). He hopes to understand intentionality and semantic content in normative terms, such as the correctness of uses or performances and the appropriateness of their circumstances and consequences. For Brandom, normativity and normative accountability provide the basis for understanding concep-

15. Strictly speaking, commitments and entitlements mark two importantly different kinds of normative status on Brandom's account, but I can overlook this difference for present purposes.

16. Although 'belief' parsed as a psychological state or disposition thus drops out of Brandom's lexicon, 'knowledge' retains an important interactive role. Use of the term 'know' expresses the hybrid normative status defeasibly invoked when one simultaneously attributes a commitment to another and undertakes the same commitment that one attributed. This interpretation of the normative and expressive significance of the *use* of the concept of knowledge thus meshes with the dynamic, deflationary account of knowing that I discussed in chapter 5 and have developed more extensively in *Engaging Science* (Rouse 1996b). Brandom and I both drop the idea that 'knowledge' refers to a theoretically coherent class of achievements, while insisting that the *use* of the concept has an important role in the ongoing assessment and development of discursive practices. Brandom's view also shares a crucial consequence with mine. In his terminology, participation in discursive practices commits one to the objective normativity of the commitments and entitlements one implicitly attributes to one another (we could all be wrong about any of the statuses we attribute or undertake), but no individual's or group's actual place within discursive practice authoritatively embodies such objectivity. In my terms, whose connection to Foucault was noted in chapter 5, that is a conception of knowing that eschews any standpoint of epistemic sovereignty.

tual content, truth and reference, and the alethic modalities of necessity and possibility (the relation between these modalities and normativity is, however, only hinted at in Brandom 1994). Brandom may then seem merely to have replaced one intractable philosophical problem (how to understand mental and linguistic representation) with another (how to understand the normative authority and force of discursive commitments).

Brandom's response to this problem begins by insisting that normative concepts are first and foremost implicit in practices. This emphasis follows from his Wittgensteinian criticism of "regulism" and "regularism," two alternative strategies that understand normativity as accord with explicit or explicable rules. Regulism (which identifies normative performances with those that are explicitly guided by rules) invokes a vicious regress: rules only specify norms if they can be followed correctly, but the correct application of the rule is a norm that must invoke a further rule, itself open to correct or incorrect application. Regularism (which identifies normative performances with those that merely exhibit regularities) runs afoul of the familiar gerrymandering problem. Any finite sequence of performances can be understood to accord with indefinitely many rules, so that a regularity alone cannot determine the norm that the performance obeys.[17] These two arguments respectively undermine intellectualist and naturalist explanations of norms in terms of non-normative performances or capacities.[18] Brandom's own aim, however, is to show how such implicitly normative practices could then be explicitly displayed, by showing how words and sentences must be used in order to express concepts. The project is thus expressive rather than explanatory, enabling discursive beings like us to articulate to ourselves what we are doing when we (already) implicitly understand ourselves as intentional. To solve in this way the problem McDowell saw in Davidson (or more generally, to resolve the dualism of nature and normativity), he must explicitly show how discursive performances are implicitly accountable to something other than just our own performances and commitments.

17. Brandom's reading of Wittgenstein on rules shows clearly why the end of regresses of justification must be construed not as a behavioral regularity (for the gerrymandering problem reemerges in grasping which regularity this is), but as a propriety. As I noted earlier, the invocation "This is what we do" in such circumstances has the rhetorical inflection of "We don't hit other children, do we?" (Wheeler 2000).

18. Thus, in terms of my distinction in the previous chapter, Brandom's arguments show why discursive practices must be understood normatively rather than as performative regularities.

Brandom's strategy is to show how the pragmatic norms governing discursive performances can account for the semantic significance of the linguistic expressions they employ. Semantic content accrues to performances (which thereby become linguistic utterances) when they are appropriately caught up in practices, not merely by being used in the right circumstances, but also by appropriately eliciting responses. In this way, the normative status of utterances and the expressions usable in such utterances replace representational relations as the basis for determining semantic content. Instead of claiming that utterances or beliefs have representational content (whether intrinsically, conventionally, or instrumentally), Brandom argues that they invoke commitments to some further linguistic moves and confer entitlements to others. Moreover, these types of status affect one another; commitment to some performances or statuses, for example, may remove one's entitlement to others.

The semantics that Brandom develops on this basis works *inferentially* to institute conceptual roles. The occasioned use of an expression acquires conceptual content from the appropriateness of when and how it is used. Specifically, the meaning of such expressions-in-use can be characterized as a function from the (appropriate) circumstances of their use to the (appropriate) consequences of that use in those circumstances. More colloquially, conceptual content is determined jointly by whatever would count as good reasons for the use of an expression, and by the subsequent statuses or performances for which that use could in turn be a good reason. Brandom therefore follows Kant and Frege in taking judgments or sentences to be the primary bearers of conceptual content, since only the assertion of sentences can both provide reasons and be reasoned for.[19] The conceptual role and content of assertional components and complexes will then be established inferentially, while the pragmatics of other speech acts will be subsumed within Brandom's larger account of the inferential articulation of action.

This inferentialist account of meaning results in a thoroughgoing semantic holism. Since meaning is a matter of inferential relations, and sentences and their conceptual components acquire much of their inferential

19. Brandom argues for the pragmatic primacy of assertion, and the semantic primacy of assertible propositions, by claiming that while a language game that consisted solely of assertions would be intelligible, a language game that contained no assertions would not; its semantic content could not be construed in normative terms alone without semantic primitives.

structure in conjunction with other background commitments not shared among participants in discursive practice, no two speakers are likely to assign the same inferential relations to any particular utterance or expression. Such semantic holism is usually rejected out of hand, as making communication impossible. Surely it is absurd, the argument goes, to think that each change in background commitments changes the meanings of every other linguistic expression. Brandom responds that inferential holism is not at all debilitating if speakers can track the evolving differences in their inferential commitments. On his view, communication is a shared *practice* of tracking participants' inferential commitments and entitlements, rather than an exchange of semantic contents mediated by a currency of conventionally shared meanings. Meaning for Brandom is not a property of expressions or utterances, but instead belongs to the whole field of discursive practice surrounding those expressions and their deployment.

The result is a dynamic account of how discursive practices sustain semantic content. A contrast with Davidson usefully highlights this point. Davidson models interpretation of another's utterances, actions, and beliefs as the construction of a truth theory for a speaker. What each interpreter thereby aims to understand (each in her own terms) are the conditions under which the speaker would (at that time) assent to various sentences.[20] As speakers make new, previously unanticipated utterances, interpreters must then implicitly construct a new truth theory to take these utterances into account. Brandom offers an alternative model of ongoing interpretation (rather than of synchronic interpretation) as "deontic scorekeeping." To participate in discursive practices is implicitly to keep track of the various commitments and entitlements accrued by participants, and to recognize the significance of their subsequent performances by changing the normative score in appropriate ways. Meaning is therefore construed as the ongoing reconfiguration of a dynamic field, rather than as a static structure. The symmetric and interactive character of interpretation is also expressed by the essentially perspectival character of the scorekeeping model. Score is kept only by participants in the practice, who

20. Davidsonian truth theories are also strongly holistic. A crucial constraint on the construction of such theories are their recursive structure, which requires the use of the subsentential components of any one sentence to be consistently aligned with their use in other sentences. This inferential interconnectedness, which Brandom's semantics wears on its sleeve, is mostly buried within Davidson's technical apparatus, although it is also a crucial aspect of his theory.

thereby attribute commitments and entitlements to themselves as well as to others. Ascriptions of commitment or entitlement thus take two distinctive forms: attributions (to others) and acknowledgments (which performatively undertake commitments and claim entitlements).

Brandom's inferential semantics extends his earlier criticism of regulism. Just as norms implicit in practice must extend further than the explicit formulation of norms as rules, so inferential norms must extend beyond explicit rules of inference. Brandom's semantics takes the materially good inference as its most basic concept. An inference is materially good if commitment or entitlement to its premises ought to confer commitment or entitlement to its conclusion.[21] The use of concepts, for example, incorporates material-inferential commitments: to believe that Middletown is south of Hartford, for example, commits me by material inference to believe that Hartford is north of Middletown. Which inferences actually are materially good, of course, is open to challenge within the practice of asking for and giving reasons. Brandom avoids vicious infinite regresses in the specification of semantic content by recognizing default entitlements, which are defeasible but not open to general skeptical challenge. Participants in linguistic practices are responsible for providing reasons sufficient to entitle them to their assertional commitments whenever legitimately challenged, but not all challenges are legitimate. Default entitlements place the burden of inferential justification upon the challenge rather than the commitment challenged. One typically acquires default entitlements by inheritance, either by taking over the same content expressed in another performance or by recognizing new content as the justified consequence of prior entitlements. The notion of 'inference' is thereby expanded to include interpersonal deference, through which I become entitled to claim *p*, by deferring to your appropriately entitled utterance of *p*.

An important virtue of Brandom's reliance upon the notion of materially good inferences is that it quite naturally yields an account of formally good inferences.[22] Such an account reverses more familiar conceptions of

21. Materially good inference plays a role in Brandom's semantics and pragmatics comparable to the role of assent and denial in Davidson's or Quine's accounts of radical interpretation or translation. Which inferences one takes a putative speaker to endorse provides the data for an interpretive assessment of whether the speaker's performances ought to be regarded as having semantic content at all, in the way that a Davidsonian or Quinean would begin a similar enterprise on the basis of what sentences one takes the speaker to assent to.

22. The general structure of the argument for this claim involves the notion of a substitution-inference (the inference from one sentence to another in which some compo-

the relation between formal logic and ordinary language use. Logic is often understood to have its own, independent justification, which then constrains what one ought to say or believe. On Brandom's account, logic would instead acquire its justification from its elegant expression of norms already implicit in discursive practices. The static structures of formal logic express a pattern of invariances within the normative dynamics of discursive practice rather than a prior constraint upon it.[23] Moreover, the priority of material-inferential norms over formal logical relations is supported by contrast to the inability to express material inferences in formal terms. One cannot readily extend a prior account of formal validity to make sense of materially good inferences, except by stipulatively denying the possibility of materially good but formally invalid inferences.[24]

Since inferential norms are embedded throughout discursive practice (which on Brandom's account is simply the "game of giving and asking for reasons"), logic plays an expressive rather than a justificatory role. The introduction of logical vocabulary to a discursive practice previously lacking it would primarily enable inferential norms themselves to be asserted explicitly, and thereby opened to explicit challenge and justification. Locutions that express conditionals and negation play an especially significant role, by enabling material inferences and material incompatibilities to be articulated as assertions for which reasons can be solicited or given. But Brandom also treats semantic vocabulary, and his own semantic and prag-

---

nent of the sentence has been replaced by a different expression). Given a prior determination of which material inferences ought to be accepted within a discursive practice, one can partition the vocabulary of the practice into two parts in various ways. Having done so, one can ask whether the systematic substitution of terms in one partial vocabulary for one another would change materially good inferences into bad ones. The validity of formal logical patterns of inference could then be expressed by the partition of discursive practices into logical and nonlogical vocabularies, such that the substitution of nonlogical terms for one another in the inferences already taken to be materially good would not turn any materially good inferences into bad ones.

23. Brandom's book indicates a strategy for how to carry out such an account rather than a working out of the technical apparatus needed to vindicate that strategy. Some of this apparatus is constructively developed by Lance and Kremer (1991, 1996) and Lance (forthcoming).

24. As Friedman notes (1996, 2000), a central issue in the philosophical exchanges between Carnap and Heidegger in the 1930s concerned where to locate relative priority in intelligibility between formal inferential relations and what Heidegger called the understanding of being. Since understanding of being is a practical, normative capacity to deal appropriately with one's surroundings (including social practices and the significations they incorporate), Brandom's Sellarsian argument for the priority of material inference to formal inference is perhaps the most important subsequent philosophical achievement that further develops Heidegger's side of that exchange.

matic theory, as expressive in this same way. In particular, Brandom shows how to accommodate in normative terms the representational locutions that play fundamental roles in other approaches to semantics. He does so by displaying in those terms just what expressive resources such locutions would add to discursive practices previously lacking them.

In the crucial case of 'true' and 'refers' (crucial, because these concepts play fundamental roles in rival semantic theories), Brandom treats the terms not as properties or relations at all, but as "anaphoric proform-forming operators." This extension of the prosentential theory of truth treats 'is true' as forming a "prosentence" (analogous to a pronoun) whose semantic content is anaphorically dependent upon another linguistic tokening. In the most familiar cases, '*p* is true' is materially equivalent to *p*, but the anaphoric construction extends the expressive role of 'is true' into quantificational and nominalized contexts that indicate a propositional content without articulating it. On this account, facts are true claims (in the sense of what can be truly claimed, not the claiming of them), but there are no distinctively semantic facts over and above the facts expressed in true claims (thereby supporting the conclusions I drew from Quine's indeterminacy thesis in chapter 3). The truth operator provides new expressive resources to articulate what one (already) does in making claims, but it thereby articulates no new facts. Brandom treats 'refers' in similar fashion as a pronoun-forming construction that enables its user to discuss what some contextually specifiable semantic tokening was about, without needing to be able to identify its referent directly.[25] He thus accounts for the familiar uses of 'true' and 'refers,' without having to posit or explicate a representational relation between terms and objects or sentences and states of affairs. These terms instead express norms implicitly governing the use of linguistic expressions in discursive practices.

25. Note that the deflationary account of 'know' from *Engaging Science,* which I briefly mentioned above in relation to Brandom's treatment of 'knowledge' as a hybrid deontic status (1994, 201–21), directly parallels Brandom's deflationary treatment of semantics. There are likewise no *epistemic* facts, no portion of the world that would constitute knowings explainable non-normatively. There are, of course, the pragmatics of discursive scorekeeping, the facts, if you will, about who attributes or acknowledges which claims, but these are perspectival assessments of a normative status within the ongoing practice of giving and asking for reasons, not objective relations between representations and things represented. Just as Brandom's pronominal account of the expressive function of 'refers' leaves no space for a causal theory of reference, so these accounts of the pragmatic, normative function of 'know' leave no space for a causal-reliabilist or other strongly non-normative theory of knowledge.

The stage is now set for Brandom's third strategic shift. So far, he has shown how the full range of what can be articulated in language can be understood in terms of inferential norms implicit within discursive practices.[26] This exposition of intentionality in terms of the pragmatic norms implicit in discursive practices only serves its purposes, however, if Brandom can show how these norms are authoritatively binding upon participants in discursive practices. This issue is especially serious, because he has characterized the normativity of discursive practice in strictly intralinguistic (inferential) terms. On Brandom's analysis, to assert a sentence is to undertake a commitment to assert other sentences under appropriate circumstances (for example, to give reasons for one's assertion if it were appropriately challenged, or to show the consequences of commitment to that assertion). Fulfilling such commitments entitles you to make still other discursive commitments. Unless the inferential relations among sentential commitments and entitlements have consequences beyond this intralinguistic network, however, it is not clear what would be at stake in making a discursive commitment in the first place. The point can be seen by an analogy that transposes Brandom's account into financial rather than discursive scorekeeping. Imagine a complex practice of tracking implicit exchanges in a newly introduced currency, in which participants were committed to some payments in this currency and entitled to others. Unless such commitments and entitlements were redeemable by something other than further transactions in this currency, however, there could be no serious meaning to (nothing at stake in) such exchanges.

This issue for Brandom is analogous to the problem McDowell uncovered in Davidson, and to the related failures Heidegger and Neurath found in Husserl's and Carnap's appeals to logical or transcendental necessity: he must make intelligible the force of logical and semantic norms in the concrete historical, material world. Brandom proposes to show that something is really at stake in discursive practices, by showing their accountability to objects independent of what one says about them. Such accountability is possible, because in addition to talking about objects, speakers

26. Strictly speaking, I have not yet discussed some important aspects of the inferential apparatus Brandom develops to express the full range of linguistic articulation in his terms. In particular, I discuss below his account of how to extend the inferential norms governing sentences to the inferential roles of subsentential expressions within discursive practices. I do so because this apparatus plays a crucial role in his account of how discursive norms have authority and force.

interact with them causally in perception and action.[27] Once again, the contrast to Davidson is illuminating. For Davidson, discursive exchanges and causal interactions run on parallel tracks. Both are going on all the time, but there is no intelligible way to connect one track with the other.[28] Brandom instead wants to show how perceptual and practical causal interactions with objects are normatively connected to discursive interactions, such that what we say is accountable to what we see and what we do (not merely, as for Davidson, to what we say and believe about what we see and do).

Brandom proposes to make the shift from representation to objective normativity in two steps. He first shows how discursive interactions with other speakers depend upon the ability to connect one another's expressions to the same "objects" (I put the term in double quotes here, because at this point in his discussion, an object is specified solely by the inferential role of singular terms in discursive interactions). He then shows how the inferential role of singular terms can be extended to encompass perceptual and practical interactions with causally efficacious objects. In the remainder of this section, I will explain the first step in his account and indicate the indispensable importance of the second step. The rest of the chapter will be devoted to exposition, criticism, and reconstruction of the second step. My reconstruction will preserve the core of Brandom's inferential-normative account of discursive practices but dramatically transform its relation to causal interaction with one's surroundings.

Brandom's initial difficulty is that the inferential norms that account for semantic content seem only applicable to whole sentences, and not to terms referring to objects. Only sentences can serve as premises or conclusions to inferences. So Brandom's immediate task is to develop an additional conceptual apparatus to express in normative terms the referential use of singular terms. He does so by showing that sentences have two dis-

27. Brandom thinks he can vindicate the normativity of discursive practices by connecting them to those objects (such as mathematical objects) whose independence from what we say about them is not causal independence. The narrow domain of mathematical (or other comparably a priori) semantic content establishable in this way could then be substantially augmented by the empirical content made accessible through perception and action. In later sections of this chapter, I show the utter failure of this aspiration to render semantic content independent of causal interaction with one's surroundings.

28. Davidson thinks the tracks are already connected, indeed overlaid one upon the other, since every belief or other normatively ascribed mental event is token-identical with a causal state. He nevertheless denies any systematic, rationally explicable relation between discursive norms and causal relations.

tinct components that work quite differently in substitution inferences. Singular terms substitute symmetrically for one another (whatever is true of Frege is true of the author of "Über Sinn und Bedeutung," and vice versa). Sentence frames, the more general structure usually identified with predicates, substitute asymmetrically (whatever is true of reptiles is true of snakes, but *not* vice versa). This distinction allows Brandom to express the relation between singular terms and objects in terms of their inferential pragmatics, rather than in terms of reference or representation.[29]

The other semantic apparatus needed to articulate the referential dimension of discursive practices are anaphora, which express the token repeatability of inferential relations. Anaphoric expressions such as pronouns enable a discursive performance to inherit the inferential commitments and entitlements of another performance without having to articulate its specific content. Such expressions are crucial to keeping track of discursive commitments, because one can use them to talk about whatever someone else is talking about, without having to understand or endorse the concepts she used to talk about it. Anaphora are the linguistic expressions that enable communication to proceed in the absence of shared meanings.

This communicative role is central to Brandom's understanding of discursive practices. The meaning of any utterance or action is determined by its inferential role. What one says and does only has such inferential significance, however, when conjoined with other collateral commitments (the beliefs and desires attributed to one another to make sense of what they say and do). But interlocutors do not share the same collateral commitments, and hence they invariably understand one another's performances differently. A central feature of the normative scorekeeping implicit in discursive practice is the need to keep two parallel sets of books. Speakers must implicitly track the differences between the inferential significance

29. This strategy for understanding the inferential role of singular terms also straightforwardly shows how to express quantificational relations. Brandom goes considerably further, however. His very long and difficult sixth chapter cleverly demonstrates that the different substitution-inferential roles of singular terms and sentence frames are a necessary condition for *explicitly* articulating inferential relations at all. Hence, he concludes, discursive practices must utilize singular terms "because it is so important to have something that means what *conditionals* mean" (1994, 404). It would be possible to engage in discursive practices that do not implicitly distinguish these two inferential roles for sentential subcomponents, and even to attribute original intentionality to one another implicitly, but it then would not be possible to articulate one's inferential norms as debatable assertions in the form of conditionals.

of someone else's utterances, given their own collateral commitments, and their significance in concert with the different collateral commitments attributable to the speaker.[30]

Thus, what singular terms and anaphora collectively contribute to discursive practices is the ability to coordinate one's own and others' utterances, by connecting them to objects in shared surroundings. Brandom understands the familiar distinction between *de re* and *de dicto* uses of referring expressions as a device that connects the inferential relations attributed to others' utterances to the objects whose characteristics and interactions are taken to govern the correctness or incorrectness of one's own utterances. Reserving the term 'of' to preface the commitments I acknowledge *de re,* and 'that' to introduce those I attribute *de dicto* to others but do not acknowledge, I can make sense of another's odd claims by taking him, for example, to believe *of* the earth *that* it is the stationary center of the universe, and *of* the sun *that* it revolves in an epicyclically circular orbit around the earth. I understand him as talking about the same things I am, but also as having rather different beliefs about them. These are familiar points, but Brandom's signal achievement is to show how to make them without needing a prior concept of reference or representation. Brandom thereby secures both the communicative and representational dimensions of discursive practice simultaneously and mutually. I can track others' performances as semantically contentful discursive moves only by understanding them as reasoning about the same objects in our shared surroundings to which my own claims and other commitments are accountable.

This communicative use of expressions to refer to objects has a fundamental normative significance. Singular terms are inferentially interchangeable. To use a singular term *de re,* therefore, is to commit oneself to endorse every other expression that is correctly substitutable for it, even those one does not yet endorse or understand. The resulting accountability to conceptual contents that exceed or even contradict one's present understanding is what Brandom calls "tactile Fregeanism."[31] The tactile metaphor empha-

30. It is not only the inferential consequences of utterances that differ perspectivally, but also the reasons that would entitle one to make the utterance in the first place. Brandom is offering a "two-way" theory of meaning, in which the meaning of a performance is jointly expressed by what would be a good reason for it, and its inferential consequences. Brandom's semantic holism and perspectivalism follow from the dependence of both aspects of meaning upon which collateral commitments are conjoined to an utterance or action.

31. Brandom's tactile Fregeanism bears interesting parallels to Derrida's insistence ([1967] 1973) upon the inevitable supplement to any signification, or his account (1988)

sizes that although we only partially grasp the concepts we use and the objects we talk about, our normative reach exceeds our grasp. In using those concepts we are responsible for all of their correct inferential connections, and thus for everything true of those objects, whether we know it or not. Although Brandom does not put it quite this way, his account of discursive practices thus subsumes them under the more general concept of responsible action: we can speak meaningfully to the extent that we can hold one another responsible for our actions (including our speech acts) and their consequences.

This first step in Brandom's treatment of discursive normativity has accounted for the indispensable role of talk *about* objects in discursive practices. Such talk is needed to triangulate one's own discursive commitments and the perspectivally different commitments of others with objects to which both are responsible. Yet what must ultimately serve this role is not talk about objects, but the objects themselves. Brandom therefore cannot, and does not, limit discursive practice to intralinguistic practice. Discursive practices cannot be just a tissue of words, but must incorporate the material, causal circumstances to which they are accountable. They thus cannot be social practices in a sense that excludes the natural world, but must be the discursive configuration of the *entire* world as a meaningful setting for action.[32]

So far, however, I have only examined Brandom's treatment of the intralinguistic aspects of discursive practices. Within those limits, I endorse what Brandom says and take it over as an important component of my own position. Where we differ is in how to understand the crucial connection between talk about shared material circumstances and those circumstances themselves. Brandom acknowledges from the outset that he confers theoretical primacy upon the inferential articulation of concepts in discourse. He places sapience, the capacity to reason and understand, at the core of original intentionality, rather than the sentience humans share with other animals or the differential responsiveness shared with thermostats and automatic door openers. Brandom is right to avoid taking sen-

of the citationality of signs. Derrida, of course, would not endorse Brandom's commitment to the explicative and constitutive priority of asserting over other discursive performances.

32. Rouse 1996b, chap. 5, provides a more general account of practices as incorporating their material setting, but that account recognizes that there is a discursive dimension to all such practices, even if no words are actually employed in them.

tience or differential responsiveness as the starting point for understanding intentionality, but there is another important possibility he has not considered: practical competence in coping with one's surroundings. My task in the remainder of this chapter will be to work out how and why to situate Brandom's account of sapient participation in discursive practices as part of a larger conception of practical competence.

Brandom is acutely aware of this concern, even though he does not include competence among the possible contrasts to sapience as the primary mark of intentionality. His response is to understand perception and action as causal interactions with objects, which must also be inferentially connected to talk about them so that the talk can be responsible to the objects. He would thereby retain sapience as the crucial aspect of intentionality, because perception and action then only figure in discursive practice as conceptually (inferentially) articulated. What matters in perception, for example, is not an awareness of qualia, or the ability to respond differentially to red and non-red things, but the ability to respond inferentially to them. One does not perceive something as red in this sense unless one also displays a capacity to infer correctly that it is therefore not green, is colored, is not in complete darkness, and perhaps even that it is not a U.S. dollar bill. Similarly, action occurs only under a description, in the sense that its status as an action depends upon attributing to the actor an inferentially articulable understanding of what she does.

Brandom's account of perception and action as both inferentially articulated and causally efficacious interaction with one's surroundings is an initially promising attempt to overcome the dualism of nature and normativity. This way of construing the inferential and causal dimensions of human life aims to meet the challenges posed by Sellars (to conjoin the manifest and scientific images of "man") and by Heidegger (to understand intentional comportment as a unified phenomenon of "being-in-the-world," without losing sight of either its materiality or its existential significance). Without question, he has succeeded in showing how object-talk can be understood without presupposing semantic or representational locutions. Thus, if his interpretation of perception and action are adequate, he can plausibly claim to have articulated the full range of discursive, practical, and perceptual engagement with the world in terms of the normative responsibilities incurred by (being correctly taken as) participating in discursively articulable practices. I therefore turn to Brandom's account of perception and action.

Perception and Action

Consider first how perception and action are related to language within discursive practices. Brandom claims at the outset (1994, 234) that while perception and action provide important enrichments of what can be said discursively, there could be discursive practices without these enrichments. Inferential and deferential authority is essential for articulating semantic content at all, but the empirical content contributed observationally and the practical content expressed in nonlinguistic action are not. This claim both emphasizes the philosophical primacy Brandom accords to intralinguistic practice and allows for the autonomous intelligibility of discursive practices like mathematics that need have no empirical content. The claim is mistaken, however. Discursive performances could express semantic content only if they could be mistaken (not just internally inconsistent, but wrong about something else). Brandom obviously thinks that mathematics is a paradigm of discursive practices whose objects have authority over what is said, without requiring empirical or practical access to those objects. Yet Brandom has made no attempt to show that such a nonempirical discursive practice could be learnable and hence transmissible without the mediation of empirically anchored assertions about material objects.[33]

Even if one initially believed that such an argument could be developed, a further consideration shows that any such autonomy for intralinguistic practice would be incoherent. Discursive practices must be public. They require mutual recognition of accountability to norms governing the correct use of linguistic expressions. But this condition requires public access to token uses of those expressions. For me to defer to and infer from what others say, I must have access to their saying of it. Even if one did not need perceptual or practical access to objects referred to by linguistic expressions, there must be such access to the uttered expressions themselves. Without such access to public significations, nothing would distinguish the intentionality of discursive practices from the intentionality of private mental states.

Perhaps one could conceive such access to others' significations on the model of telepathy, or of electronic communications between computers. Yet familiar private language concerns then arise: there must be some pub-

33. Recall the importance of the conditions for language learning in establishing Quinean indeterminacy of meaning according to the line of argument developed in chapter 3.

lic, mutually accessible check upon each participant's understanding of which linguistic expression was actually uttered. The perceptual apparatus that would provide such checks need not resemble yours or mine, but it would have to be genuinely perceptual in the sense of being reliably responsive to something that is not just an internally accessible state. The upshot of this argument is that perceptual and practical engagement with the world is integral to Brandom's account of intentionality and language, and not merely a powerful expressive addition to an essential but minimal linguistic core.

So how does Brandom understand the place of perception and action within discursive practices? Brandom follows Sellars in understanding perception and action as language-entry and language-exit moves that complement the intralinguistic moves of inference and deference. If his account is not to reinstate an unintelligible dualism of norms and causes, however, these entry and exit moves must at a minimum have both a normative significance and a causal nexus. Brandom thus understands perception as having two components, a causal elicitation and an inferential articulation. Its elicitation is understood in externalist terms of (causal) reliability, while inferential articulation is internal to the discursive game of giving and asking for reasons. Yet Brandom takes these two components to be intertwined normatively, because he takes the relevant sense of 'reliability' to be the correct attribution of reliability rather than just as the outcome of a contextually reliable causal chain.[34] In practice (which always comes first in Brandom's analysis), perception enters into discursive interactions through implicit deference to observational reporting. One "implicitly endorses the propriety of a permissive inference from the informant's claiming that *p* to *p*" (1994, 218), by relying upon that inference in practice, and only subsequently by explicitly claiming that the informant is a reliable reporter. What matters to the propriety of the inference is only the reliability of the informant's differential responsiveness to contextually relevant features of the world. Hence the causally interactive contribution of per-

34. Brandom's reasoning for the primacy of normative attribution of perceptual reliability over a strict reliabilism involves the familiar gerrymandering problem that applies to any regularist account of normativity. Here what is gerrymandered is the reference class within which one assesses perceptual reliability. As Brandom concludes, "in the case where the claim one is making is true, there are always *some* reference classes with respect to which one is reliable" (1994, 211).

ception can be supplied by machines or nondiscursive organisms as readily as by fellow participants in discursive practices.[35]

Brandom treats nonlinguistic action by analogy to perception: action is the correctly attributed reliable disposition to *fulfill* acknowledged commitments, whereas perception is a similarly attributed disposition to *acquire* them. One is only an actor if one can recognize practical inferential relations: for example, that going to the beach is incompatible with working on my paper, while a promise to drive my friend to the beach commits me to going to the beach. Actions may involve more than the actors acknowledge, however. In flipping the switch, I may also start the fire, even if I am unaware of the short in the wire next to the flammable rubbish. Agency thus incorporates any and all reliable pathways for bringing about consequences, even if the responsible agent does not herself acknowledge the reliable connection.[36] In this way, the normative, inferential significance of action (and perception) could encompass causal connections that outrun one's understanding. Indeed, the normatively configured world of discursive practitioners extends to incorporate all the links in causal chains that are inferentially relevant to perception and action.

'Perception' and 'action' (or 'intention') thus parallel 'belief' in the ambiguity between acknowledged and unacknowledged commitments: in the narrow sense, I only perceive, believe, intend, or do what I (perhaps implicitly) acknowledge myself to perceive, believe, intend or do, but there is also a broader sense of perceptual, doxastic, intended, and practical content that incorporates what is correctly attributable to me regardless of what I would acknowledge.[37] Here also, Brandom removes the ambiguity by

35. Brandom notes that the hybrid deontic status of 'knowledge' can only be ascribed to those who count as participants in discursive practice, since attributing knowledge requires attributing (as well as acknowledging) an inferentially articulable conceptual content to the knower. But Brandom still recognizes an externalist dimension to knowing: the knower must be "in the space of reasons," but need not herself endorse (even implicitly) the inference of reliability. If others correctly take me to be a reliable reporter of the presence or absence of positron tracks or of 1965 Mustang convertibles, then I can *know* whether they are present, even if I do not believe that I know it.

36. The parallels to perception are sustained here: my capacity to discriminate objects or features reliably may outrun my recognition of my reliability, just as does my capacity to bring about states of affairs reliably. Moreover, in both cases, reliable differential perceptual discrimination and reliable differentially consequential agency may be mediated by the capacities and performances of other speakers and things, even though Brandom only attributes knowledge and agency to those beings that sufficiently recognize (in practice) the inferential significance of perception and action.

37. It is of interest in light of my subsequent arguments that Brandom explicitly notes the ambiguity of 'belief' (1994, 195) and 'intention' (1994, 261), but does not correspond-

treating both senses in strictly public terms of acknowledgments and attributions, with no mentalistic residue: the normative status of acknowledging a commitment or entitlement (perhaps only implicit in practice) replaces the mental state of awareness. Sapience, not sentience (or consciousness more generally), is the crucial dimension of perception and action for Brandom, and sapience is construed as emerging from discursive performances rather than mental events.

Although Brandom draws a close analogy between perception and action in their indispensable role connecting causes to norms, he explicitly notes three differences that disrupt any simple parallels. First, the causal reliability they involve have different significance: perceptual reliability yields truth, while practical reliability yields the achievement of practical goals. Second, perception and action have reciprocal temporal limitations: I cannot perceptually discern whether a car will be painted green in the future, whereas I cannot practically bring about that it was painted green in the past. Finally, but most important, Brandom argues that the testimonial significance of observational reporting has no adequate practical analogue.[38]

This last point, the possibility of testimonial deference to the claims of others, marks for Brandom "a fundamental asymmetry between expressing a belief by making a claim and expressing it by performing an action" (1994, 239). He takes the ability to inherit entitlements from testimony to be especially crucial for empirical practice, since testimony is the means for coordinating and tracking one another's responsiveness to objects in shared surroundings. My argument at the beginning of this section raises the stakes on this point, because the intelligibility of all discursive practice (not just its "empirical content") now depends upon the public and mutual accessibility of objects. Yet testimonial inheritance is precisely where the doxastic and practical dimensions of discursive practice diverge for Brandom. Your having reason to take a claim as true is, apart from the compli-

---

ingly characterize 'perceive' as ambiguous, even though the externalist dimension of perception figures prominently in his view. I suspect that too much of Brandom's differential treatment of perception turns on that ambiguity for him to acknowledge it so readily and explicitly.

38. Brandom does allow for a limited practical analogue to inheritance from perceptual testimony in cases of practical subordination to the commands of institutional superiors. When I argue below against Brandom's contrasting of perception and action, I shall also show why subordination of one's intentions to the commands of others is the wrong analogue to testimonial deference to others' observation reports.

cations of differing collateral commitments, a reason for me to take it as true. Indeed, Brandom argues, the default and challenge structure of asserting is only intelligible if testimony (most notably observational reporting) grants "a reassertion license to do just what the asserter did" (that carries "just the same authority as the status or performance that authorized it" (242, 243).

Your reasons for action do not similarly provide me with reasons to act, for I may not share your desires and interests. This difference, Brandom argues, has far-reaching significance:

> Differences among agents as to desires and preferences (or institutional roles) need not have the significance of indications of normative failures. Whenever two believers disagree, [by contrast,] a diagnosis of error or ignorance is appropriate for at least one of them. . . . We come with different bodies, and that by itself ensures different desires. . . . Our different bodies give us different perceptual perspective on the world as well, but belief as taking-true incorporates an implicit norm of commonality . . . aiming at a common set of beliefs that are equally good for all. (1994, 240)

As Brandom notes, this normative contrast between perception and action can be "colorfully" expressed by saying that "the Truth is One" but "the Good is not in the same way One" (240).[39] Brandom happily endorses the less colorful, deflationary expression, which accounts for a crucial difference between truth-reliability and success-reliability: the latter differs perspectivally according to one's aims, while the former does not. Significantly, however, Brandom offers no further account of the temporal polarity he ascribes to perception and action, which simply function as brute data in his discussion.

The Intertwining of Perception and Action

I can now begin to assess Brandom's understanding of perception and action as the sites where causes and norms come to bear upon one another.[40]

39. In his chapter 5, Brandom notes that claims like "the Truth is One" are "one sort of truth and reference talk that is not recoverable on a prosentential rendering of 'true'" (1994, 323). If my arguments are correct, this recognition did not prevent him from having been misled by such talk in his chapter 4.

40. A similar assessment could also be developed for McDowell's account (1994) of how the exercise of conceptual spontaneity (his analogue to what Brandom calls the inferential normativity of discursive practices) must always be accountable to conceptually articulated experience. Like Brandom, McDowell seeks to make conceptual norms objectively

What is at issue in this assessment is not just the extent of the parallel between perception and action, however. In showing Brandom's distinction between perception and action to be untenable, and reconceiving perception and action as intertwined, I also fundamentally reconfigure Brandom's account. Brandom conceived the intralinguistic aspects of discursive practice as the core of intentionality and understanding. Perception and action are then tacked on to either end of language and thought, as complements that merely extend one's conceptual capacities. Properly reconceived, I argue, practical/perceptual coping with one's surroundings should instead be placed at the center of human intentionality, with conceptually articulated discourse as a powerful extension of these integrated *bodily* capacities. Instead of asking how mind (Brandom's discursive sapience) reaches out to the world, I consider how bodily responsiveness to one's surroundings acquires conceptual articulation through the ability to speak and understand language. The potential importance of this shift for philosophical naturalism should be clear: causal interaction with the world as active, responsive bodies incorporates and expresses conceptual understanding.

Consider first an important lacuna in his discussion of perception and action. While Brandom very carefully shows how to understand in practical-normative terms the referential use of linguistic expressions, he leaves unexplicated the *perceptual* achievement of inferentially articulable "representations" and the *practical* accomplishment of inferentially significant actions. His explication of intentionality halts at the point where participants in discursive practices, when properly "placed," are appropriately taken as reliable reporters of conceptually articulated facts about objects and reliable performers of actions that fulfill their commitments.[41]

---

accountable by showing how they incorporate perceptual encounters with objects. McDowell's account would nevertheless need to be developed further in three important respects before these issues could even be joined. His emphasis upon experience would need to be expanded to encompass action, both would need to be considered as causal interactions with the world rather than as experience, and his understanding of conceptual spontaneity as mental rather than as discursive practice would need to be revised. These revisions are especially important to my concerns, because without them, McDowell could not do justice to the practical, material, and discursive aspects of scientific practices. To develop parallel arguments against McDowell to the same extent that I do against Brandom would therefore seriously distract from my main concern.

41. "Proper" placement includes not merely spatiotemporal location and orientation, but also the development of discriminatory capacities that make the perceiver sensitive to relevant differences, whether among colors, words, moods, hornbeams, or mu-mesons. In each case, however, Brandom attends to the "downstream" inferential implications of ap-

In Brandom's defense, he is primarily interested in understanding the linguistic dimension of intentionality, and for that purpose, the achievement of reportorial reliability and practical competences might justifiably be taken for granted. This rejoinder presumes, however, that a more fine-grained account of the practical normativity of perceiving and acting would be compatible with the remainder of Brandom's account of discursive practices. As I shall show, this presumption cannot be sustained, so that further reflection upon perception and action requires fundamental revisions in Brandom's understanding of discursive practice. I should note, however, that at issue are the practical skills involved in reliable observational reporting and competent action, skills that are publicly expressed and open to normative correction. I am not denying Brandom's entitlement to take for granted perceptual and motor physiology; we can usually discriminate reliable and unreliable reporting, or competent and incompetent action, without having to investigate the physiological mechanisms enabling them. I only challenge Brandom's entitlement to leave implicit the practical, corrigible skills involved in reliable observational reporting and practical competence.

Let us return to Brandom's distinctions between the inferential roles of perception and action. Recall first that, on Brandom's view, perception is not merely a reliable differential responsiveness to features of the world, but a conceptual grasp of objects.[42] Moreover, he recognizes that a skillful, and hence normatively articulable achievement underlies the display of such a conceptual grasp in reliable reporting. The ability of an accelerator technician to recognize the tracks of mu-mesons, of a naturalist to identify hornbeams, or of a decorator to discriminate color compatibilities requires the acquisition of distinctive perceptual skills, where some grasp of conceptual significance is embedded in perception (I needn't have mastered all of the inferential relations invoked in recognizing a mu-meson, but if I have no practical grasp of them, I cannot have *perceived* a mu-meson.) These are acquired capacities not shared by all otherwise competent observers.

propriate discrimination, without considering the "upstream" perceptual achievement of discriminatory capacities (or the further "downstream" achievement of different practical capacities).

42. Perception is conceptual, for Brandom, if one can recognize and respond appropriately to the inferential significance of one's differential responsiveness. One perceives objects if one's inferential resources include the ability to recognize appropriate symmetric substitution inferences, and thereby to identify two distinct token performances as inferentially interchangeable expressions about the same object.

Brandom also acknowledges the possibility of technological extensions of perceptual skills and capacities: unaided, human beings cannot reliably report the presence of x-rays, but some now can do so with the aid of properly functioning apparatus. Indeed, although Brandom does not explicitly note this point, the ability of language users to discriminate sentences, words, and the phonemes or marks that compose them is a necessary condition for the possibility of participation in discursive practices.[43] Perceptual reliability is hence a means of language entry in a much stronger sense than Brandom acknowledges; moreover, such entry requires a reliable ability to produce phonemes and marks that mesh appropriately with other speakers' capacities for perceptual discrimination.

Perception in Brandom's conceptually articulated sense also involves actions whose reliability conditions are practical. The practical dimension of perceiving is often ignored by attending only to visual and auditory examples, whose implicit practical dimension is unobtrusive. Philosophers now more readily reject a naive empiricism by recognizing the *conceptual* activity involved in perceiving, but they less frequently attend to the *bodily* activity that is also required. Consider first, then, the tactile perception that contributes to reliable reporting about textured surfaces. Optimum attentiveness to texture requires a successful movement of the hand across the surface, without which one cannot adequately discriminate the texture.

Now consider what happens perceptually in cases of practical failure. To take a different example, if while anticipating a sip of milk, one actually imbibes grapefruit juice, one can readily recognize the failure ("Yuck, that's not milk!"), but is at first poorly positioned to report reliably what one did taste. Only after "resetting" oneself to explore the unexpected taste can one's discriminatory capacities come back into play.[44] The exploratory activities that are carefully cultivated in a more specialized discriminatory activity like wine tasting are thus an ordinary but implicit part of everyday perceptual activity. Moreover, visual and auditory perception are not fundamentally different. Successfully seeing something obviously involves getting oneself to the right place and turning in the right direction. But it also requires appropriate scanning, focusing, tracking, and registering, which

43. The perceptual discrimination of phonemes and written symbols is a good example of normatively adjudicated facts. Physical regularities notoriously do not account for what does or does not count as the successful articulation of a linguistic expression, or of which expression is articulated.

44. I borrow this example, slightly modified, from Dreyfus 1979, 242.

are ways of actively directing one's body. Sounds, of course, often intrude quite involuntarily, but just as in the example of unprepared tasting, we are poorly positioned to register and recognize what we hear when startled by a noise until we reorient ourselves to listen.[45] Here, too, we often have to turn, move, or set ourselves properly in order to discriminate sounds, and thus to hear. Looking and listening are indispensable aspects of seeing and hearing. Since Brandom's account of perception emphasizes its inferential articulation, external reliability, and discursive corrigibility, I should note that these features also characterize perceptual *activity*. Perceptual activity is publicly corrigible independent of any assessment of the resulting observation reports. One can see that someone is not listening, or failing to keep her eyes on the ball, without having to wait to be told that she didn't hear or didn't see. Perceiving includes corporally active, publicly corrigible comportments that involve practical success or failure.

The practical dimension of reliable reporting becomes more rather than less prominent in moving from everyday attentiveness to one's surroundings into the more disciplined observational context of scientific practices. Learning to be a reliable reporter in these settings typically involves learning how to use instruments properly, and often requires preparing and manipulating what one is trying to see. In many contexts, the relevant empirical inputs bypass human observational capacities altogether, as machines produce empirical data in readable rather than observable forms. Here, what is involved in reliable reporting is constructing and utilizing the right technological extensions of one's naked perceptual abilities, and being attentive to their proper functioning. Such considerations led Ian Hacking (1983, chap. 10) to conclude, rightly, that in philosophical reflection upon science, "observation is over-rated" compared to the importance of being *observant,* attentive to the practically successful functioning of active intervention in the world.

In challenging Brandom's Sellarsian mapping of perception and action as language-entry and language-exit moves, it becomes evident how the priority Brandom ascribes to intralinguistic practices still retains a vestige of a representationalist conception of mind. As John Haugeland noted,

45. Todes (2001, 116) takes the experience of being startled to exemplify a significant structural feature of perceptual practice. When startled, we recognize *where* something is coming to pass, but not *what* it is, and consequently see the need for appropriate bodily anticipation in order to determine what is happening.

> the very distinction between perception and action is itself artificially emphasized and sharpened by the image of a central processor working *between* them, receiving "input" from the one and then (later) sending "output" to the other. The primary instance is rather *interaction,* which is simultaneously perceptive and active, richly integrated in real time. (1998, 221)

This failing is not substantially relieved by Brandom's reconception of the central processor as tracking inferential commitments rather than mental representations.

To undermine this vestigial representationalism, first recall the priority Brandom ascribes to material inferences. The autonomy of material inference allows the possibility of perfectly good inferences from one sentence to another, in the absence of the intervening premises needed to make the inference formally valid. Material practical inferences go further, from premises such as 'it is raining' to a commitment to take an umbrella. The next step extends Brandom's conception to recognize that perception and action can be inferentially related to one another without needing any intralinguistic moves at all. For example, the material practical inference from seeing a hanging curve ball to swinging for the fences is perfectly in order without it enthymematically presupposing an observation report about the pitch.[46] Furthermore, the same interaction with the world can serve both as a perceptual entry into empirically contentful inferences and as the conclusion to practical inferences. The undertaking that fulfills my inferentially acquired practical commitment to look through the crowd for my sister, for example, also entitles me to the perceptually acquired claim that she is here.

Such entanglements are pervasive, so that they obliterate the artificial distinction that Brandom has made between perceptual language entries and practical language exits. I have emphasized the ineliminable practical dimension of perception, but the reverse is true of most actions. Whether I am driving a car, flipping a switch, dribbling a basketball, or dancing, successful practical action is inextricably connected to a conceptually rich

46. It may sound strange to call such practical responsiveness an inference, but doing so is a natural extension of Brandom's claims that inferential norms are implicit in practice, and that action is inferentially incorporated into discursive practices. Once inferences can be made explicit, the implicit inferential normativity of actions becomes identifiable. In choosing this example, I acknowledge Haugeland (1998, chap. 7) as the authoritative source claiming an essential conceptual link between intentionality and baseball.

perceptual responsiveness to my surroundings, without involving explicit linguistic mediation. Discursive practice is therefore not a matter of intralinguistic moves situated between distinctive perceptual entries and practical exits. The normative status of semantic explicitness instead accrues to some practical/perceptual interactions with the world, against the background of other responsive interactions whose inferential significance is thereby recognizable as implicit within them.

Even if these arguments are sound, however, Brandom might still insist that the norms governing the testimonial inheritance of assertional entitlements *establish* a difference between the doxastic significance of perception and the practical significance of action. But is Brandom entitled to this commitment to the differential significance of testimonial inheritance? Consider first a telling but unobtrusive shift in how Brandom talks about perception and action. Brandom starts with a tripartite classification of perceptual language entries, intralinguistic moves, and practical language exits. He then unobtrusively substitutes a dual distinction between doxastic and practical commitments that result from these three kinds of performance. Only in passing does he note his assimilation of perception to intralinguistic moves, as doxastically rather than practically normative.[47] But the assimilation is crucial to his argument, since the differential significance of testimony concerns, not perception and action, but doxastic and practical commitments.

Brandom claims that reliable reporting entitles others to reassert the *same* content asserted in the original report. This claim, however, is already substantially altered by the qualification that Brandom takes conceptual content to be essentially perspectival:

> Conceptual contents, paradigmatically propositional ones, can genuinely be shared, but their perspectival nature means that doing so is mastering the coordinated system of scorekeeping perspectives, not passing something nonperspectival from hand to hand (or mouth to mouth). (1994, 590)

To talk about sameness of content is thus to bracket the pervasive and ineliminable differences in conceptual perspective that result from the in-

47. Brandom never explicitly says that perception should be conjoined with assertion as involving doxastic rather than practical commitments, but he does speak of "my noninferentially acquired doxastic commitment to the effect that there is a red thing in front of me" in contrast to a practical commitment to raise my arm (1994, 236).

ferential significance of differing collateral commitments and different embodied locations.[48] It might be more natural to say that, on Brandom's account, one could only inherit a perspectivally shifted conceptual content from others' observation reports.[49] In what follows, I shall try to avoid confusion by reserving the phrase "same content" for perspectivally invariant commonalities and develop other locutions to express what connects two irreducibly different "perspectives."[50]

Suppose for the moment that one could hold such collateral propositional commitments constant. Would there remain significant perspectival variations in what is inherited from observation reports? In the prototype face-to-face communication, I must still reorient my interlocutor's reports of what is to the right or the left, and the front or the back. The bodily asymmetries of human perceivers introduce at least this much nonheritable perspectival variation into our practical/perceptual interactions with things. But these are not the only variations in human bodies that introduce differences into communicative practice. Someone who, like me, is nearly two meters tall gradually learns to adjust what others take to be within reach, readily visible, or overhead. Others, of differing strength or stature, correct for variance in what is lightweight or portable. Women learn that men's reports of where it is easy to walk must be revised by considerations

48. Both Donna Haraway (1991, chap. 9) and Peter Galison (1997) argue that understanding scientific practice requires abandoning the metaphor of translation as a sharing of common meanings. Galison argues that communication between experimentalists and theorists (the exemplary scientific case of observational testimony) should be understood as a practical "coordination of action and belief" rather than as meanings shared through translation. Such coordination uses scientific correlates to pidgins and creoles, more limited idioms without the expressive resources needed for fuller understanding. Haraway insists that "translation is always interpretative, critical, and partial" (1991, 195), even in one's own case: "These points [about vision] also apply to testimony from the position of 'oneself.' We are not immediately present to ourselves. Self-knowledge requires a semiotic-material technology linking meanings and bodies. Self-identity is a bad visual system. Fusion is a bad strategy of positioning" (192).

49. There is a formal sense in which sameness of conceptual content is preserved despite different collateral commitments and their divergent inferential significance. If the conceptual content of an expression is a function from the circumstances of its appropriate application to the appropriate consequences of its correct applicability, then one can mark sameness of conceptual content as expressed in the claim that this function is an eigenfunction. What marks the "sameness," however, is a trackable isomorphism among inferential roles revealed by the transformations that different discursive scorekeeping perspectives make upon one another, not any substantially shared content.

50. For now, I follow Brandom in speaking of perspectival variations in conceptual contents, perceptual responses, and practical capacities and configurations, but as I noted at the outset of this section, I shall eventually (in chapter 7) point out some fundamental problems implicit in this way of speaking.

of safety. Gays and lesbians reorient straight people's claims about who is attractive and where are good places to meet people, among many others. Those who are deaf, unable to read the predominant language, or allergic to onions recognize the need for other and sometimes more substantial adjustments,[51] while the prosthetic embodiment offered or imposed by automobiles, wheelchairs, or well-equipped laboratories call for their own corrections. Some bodies are marked by the comparative invisibility of their unmarked interlocutors' reciprocal need to correct for their perspectival differences. The ability to make one's way in the world with the implicit presumption of normal embodiment displays the privileged position of those for whom others are called on to adjust.[52]

Yet one should not assume that perspectival variation results only from encounters with others. The inheritance of perceptually acquired entitlements begins at home. Remember that a crucial aspect of the skill needed to be a reliable perceptual reporter is the ability to recognize what one encounters as appearances of conceptually articulated objects. Brandom emphasizes the conceptual, inferential dimensions of this ability. These include the conceptual roles of singular terms and anaphora as articulating the inferential role of the mutually accessible objects that triangulate communication. But there is also a practical, perceptual dimension to this ability, namely, tracking the same object over time and recognizing it as the same in its multiple appearances (think of the ability to recognize a person you have seen before, or a familiar object from an unfamiliar point of view). The exercise of this capacity is so deeply implicit in everyday bodily comportment that the underlying complexity revealed by its explicit articulation may be surprising. Consider Samuel Todes's discussion of what is involved in keeping track of the perceptual orientation of an object as one passes by it:

51. Star (1991) effectively articulates the practical/perceptual reorientation required by her allergy to onions as illustrating the epistemic contribution of marginality and partial perspective.

52. These examples are all drawn from differences in human embodiment. Brandom's philosophical aspiration extends rather further, to consider what would have to be true of quite differently embodied beings for them to count as one of "us" who can at least implicitly say "we" by adopting the intentional stance. Interpreting dolphins, computers that interact with their environment, or gaseous extraterrestials would require rather more substantial reorientations of perceptual reports. One need not understand what it would be like to be one of those beings in Nagel's sense of a first-person view (1974), but one would need to be able to take pragmatic account of how they get a different bodily grip upon things.

> When we approach an object, pass it, and leave it behind, . . . *each part of the object passes through all four orientational quadrants.* What was at first actually perceived as facing is finally virtually sensed as facing-away, and vice versa. What was at first actually perceived on the right is finally virtually perceived as on the left and vice versa. . . . The passed object is thus a compound mirror image of the object before us. If we stand between an object and a mirror, the left-right distinction in the object is reversed in the mirror. But the near-far distinction remains invariant. If we place an object between ourself and a mirror, the near-far distinction in the object is reversed in the reflection. But the left-right distinction remains invariant. Compound mirroring is peculiar to perceived motion. It is the perceptual work of motion through time. (2001, 121)

Making explicit the transformations implicit in this ongoing perceptual work further qualifies Brandom's claims about the norm of commonality that is supposed to permit the inheritance of entitlement to claims about perceivable objects. As Todes notes:

> By being motion-mirrored a passed object has a kind of concrete unity which a merely anticipated object cannot have. . . . The *variation* of orientation . . . integrates the *opposite* sides into parts of the *same* object by demonstrating the *interchangeability* of these sides in a single orientational field. (122)

Thus, just as grasping semantic content requires tracking and preserving perspectival differences among speakers who recognize one another as belonging to the same discursive practice, conceptually articulable perception of objects requires tracking and preserving perspectival differences among agents who recognize one another as belonging to the same practical-perceptual field.

This recognition of the object-constitutive role of perspectival difference does not mean Brandom is mistaken in seeking norms of mutuality within discursive practice. The mistake is to identify those norms as "aiming at a common set of beliefs equally good for all" (1994, 240), and to attribute them to the deferential inheritance of the same content and authority expressed in others' observation reports and other testimonial acknowledgments.[53] We appropriately inherit not the same substantial con-

53. Remember that Brandom recognizes that what is shared among discursive interlocutors is not a *substantially* identical conceptual content, but a shared practice of tracking

tent, but its perspectival transform; the underlying norm is not common beliefs, but materially compatible ones.[54] We share the same world, not the same doxastic obligations.[55] To equate the two would be to presume an obligation to acknowledge all expressible truths about the world, including those employing not yet invented vocabularies, connected by unfamiliar "styles of reasoning," or relevant to novel purposes.[56] Such an obligation may not even be coherent if, as seems plausible, there are practical barriers to the simultaneous use of some vocabularies or styles of reasoning. I can (indeed must) understand practitioners of phlogiston chemistry as expressing (some) truths if I am to understand them at all, but my engagement with modern chemical practice precludes my acknowledging those truths through the practical use of mid-eighteenth-century vocabulary and material inferences.[57]

Neither acknowledging irreducible perspectival differences in perceptual/practical interaction with things nor interpreting testimonial inheritance in terms of materially compatible (rather than identical) commitments reinstitutes Davidson's third dogma of empiricism, the scheme/content distinction. Acknowledging perspectival differences does not invoke a prior manifold of appearances, which are then synthesized into objects; there are only multiple manifestations of objects, whose identification

---

and coordinating differences in conceptual perspective. I insist upon distinguishing "same content" from "perspectivally variant contents" because I believe that Brandom's use of the former to denote the latter eventually leads him astray.

54. The inheritance of authority is more complicated. On the one hand, my entitlement to what is expressed by the (perspectivally adjusted) observational report of another only licenses others to attribute the same authority to my reassertion of it if I am (taken to be) a reliable reporter. There is some appropriate presumption of a decay of entitlement upon reassertion. On the other hand, equality of authority would allow no augmentation of a report's authority accruing from my endorsement of it. For an extended discussion of why reassertion of sentences (like the adoption of tools) prima facie enhances their authority, see Latour 1987, part 2. Note well, however, that such augmentation does not fix the content of what is thereby more authoritatively expressed. An adequate account of authority within communicative transactions in the end presupposes a much more extensive account of power than can be provided here. For the beginnings of such an account and its relation to epistemic authority, see Rouse 1996b, chaps. 7–8.

55. For further discussion of the difference between sharing a situation and sharing beliefs about it, see Rouse 1996b, chap. 6.

56. Hacking (1982, 1992) discusses how styles of reasoning are implicated in the articulation of semantic content, and how differences in styles of reasoning institute not differences in what is true (or what is false), but in what is "true-or-false," an intelligible *candidate* for truth, and hence also for falsity.

57. For further discussion of pragmatic incompatibilities among such discursive commitments, see Kuhn 1987 and Hacking 1993.

as the same or different is a normatively responsive practical skill. Understanding the perspectival transformations in testimonial inheritance does not appeal to differences in conceptual scheme, but only to conceptually articulable differences in how we engage the world practically. The principle of charity that undercuts conceptual scheming does not require interpreting others as sharing (our) beliefs, but only as expressing mostly truths about the objects around us. These two categories would be extensionally equivalent only if one ought to undertake commitments to all expressible truths.

Brandom might still insist that, despite these modifications in his account of testimonial inheritance from observation reports, there is still no analogous inheritance of practical entitlements from the actions and intentions of others, and hence there remains a residual functional difference between perception and action. This claim would also be mistaken, however. Brandom is right that we do not normally inherit reasons for an action from others' intentions to perform the same action. But we do not inherit substantially the same content from observation reports either (even from our own reports at different times).

Moreover, just as we must perspectivally reorient our perceptual-testimonial inheritances from others' observation reports, others' actions and intentions also bequeath us perspectivally transformed reasons for action. Your expressed intention not to come to campus today does not directly give me a reason not to come. It does, however, remove my entitlement to intend to meet you there for lunch. To the extent that the success of my (intended) actions depends upon appropriate supporting roles played by others, I gain or lose entitlements from the testimonial authority of others' expressed intentions. Such roles for others' actions and intentions in shaping my own practical commitments and entitlements are pervasive. They extend beyond particular interactions such as meetings, joint ventures, or shared transport, to encompass the social roles and other patterns of action that my various actions fulfill. My entitlement to intend to continue teaching in a private university, for example, depends upon the commitments of students to attend, which in turn inherit practical entitlements from the intentions of parents to pay tuition, employers to respect credentials, and administrators to make appropriate institutional arrangements for them and for me. Scientists' entitlement to undertake a line of research as a potential contribution to a discipline or research program likewise depends upon the expressed or implicit intentions of their colleagues. Once again, this pervasive phenomenon of practical entitlement inheritance is typically overlooked because it is largely implicit and taken for granted.

Power is also a central issue in the inheritance of practical entitlements. Relatively disempowered agents know well that in practice, their entitlement to various possible intentions is substantially dependent upon the expressed or implicit intentions of others. Teachers, parole officers, employers, peer reviewers, bureaucrats, "breadwinners," creditors, editors, parents, and many others occupy roles or positions that effectively bequeath licenses or obstructions to the intentions of others subordinate to them. The field of possible action of these relatively disempowered subordinates is substantially shaped by what these more powerful others do, or would do under different circumstances. Power, then, is the situated ability to act upon the field of possible actions of others.[58]

Brandom does recognize the relevance of power in this context when he cites an institutional superior's explicit orders or permissions to a subordinate as a partial practical analogue to the inheritance of observational authority. But his comparison makes at least three significant mistakes. First, the transmission of the same practical content from superior to subordinate is an analogue to inheritance from perceptual stand-ins rather than from perceptual interlocutors: the reliance of the blind Oedipus upon the observational testimony of Antigone is *not* typical of testimonial inheritance.[59] The reason, as argued above, is that normally one does not inherit the same content either from observation reports or from explicit orders to act. Second, the power relations that do license or obstruct inheritance of practical entitlements need be neither explicit nor institutionalized, as

58. This way of characterizing power will be familiar to many readers from Foucault (1977, 1980), but it is often misunderstood (for a more extended exposition of Foucault's conception, see Rouse 1994). Recall my discussion from chapter 5 of why power is not a property or capacity possessed or exercised by agents, but a feature of the dynamics of the situations in which those agents act. In chapter 7, I say more about the expressive role of the concept of power: the term expresses the *normative* significance of *causal* capacities, not because might makes right, but because the effects of the exercise of causal capacities also reconfigure the meaningful situations in which various courses of action are intelligible possibilities for the agents within them. Recognizing that power is always involved in discursive practices is another consequence of taking seriously my argument against Brandom that discursive practices always belong to larger patterns of practical/perceptual intra-action.

59. As Oedipus says, "Friend, my daughter's eyes serve for my own" (Sophocles 1969, 83). Note that the difference between interlocutors and perceptual stand-ins is easily overlooked if one does not distinguish the two different senses in which Brandom speaks of different performances having the "same conceptual content."

Brandom claims (1994, 241). The path-blocking step by a menacing figure on a dark, deserted street removes my practical entitlement to intend to pass safely. Or, illustrating the intertwining of testimonial and practical inheritance, a shared understanding of whose word will be taken as credible may compel one person to defer to another's implicit intentions. The circumstances in which others would likely defer to a woman's "observation report" about rape can differentially shape the practical entitlements of men and women without ever needing to be formally institutionalized or even articulated.[60] Finally, Brandom's implicit conception of power and practical authority is too static and one-dimensional. It ignores the possibilities for resistance to, realignment of, and freedom within the practical situations in which the intentions and actions of the powerful transform the parameters within which relatively disempowered agents acquire practical entitlements and obligations.[61]

So what should we make of the more "colorful" claim that "the True is One but the Good is Many"? What I have shown so far are different senses in which the true is both many and one. Different bodies as well as different collateral commitments give us each an irreducibly different practical, perceptual, and conceptual hold upon the world.[62] Yet the normative significance of Brandom's "tactile Fregeanism" commits us to the material compatibility of all the perspectivally varied commitments anyone is actually entitled to. Participants in discursive practices are responsible

60. I am grateful to Linda Alcoff for reminding me of the central epistemic and political significance of this example of what Brandom would call the testimonial inheritance of a reassertion license. Brown (1995) shows why it is politically crucial to characterize this issue in terms of the different practical configuration of women's and men's situations, rather than of the authoritative *experience* underlying testimony. Such a shift from first-person experience to public, practical significance is, of course, more generally characteristic of Brandom's approach to intentionality.

61. For a more adequate and dynamic conception of power, see Wartenburg 1990. For an extension of that conception that recognizes the mutual entanglement of doxastic and practical reasoning, see Rouse 1996b, chaps. 5–8. In general, the almost complete lack of any explicit reflection upon the role of power in discursive practices is a glaring omission in Brandom's otherwise remarkably comprehensive account. This omission may be compensated by the happy fact, if I am right, that Brandom's discursive pragmatics is almost completely compatible with the best available philosophical conceptualizations of power. See the discussions of power as dynamic in chapter 5, and of the expressive role of the concept of power in chapter 7. For a critical reflection upon the concept of resistance as an expression of the incomplete hegemony of power relations, see Brown 1995.

62. In chapter 7, I amend this claim to incorporate the related ways in which the world has a normative grip upon us. The metaphor of visual perspective does not take adequate account of how understanding is located within the world, and subject to it.

to what others truly express or imply. But that responsibility does not extend to adopting the bodily comportments, doxastic commitments, or conceptual resources through which they express it. Brandom's mistaken insistence upon shared belief and aperspectival perceptual inheritance reinforces the pervasive normalization of unmarked bodies (typically coded white, male, straight, or abled) that powerfully affects the ways other bodies can inhabit the world, and deflects efforts to challenge the domination thereby effected.[63] Yet Brandom's tactile Fregeanism and his expressivist prosentential conception of truth are also powerful resources to challenge such forms of domination. His account thereby suggests how a differently embodied engagement with the world can be articulated with the normative authority and force of claims that others ought to endorse, that is, of claims that are true. With an appropriate revision of his views about perception and action, then, Brandom makes explicit a pragmatics and semantics complementing Donna Haraway's account of "situated knowledges" (1991, chap. 9) that express the objectivity of partial perspective.

In this idiom, the good is both one and many in much the same way as the true. Brandom cites differences in bodies and desires as accounting for the familiar sense in which the good is many (1994, 240). Yet Brandom recognizes that endorsing the satisfaction of my desires also commits me to endorse whatever means must be employed to fulfill them, and whatever collateral consequences would follow from their fulfillment. Since successful action (including the successful enactment of social roles and life patterns) inevitably presupposes the cooperative actions and other responses of people and things, there is an important sense in which my intentions and desires, in ways largely unbeknownst to me, must be reconciled with the success or failure of others' desires and intentions in the same world.[64] After all, Brandom's tactile Fregeanism commits us each to desiring not just some particular outcome, but the (possible) world in which that outcome would be realized. Within the game of giving and asking for reasons, the concept of the good expresses the responsibility to vindicate my entitlement to those desires. So the good may be said to be one, I think, in the

63. Brown (1995) persuasively argues against familiar characterizations of such forms of domination as "silencing."

64. MacIntyre (1981, chap. 15) offers a helpful account of why any conception of the good must extend to the understanding of social roles and ways of life.

sense that the intentions and desires to which I am rationally entitled must be materially compatible with the comparably entitled intentions and desires of others. But this sense is exactly parallel to the sense in which the true is rightly regarded as one: the materially compatible unity of the good results from the need to realize our divergent perspectives on what is good within the same world.

This parallel may be overlooked because reflections on the true and on the good have characteristically different temporal orientations. 'Truth' is usually invoked in the past or present tense about facts already determined; one considers future truths in the future perfect tense that projects a standpoint from which the future in question *will have been past.* 'Good,' by contrast, is usually spoken as futural, and is directed to the past in the past subjunctive tense that projects a standpoint from which the past in question *would not yet have happened.* Brandom's brief, unanalyzed discussion of the temporality of perception and action fits into this mode. I cannot (now) perceive whether an automobile *will* be painted green, and cannot (now) bring it about that it once *was;* however, I *will* be able to do the former, and *could have* done the latter, respectively.

These familiar orientations, however, mask the full temporal horizons of both the true and the good, understood in terms of the material compatibility of entitled perspectives. Suppose (rightly, I think) that we understand truths about the future as not yet determined, and not merely as unknown. We could then say, if such idioms were our wont, that truth is many, in the sense that there are many different ways things may turn out that are materially compatible with what has already happened.[65] This plurality applies just as well to observationally elicited truths: there are many different observation reports I might make in the future that are materially compatible with my past and present circumstances (i.e., with all the determinate facts!), depending upon where I go and what I do. Following out the parallel, in this colorful idiom, the good is one in the sense that the material reconciliation of past differences over how the world ought to be is already settled. We may now, of course, adopt different evaluative perspectives upon that past settlement. But as Marx knew well, the point of denouncing the past is to make the future differently. Moreover, the significance of the

65. A similar point of view about the futural temporality of scientific knowledge is articulated, in a less philosophically fancy idiom, in Rouse 1996b, chap. 6.

past and present is always to this extent up for grabs: it is not yet settled what future they are the past and present *of*.[66]

Discursive Practices as Practical/Perceptual Competence

Reviewing this extended criticism of Brandom's account of the integration of perception and action within discursive practices, I want to emphasize two interconnected conclusions. First, discursive practices are practical-perceptual activities. Discursive practitioners are bodily beings, whose discursive performances materially affect and are affected by their circumstances. Even the specifically linguistic components of these practices (uttering and responding appropriately to sentences) crucially draw upon practical/perceptual capacities. Moreover, these narrowly linguistic capacities do not float relatively free from other practical involvements, but instead only acquire normative accountability and meaningful content as part of a larger pattern of practical intra-action with the world. I do not deny the familiar Quinean recognition that the interanimation of sentences can free particular utterances from close inferential connection to any particular experiential or practical involvements. I just claim that the discursive normativity that makes possible meaningful utterances can only get a grip upon agents who make, track, and differentially respond to material transformations of the world.[67] My insistence upon the materiality of discursive normativity will take on further significance in later chapters when I discuss causality and ascribe priority to causal intra-actions over any supposed governing laws or exhibited regularities. Indeed, I shall point out parallels to Brandom's criticisms of regulist and regularist reductions of normativity in my criticisms of the familiar regulist and regularist conceptions of causality.

66. Compare Hans-Jörg Rheinberger's elegant conjoining of scientific and political examples of this point: "Early in 1991 a group of historians of biology assembled at the Natural History Museum in what had been East Berlin less than two years before. A paleontologist who had spent his whole professional life ordering and reconstructing fossil material pointed at the famous Solnhafen Archaeopterix of the Berlin collection and summarized the experience of forty years of work: . . . 'At the point of the emergence of the new, the new is not the new' " (1994, 66).

67. Once this perspective is assimilated, it is clear that the Quinean interanimation of sentences is itself a substantial material effect. A pervasive feature of the practical/perceptual environment in which most human beings live is its *wordiness;* indeed, words and sentences are also highly salient features of most human surroundings. Yet the point of my arguments is that they could not function in this way (as contentful sentences and subsentential expressions) except in concert with an extensive practical-perceptual responsiveness to other nonverbal (albeit inferentially significant) features of one's surroundings.

The second point I want to emphasize is that all discursive practices have an ethicopolitical dimension. The reason that this dimension of discursive practices is not fully separable from their semantic/epistemic normativity is precisely the materiality of discursive practices: talk and response to talk are part of a larger pattern of activity that continually reshapes the situations in which agents live and understand themselves. Discursive practices always involve actions with material consequences, and the consequences are always open to assessment in terms of goods as well as truths. These consequences are not merely the direct effects of actions upon other actions, but their effects upon the pattern of commitments and entitlements within which subsequent sayings and other doings are intelligible. Discursive practices, that is, are always power-laden, where the concept of power expresses how material effects have normative significance.

It would be easy to misunderstand my claim that discursive normativity is an expressive extension of practical/perceptual competence in one's surroundings as a simple inversion of Brandom's failed attempt to treat discursive norms as autonomous. Brandom tried to extend an account of autonomous discursive norms to incorporate perception and action. An inversion of this strategy would claim that an autonomous, prediscursive domain of practical/perceptual intentionality can be extended to incorporate discursive articulation as an expressive addition. John Haugeland's attempt (1998) to understand conceptual normativity in terms of practical/perceptual skills implicitly seeks to invert Brandom's strategy in just this way. A brief discussion of how my conclusion differs from Brandom's and Haugeland's projects should help clarify my position.

Haugeland provides an extended discussion of chess, as a complicated normative practice that might illuminate intentionality and norms more generally. Chess requires skillful practical/perceptual discriminations and vigilant compliance with constitutive norms, not just in order to play well, but to play at all. Haugeland characterizes these skills as involving a practical/perceptual normative accountability, whose functioning need not depend upon the discursive articulation of its norms.[68] Chess players must be able to tell the difference between various pieces and moves, and must

68. Haugeland recognizes that in practice, chess is learned and played with considerable discursive interaction as part of the game. He suggests, however, that it is at least conceivable that "super-monkeys" could discernibly play chess without being able to speak a language.

not only tell legal from illegal moves, but must not tolerate the illegal ones. Moreover, the objects to which such skillful and vigilant performances are accountable are not simply physical objects (since the same rook in a continuing game might be instantiated by very different physical tokens, or even by symbols in a chess notation). In principle, he argues, one could play chess without having a language in which to talk about playing chess, yet chess is clearly a conceptually sophisticated intentional comportment. To vindicate fully an inversion of Brandom's strategy, of course, Haugeland would ultimately need to show how the practical/perceptual norms governing the appropriate use of words and sentences could then be understood on the model of the constitutive standards governing a complex practice such as chess.[69]

The difference between Brandom's and Haugeland's strategies for understanding conceptual normativity may seem to exemplify a sharp philosophical divide among those philosophers who follow Sellars and Davidson in rejecting empiricist strategies for explicating intentionality.[70] Whereas some philosophers have argued that language presupposes a bodily or practical background intentionality that cannot be made explicit in language, others have claimed that all understanding (or cognitive content) must be expressible in language.[71] Haugeland might seem to exem-

69. Haugeland does not actually discuss discursive practices or norms directly. He begins by considering how practical skills for coping with objects could constitute modally robust "excluded zones" of impossibility. Such exclusions supposedly establish the independence that allows objects to be bindingly authoritative over what one does. Although he never quite says so, to account for discursive practices, his extended treatment of supposedly conventional practices like chess-playing in these terms would need to be further developed to make sense of skillful capacities for using and combining words in relation to the skills needed to cope with other kinds of thing. Just as Haugeland takes chess pieces to be not actual figurines but real patterns in the world that are only contingently manifest through practical abilities to keep track of the various objects that might instantiate them, so words, and the beliefs and desires they can express, must be patterns in the world that are only contingently situated in the production and recognition of particular phonemes and marks as semantically significant syntactic types.

70. Empiricism here denotes any position that would account for the intentional content of thoughts and utterances on the basis of what is "given" through sense experience, perceptual appearances, surface irritations, or some other extralinguistic presentation of content.

71. Heidegger, Wittgenstein, and Merleau-Ponty are often interpreted as denying the possibility of making understanding fully explicit, and therefore as endorsing some kind of "practical understanding" inarticulable in words. Polanyi 1958, Dreyfus 1979, Searle 1983, and Bourdieu [1980] 1990 are among philosophers who explicitly advocate some version of what one might call the inarticulability hypothesis. Carnap (1959a) may be the clearest advocate of its antithesis, which would identify understanding or cognition with linguistic articulability, although many defenders of the inarticulability hypothesis attribute the antithesis to a wide range of philosophers.

plify the former strategy, and Brandom the latter. I am instead denying that either language or prelinguistic practical skills take precedence in this way. When properly understood, I argue, these two starting points for a philosophical account of intentional normativity are interchangeable. One should be able to arrive at my conception of discursive practices as themselves practical/perceptual skills by either route.

One reason these two starting points may seem to be irreconcilably opposed, I think, is a residual contrast to what Brandom (and others) have revealed to be an impossible conception of what it is to make something explicit.[72] Those who argue for the essential inarticulability of practical coping have in mind, I believe, a conception of explicit articulation as something like the immediate presence of a conceptual content: examples of such conceptions might include Husserl's notion of expression *(Ausdruck)* as a presence *to consciousness* that confers content upon mere indicative signs, or the concept of a program or algorithm whose correct execution fully articulates what the program "says." By such criteria, however, nothing can be explicit. An algorithm cannot express its own correct execution, and the expressive mental voice cannot be authoritative over what it actually says.[73] Brandom offers an alternative, conceivable account of making explicit, as saying something in words. Making explicit is never exhaustive, for there is always more to say, further distinctions and articulations that one could make. Making explicit is never fully authoritative, for the commitments one thereby takes oneself to acknowledge may be supplemented or corrected by those that others attribute. Making explicit is never fully accomplished, because *what* has been said depends upon the different collateral commitments with which it might have inferential significance. Moreover, whether something *has been* made explicit is itself a normative, explicable determination. Various utterances, marks, or gestures do not magically acquire the status of words and sentences, but are only constituted as discursive by resiliently skillful capacities for semantically appropriate responses to them.

A second reason why Brandom's and Haugeland's strategies may not seem to converge turns on a subtle difference in the locus of intentional

72. This impossible conception of making explicit is what Samuel Wheeler calls a "magic language" account of meaning, for which "representing tokens have natural semantic natures" (1991, 200).

73. These claims correspond to Brandom's (and others') Wittgensteinian criticisms of regulism (1994, 18–26).

normativity. This difference correlates with a difference between semantic theories that explicate reference through its contribution to truth and those that explicate truth in terms of reference. Nowadays, most philosophers who ascribe priority to linguistic intentionality, including Brandom, undertake the former strategy, giving priority to sentences rather than names. Those who ascribe priority to practical/perceptual intentionality might well be tempted to give priority instead to interaction with objects (the practical/perceptual correlates of names) rather than phenomena (or situations, states of affairs, or whatever else one takes to be the practical/perceptual correlates of sentences). Haugeland's own discussion is tellingly ambiguous on this point. Sometimes, Haugeland argues that the constitutive standards and skills of practices like chess govern the phenomena that (can) occur within the game.[74] At other times, however, he speaks of standards and skills as making constitutive contact with objects.[75] In the former case, for example, the constitutive standards and skills of chess would rule out the phenomenon of a-rook-moving-along-a-diagonal. In the latter case, the constitutive standards and skills rule out the identification of a rook with a diagonal-mover. Haugeland is ultimately drawn to the latter strategy because he aspires to link the objective accountability of practical norms to a modally robust practice-independence of objects.

74. This locution appears most prominently in section 7 of Haugeland's "Truth and Rule-Following" (1998, chap. 13), where he introduces the concept of constitutive standards, but also in section 12 where he describes the importance of establishing an excluded zone of conceivable but unacceptable phenomena.

75. Haugeland offers one explicit reason for beginning with the practical recognition of objects rather than of phenomena. In chess, for example, he argues that "the skills for recognizing pieces and moves must be separate and exercisable independently of one another. To recognize a piece of a certain type moving in a certain way involves (at least) three recognitive abilities: for the type of piece, for the way it's moving, and for, so to speak, their combination (in the *same* piece)" (1998, 332). Haugeland's argument for this claim is insufficient, however. It is certainly true that "recognizing that something is a rook cannot amount simply to recognizing that it is presently moving along a rank or file" (331). That truth by itself, however, does not determine how best to account for it. One might, as Haugeland suggests, proceed from object-recognition skills, but one might equally well proceed from Brandomesque practical-inferential capacities for coping with phenomena. Recognizing something moving along a rank does not commit me to its being a rook, but it removes my entitlement to regard it as a bishop, while preserving whatever entitlement I had to take it as a rook. Haugeland's suggested strategy of distinguishing three recognitive abilities has this underlying problem: how does the third recognitive ability actually connect in practice to the objects recognized by the other two, rather than just adding an additional, independent "objectification" of the situation. If one *needs* the explicit practical recognition of combination-in-an-object to bring together the first two, it is unclear how that need could be satisfied by an additional practical recognition skill, whose relation to the first two would seem to present the same problem that they present to one another.

I call attention to this issue now, because I think that following Haugeland in this latter direction would be a serious mistake. On the one hand, giving priority to objects rather than phenomena would block an adequate account of discursive normativity as practical/perceptual. Brandom's Kantian and Fregean arguments for the semantic priority of sentences within a pragmatist conception of language are compelling.[76] It is instructive on this point that Haugeland has not yet fully confronted the difficulties of proceeding from object-recognition skills to judgments, because he has not actually undertaken the burden of showing how to account for propositional contents in his terms.[77] An even more fundamental reason for rejecting the priority of objects, however, emerges in chapters 8 and 9. There I will argue that an adequately naturalistic understanding of causality, including the causal dimension of perception and action, requires that causal capacities inhere primarily in phenomena rather than objects. A third, related consideration involves the role of alethic-modal concepts in explicating normativity. Kripke (1980) has compellingly argued that object identity depends upon modal properties, and the priority Haugeland gives to the normative constitution of objects is closely connected to the modal significance he ascribes to what object-constitutive standards and skills render impossible. Both my initial arguments in chapter 1 and my discussion of alethic modality and normativity in chapter 9 below, however, suggest that alethic modalities should be explicated in normative terms rather than the reverse.

How do these considerations then affect the interpretation of my claim that discursive practices should be understood as forms of practical/perceptual competence? I am coupling three claims: first, that discursive practices themselves always involve embodied, practical/perceptual skills for

76. The basic line of argument is laid out in Brandom 1994, 79–85, but the whole book further develops the underlying themes.

77. Haugeland explicitly makes a promissory acknowledgment that, even to account for the practical/perceptual skills involved in practices like chess, he needs to develop an account of the practical correlate of judgment as the recognition of combination-in-an-object (1998, 358, 359 n. 21). The difficulty cuts even more deeply than he acknowledges, however. Haugeland's own account of the normative constitution of objective accountability gives a central role to the establishment of an excluded zone of impossible but conceivable and recognizable phenomena. If he remains committed to the explicative priority of objects over combination-in-an-object, he would need to rework that account to show how excluded zones could be established in terms of excluded object-identifications rather than excluded phenomena, and how those excluded object-identifications could be what are disclosed by constitutive skills.

recognizing, responding to, and producing inferentially interconnected performances; second, that these skills also depend upon other, nonverbal practical/perceptual capacities; but third, that once discursive articulation has been mastered, there is no residual prelinguistic or preconceptual domain of experience or competence that can be disentangled from the discursively articulated capacities it makes possible. To be sure, other organisms do interact practically and perceptually with their surroundings in the absence of any discernible discursive practices. But that does not entail that human perception and practice incorporates an autonomous, "prediscursive" component that retains its integrity even when conjoined with discursive practices. Discursive articulation might seem merely to add further performances that express what can thereby be explicitly recognized as having been implicit in the prior cases (and the articulation of logical vocabulary and norms, in turn, allows one to express what one was already doing in other discursive performances). But these expressive capacities, and the more fine-grained discriminations they make possible, affect everything we do, even though they are only possible against a nonverbal practical/perceptual background. The explicit expression of such norms enables different responses to one's prior capacities, using them in new ways. Practical/perceptual interaction with surroundings is then transformed by its interconnection with the more fine-grained articulation made possible by explicitly inferential norms. Indeed, as I argued earlier in this chapter, once logical vocabulary has been articulated, it can be appropriately applied to express the interconnections among nonverbal performances as themselves implicitly inferential (such as the implicit material practical inference from seeing a hanging curve ball to swinging for the fences).

The result of these conjoined claims is that discursive practices and practical/perceptual responsiveness are not separable domains or components within the world, but are instead dimensions of the entire world as a meaningfully configured practical field.[78] That is another way to express

78. Haugeland (1998, chap. 9) has made an important distinction that usefully applies here, although he himself does not use it in this context. Instead of claiming that discursive articulation and practical/perceptual interaction are "interrelated" components of the world that only encounter one another at a well-defined interface (Brandom, for example, conceives of observational testimony as the interface between discursive practices and perception), I am insisting upon the "intimacy" of their interaction. The interaction between what one says or hears verbally, and other practical/perceptual performances, is not a narrow bandwidth transduction, but a broad bandwidth exchange that cannot be readily localized. This is also another way of expressing the point Karen Barad (1996) makes by speaking of intra-action rather than interaction.

my reading of Quinean indeterminacy, which denies that meanings comprise a distinct domain of semantic facts (even an *empty* domain). This indistinguishability of linguistic and nonlinguistic domains can be seen from either direction. If discourse is whatever is semantically contentful, then on Brandom's account, speakers' practical engagement with the world is discursive "all the way down." Semantic contentfulness is expressed inferentially, and nonlinguistic perceptual/practical engagement with one's surroundings is fully incorporated within the Sellarsian game of giving and asking for reasons. Of course, the inferences involved are material inferences. Such inferences can always be expressed in language as conditional claims, but the inferences are not enthymematic in the absence of such expression: in particular, material practical inferences from what one sees to what one does are perfectly good inferences without being expressed in words (even though they are always thus expressible).

The inseparability of discursive and nondiscursive domains can also be seen starting from practical/perceptual engagement with the world. Just as the practical/perceptual activity of discursive practitioners is always inferentially significant, explicitly discursive articulation must also always be practical/perceptual. Recall that discourse presupposes the practical/perceptual capacity to discriminate and produce sounds, marks, gestures, or other publicly accessible performances. Otherwise, discursive performances would not be publicly corrigible, and therefore could not be inferentially articulable and semantically contentful in the first place. That is why, when properly understood, Brandom's strategy of explicating practical/perceptual coping as inferentially significant components of discursive practice and Haugeland's strategy of explicating semantic contentfulness as constituted by resilient practical/perceptual skills ought to converge in their outcome.

In chapter 7, I further explore the bodily character of discursive normativity, as well as its importance for understanding how norms are bindingly authoritative. The conception of human agents as normatively accountable subjects confronting a world of normatively inert objects has been integral to the dualism of nature and normativity. By taking more seriously the situated embodiment of discursively articulate agency, it becomes clearer how to grasp the normativity of human understanding and agency as an extended bodily competence in responding to circumstances not fully subject to practical or conceptual control.

SEVEN

Desires, Bodies, and Normative Force

I HAVE PROPOSED turning Brandom's conception of discursive practices inside out, arguing that semantically articulated language should be understood as an extension of practical/perceptual competence in dealing with material surroundings. Instead of treating conceptual norms as reaching out to incorporate causal interaction in perception and action, my interpretation emphasizes the practical/perceptual character of speaking and understanding language. I thereby hope to understand how causal interaction with our surroundings is authoritative over what people say and do.[1] Before I can go on to carry out this shift, however, I need to address a second problem confronting Brandom's aspiration to understand semantic articulation in terms of practical normativity rather than objective representation. Any normative authority supposedly conferred upon practical/perceptual causal interaction would be hollow unless one can also understand its normative force (how its supposed authority is actually binding upon participants in discursive practices). Why should we fulfill our discursive commitments? How and why does an entitlement to such commitments matter to us? These questions in turn may lead us to ask who "we" are, to whom such commitments are attributed, and by whom they must be fulfilled.

Brandom's expressivist strategy suggests that we should ask these questions from the midst of our own involvement in discursive practices. He claims that discursive practitioners attribute desires to speakers, to express what is at stake for them in the other commitments attributed to them to make sense of what they say and do. 'Desire' does double duty for Bran-

1. Although chapter 6 presented this inversion of Brandom's program as a strategy for understanding how causal interaction is authoritative over discursive practices, I have not yet carried out that strategy. Crucial parts of the argument will be developed in chapters 8 and 9.

dom, however. In addition to this role expressing what binds us to our discursive and other practical commitments, the concept also accounts for the perspectival differences in conceptual norms that discursive practices seek to bridge in the first place. We have different conceptual perspectives because of our underlying differences in bodies and desires. In the first part of the chapter, I show why this dual role for 'desire' cannot be fulfilled. Even John Haugeland's appeal to more stringent existential commitments rather than the merely deontic commitments invoked by Brandom fails to account for how discursive and other practical norms are binding upon us. Reflection upon Haugeland also reveals additional difficulties. First, an adequate reconciliation of discursive and practical/perceptual normativity turns out to challenge Haugeland's and Brandom's construal of such normativity as accountability to objects. Second, Haugeland's associated distinction of the objective normativity of natural science from the merely instituted normativity of social practices also fails. Most fundamentally, however, what the comparison of Brandom and Haugeland reveals is the mistake of conceiving practical and discursive normativity in terms of commitments undertaken by subjects that would confer normative authority upon otherwise anormative objects. Hence the need for an alternative understanding of what it is to be a responsible participant in discursive practices, someone accountable for what she says and does.

In the second section of the chapter, I begin working toward such an alternative by reconsidering Brandom's account of discursive practice as perspectival. Brandom cites differences in bodies and desires as the ground for differences in conceptual perspective. I first consider bodies, which cannot simply be objects, but should instead be identified through their situated practical capacities. This point suggests a subtle but important revision to Brandom's insistence that discursive practitioners occupy different conceptual perspectives. Perspectival locations are often conceived as subject positions, external to the situations upon which they provide a perspective. The notion of perspectival differences may then seem to require some aperspectival grounds that differentiate them. Both of these points turn out to be misleading. The relevant sense of 'perspective' instead refers to a configuration of an agent's situation, rather than to an external viewpoint upon that situation.

The final section of the chapter draws from these criticisms of Brandom and Haugeland an initial formulation of an alternative conception of normativity, which concerns how one is situated in the world such that something is at stake in how one responds to that situation. This conception may challenge familiar and deeply entrenched conceptions of the anormativity of nature as revealed through modern science. This concluding section thus points toward the final two chapters of the book by raising questions about how scientific practices could disclose the normative significance of nature and natural necessity. These chapters in turn come full circle to my central question of what it can and should mean to be a philosophical naturalist.

Desires and Commitments

To understand Brandom's difficulties in accounting for normative force, consider first how critical a role the attribution of desires plays in his account.[2] Discursive practices involve the mutual attribution of both implicit and explicitly avowed beliefs. What is thereby attributed are not private mental states, but public commitments (and entitlements). These attributed commitments must therefore be accountable first and foremost to what the speaker does (including what she says). But it is only attributed desires that "bridge the gap between what one believes and what one decides to do" (Brandom 1994, 247), by making explicit the pattern of practical inference from beliefs to action (or the propriety of action). The ability to bridge that gap is what justifies the attribution of beliefs and discursive avowals in the first place. Thus, in Brandom's terms, *the normative status of being a (scorekeeping) participant in discursive practices is ascribed by attributing desires, and thereby taking the other commitments one attributes to be binding.*[3] That is, the attribution of desires expresses that the attributee

2. Brandom sometimes uses 'desire' as a species of the genus that also includes wants, preferences, and other evaluative attitudes and sometimes to encompass the entire genus. Since these specific differences do not matter to my argument, I shall use the term more generically.

3. Assigning the normative status of having desires thus plays the role in Brandom's normative conception of the intentional-attributive stance (1994, 55–58) that the hypothetical assumption of the substantive rationality of a system plays in Dennett's naturalistic account of the intentional stance as an explanatory strategy: the correctness (normative appropriateness for Brandom, predictive success for Dennett) of that assignment or assumption would be what enables one's discursive scorekeeping (Brandom) or intentional attribution (Dennett) to track successfully the public performances of the system one takes to be a fellow discursive scorekeeper or an intentional system, thereby vindicating that interpretive stance. Brandom takes his account of intentionality to be irreducibly normative and Dennett's as cashed out naturalistically in terms of the predictive power of intentional attribution. This difference can only be sustained if Brandom's account of desire turns out to be genuinely normative.

ought to fulfill her commitments, and thereby live up to her attributed status as an equally authoritative participant in discursive practices.

There are, however, two levels of ambiguity in Brandom's concept of desire. First, like 'believe,' 'intend,' or 'perceive,' 'desire' is ambiguous between acknowledged and attributed commitments. I exploited this ambiguity in the preceding chapter to explicate a sense in which the good is one: my acknowledged desires can commit me to other desires that I may not acknowledge, which together highlight the need to reconcile my desires with those of others in a shared world. But there is also a second, more troubling ambiguity in Brandom's notion of desire. The difference between attributed and acknowledged commitments itself depends upon distinct scorekeeping perspectives. The perspectival character of discursive practice is indispensable in Brandom's account, since he only establishes the normative authority of objects through their communicative role in mediating such perspectival differences. That is why he repeatedly insists that "the conceptual contents employed in *monological* reasoning, in which all the premises and conclusions are potential commitments of one individual, are parasitic on and intelligible *only* in terms of the conceptual contents conferred by *dialogical* reasoning, which . . . essentially involves assessments from the different social perspectives of interlocutors with different background commitments" (Brandom 1994, 497). But those differences of social perspective cash out for Brandom in terms of speakers having different bodies and different desires (and hence different patterns of entitlement from the same commitments).[4] Hence, 'desires' are simultaneously patterns arising from the divergence of practical/conceptual perspectives and the ground for the differentiation of one perspective from another.

Given the crucial role that desire plays in his view, Brandom has remarkably little to say about it, but what he does say is worrisome. He introduces the concept as an analogue to what Davidson calls pro-attitudes, which together with beliefs constitute primary reasons for action

4. It would take a long, complicated textual exegesis and argument to show that Brandom is committed to the claim that distinct scorekeeping perspectives are constituted by differences in bodies and desires, and not also by differences in doxastic commitments. Here I will only give two indications of where such an argument would draw from: first, the passage noted above in which Brandom appeals to differences in bodies and desires to account for the irreducibly perspectival plurality of the good in contrast to the unity of the true; second, Brandom's Davidsonian commitment to postulating massive coincidence in beliefs (doxastic commitments) as a condition for the possibility of interpreting anyone, oneself included, as among the original-intentional 'We' that makes possible the monological perspectives that constitute 'I' and 'Thou.'

that are prima facie binding upon the agent. The word 'attitude' has very different significance for Davidson and for Brandom, however. For Davidson, pro-attitudes are psychological states that account for the causal powers of reasons for action. For Brandom, attitudes are instead public performances and dispositions to perform that *take* something as something: normative attitudes, which defeasibly *attribute* commitments, institute the normative status of *being* committed. Brandom clearly cannot accept Davidson's interpretation of pro-attitudes: that would be to abandon his project of accounting for intentionality in normative terms, because the intentionality of desires as pro-attitudes *toward* something would be simply a given fact. His account would then collapse into a regularist conception of discursive normativity, which he has already shown to be inadequate.[5] But the troubling ambiguity about desires can now be more precisely characterized. Normative statuses are supposedly instituted by normative attitudes (of which desires are now clearly the preeminent case). Normative attitudes, though, must either be Davidsonian psychological states or dispositions or else must themselves have a normative status (whether a set of events genuinely constitute a taking of something as normatively bound is a normative rather than factual question).

The initial difficulty with Brandom's conception is that desires, understood as attributed normative statuses, cannot do the work required of them. Merely attributed desires have no binding authority. If I do in fact have the desires attributed to me, they can explain why I act as I do, but they cannot bind me to act in accord with other acknowledged or attributed commitments. Indeed, having those desires cannot sustain the normative authority of commitments at all, since nothing commits me to continue to have those desires. This point underlies one of Haugeland's most significant criticisms of Brandom. Haugeland (1998) rejects Brandom's appeal to deontic commitments and entitlements as the most basic expressions of discursive normativity. He argues instead that the normative authority of objects, without which discourse cannot be semantically contentful even according to Brandom, must be secured by a more fundamen-

5. If Brandom were to adopt the Davidsonian reading of attitudes, that is, he would be taking toward desires the stance Searle 1983 takes toward all intentional states, namely, that intentionality is a natural (perhaps scientifically explicable) property. In another idiom, the failure of such a move would appear even more stark. Having started in Brandom 1979 by trying to explicate intentionality in terms of the Kantian notion of freedom as autonomy (via a Hegelian account of the social constitution of norms), he would have ended by explicating intentionality in terms of the heteronomy of de facto desires.

tal kind of existential commitment, which "is no sort of obligation but something more like a dedicated or even a devoted way of living, a determination to maintain and carry on" (1998, 341).

Seeing the force of Haugeland's concern requires further exploration of what seems problematic in Brandom's appeal to desires to explicate the binding character of deontic commitments. The concern first arises upon recognition that mere desires, whether attributed or acknowledged, are not self-sustaining. Consider Haugeland's central example of chess, a practice whose intelligibility strongly depends upon players' commitment to abide by its constitutive norms. My desire to play chess impels me to hold myself, other players, and the requisite chess paraphernalia accountable to the constitutive standards of chess, but only so long as I continue to have that desire (or desires that impel me to accede to other players' desire to play). If the requisite desires were all to disappear (such that no one cared about the difference between correct and incorrect play), the practice of chess would disappear with them. Indeed, Haugeland notes that such failure would result even if the requisite desires slacken sufficiently; the commitments that would sustain genuine normative accountability must be both "resolute" and "resilient."[6]

Although he uses chess as an illustrative example, Haugeland is ultimately most worried that the normative authority of objects over the practices of empirical science would not be genuinely authoritative if it depended upon less resolute commitments, let alone anything as fleeting as agents' de facto desires. If Haugeland is right, the problem cuts even deeper for Brandom, for the semantic contentfulness of discursive practices more generally depends upon the authoritativeness of objects; we could not count *as* discursive practitioners, and therefore would not *be* discursive on Brandom's account, if the binding force of semantic and epistemic norms

6. A commitment is resilient if it has "a kind of perseverance born simultaneously of adaptability and self-assurance [such that] one is not turned aside or discouraged at the first, or even the second, sign of recalcitrance," without persevering to the point of "complete obstinacy or rigidity"; it is resolute if it "relies and is prepared to insist on that which is constitutive of its own possibility [and] intelligibility" (Haugeland 1998, 322, 334, 341). Although Haugeland doesn't take note of this issue, Brandom's conception of deontic scorekeeping and Davidson's parallel conception of radical interpretation illustrate well the need for resilience and resoluteness. Why shouldn't deontic scorekeepers and radical interpreters give up on the presumption that their interlocutor is one of us language-users at the first or second sign of interpretive difficulty? Why do (or rather, should) we persist in trying to make sense of seemingly obscure, obtuse, or bizarre performances by others, if we could appropriately satisfy our interpretive obligations by just settling for the failure of initial efforts to make sense of them?

were subject to whimsy or irresolution. Haugeland's response is to claim that objective normativity must be rooted in a kind of participatory involvement. Such involvement would counter the potentially fleeting character of de facto desires with an insistent holding-oneself-and-others-accountable to constitutive standards. Its insistent binding of oneself along with others is what would differentiate it from a mere persistent conformity to norms, and it supposedly contrasts to Davidson's or Brandom's models of interpretation by this "return from detachment to personal involvement, from the position of the spectator to that of the participant(s)" (1998, 339–40).[7]

The issue is more complicated, however, because Haugeland does not acknowledge the extent to which Brandom's model of normativity also invokes a kind of participatory involvement. Brandom shows his own commitment to the involved character of normativity in two ways. First, his claim that discursive scorekeeping is essentially perspectival establishes this point: scorekeeping can only be done by participants in discursive practices, for whom the score matters, and who must maintain a score of their own. Second, the role of participatory involvement in constituting discursive practices is reflected in the fundamental distinction Brandom draws between attributing and acknowledging commitments. The discursive norms that institute semantic contentfulness would indeed be self-defeatingly ephemeral if the normative status of being a desire were only instituted by attribution. Desires are attributed to explicate reasons for action; in an exemplary case, my continuing to play chess would be prima facie grounds for someone to attribute to me the desire to do so. Attributed desires cannot by themselves confer or institute a reason to play, but can only answer to what already has the capacity to bind action to norms. In acknowledging a commitment, however, I take it upon myself.[8] What

7. Rebecca Kukla (2000a) argues persuasively that Haugeland mistakenly conflates the distinction between involvement and detachment with that between first- and third-person positions. I would take Kukla's point slightly further, and suggest that this conflation invokes a philosophical fiction of what it would be to be a spectator or a radical interpreter. Being a spectator, a radical interpreter, or (in the one case for which Haugeland himself recognizes this point) a scientific observer must instead be a kind of "involved participation" in what one sees. After all, as Brandom has insisted, perceiving differs from mere differential responsiveness in its normative responsibility, and this responsibility is to one another as coparticipants in discursive practices.

8. Note that while the practical capacity to take a commitment upon oneself is what is expressed explicitly by *saying* that one is committed, such a statement is only contentful against a background of mostly fulfilled implicit commitments. My uttering various noises would not amount to saying that I am committed unless that interpretation was vindicated

Haugeland's concern initially shows about Brandom, then, is that acknowledging a commitment (at least in the case of desires) cannot just be a kind of self-attribution, a "discovery" that I seem to be impelled to do something.[9]

Brandom does not explicitly assign this significance to the difference between acknowledged and merely attributed commitments: he accords symmetry to the first- and second-person standpoints in discursive practices. Suppose the acknowledgment of discursive commitments were reconceived along Haugeland's lines, however. He thinks what is required is not merely a socially responsive obligation, but an insistent holding of oneself, others, and one's surroundings to the constitutive standards of a practice. The merely deontic commitments that might be imposed by such insistence on others' part would only be derivative from their own insistence in holding me to my obligations. For Haugeland, such existential commitment is

> not a communal status at all but a resilient and resolute first-personal *stance*. Marriage and monastic vows may be deontic commitments; love and faith are existential. Such commitment is not "to" other players or people, or even to oneself, but rather to an ongoing, concrete game, project, or life. It is no more a psychological or an intentional state than it is a communal status; rather it is a *way*, a *style*, a *mode* of playing, working, or living—a way that relies and is prepared to insist on that which is constitutive of its own possibility, the conditions of its intelligibility. (1998, 341)

Such commitments, Haugeland believes, also invoke existential responsibility for sustaining their conditions.

> One can be deontically responsible only for the character and consequences of one's *own* behavior (and omissions). By contrast, a chess player as such is committed not only to following the rules of the game, but also to seeing that the rooks and bishops themselves abide by the constitutive standards. (342)

by its place in a larger pattern of practice on my part, a pattern that I must also effectively acknowledge in practice.

9. My suspicion is that here is a point at which Davidson and Brandom crucially diverge. For Davidson, I suspect, acknowledging a belief or desire *is* equivalent to attributing it to oneself in homebound radical interpretation, and thus taking pro-attitudes to be de facto psychological states, however anomalous. As the following discussion will show, that cannot be Brandom's view.

Existential commitments are thus self-generated and self-binding, but in a peculiar way. Not to live up to them would not be a failure *by* oneself, but a failure *to be* oneself. In the extreme case, in which one could sustain *no* existential commitments, there would perhaps be no self, or at least no "authentic" self.[10] But existential success or failure is not solely a matter of one's own self-constitutive resoluteness. The conditions for the intelligibility of one's life might not hold together despite one's best efforts, given their dependence upon other people and things being accountable to constitutive standards that they might not meet.

Existential commitments are supposed to be binding because through their resilient insistence, they allow something else to be genuinely authoritative over one's performances. They do so by establishing an "excluded zone" of conceivable but unacceptable phenomena; the normative force of these commitments is expressed by the practical inviolability of the excluded zone. A rook moving diagonally, for example, is simply unacceptable for a chess player. To allow such an unacceptable move to pass is not merely to default upon a specific obligation, but to abandon the practice of chess altogether. The maintenance of such exclusions thus simultaneously constitutes rooks as objects and institutes the practice of chess within which the standards governing rooks are authoritative.

At one level, any practice in which one could seriously engage would thereby constitute objects as authoritative over one's performances. The game of chess exemplifies this level of commitment for Haugeland. Presumably ordinary discursive practice, which Brandom models as "the game of giving and asking for reasons," could also play this role. The constitutive standards of communicative exchange are rather looser than those of chess, but to allow oneself or others to detach utterances from circumstantially appropriate inferential relations to other utterances and performances would likewise be to give up the entire practice of speaking and understanding.[11] Yet Haugeland would argue that such practices are not fully

10. Haugeland explicitly alludes to Heidegger's notion of authentic resoluteness as closely allied to his conception of existential commitment. Presumably, Haugeland would recognize the possibility of an "inauthentic" self, which sustains commitments sufficiently to be intelligible in its particular local activities, but not with sufficient insistence to be intelligible as a whole, as one's own person. Such irresolute agents, on Haugeland's view, need to get a life. Most of us, however, may well live in such average, everyday irresolution most of the time.

11. In both chess and ordinary discourse, it is not the illegal moves that are excluded, but the willingness to countenance such moves. One can move rooks diagonally by mistake, and one can likewise make irrational or nonsensical utterances. What is excluded is irre-

normative. In chess, the performances of players and equipment are indeed bound by an existential commitment to the constitutive standards of the game. These constitutive standards themselves, however, are not bound by or to anything at all; they are neither correct nor incorrect, appropriate nor inappropriate. The relevant standards simply are whatever the chess-playing community is willing to accept. As a result, nothing beyond the contingent continuity of that community's commitments binds them to those standards.

Haugeland then turns to empirical science to discover a more fundamental normative beholdenness to objects. In the contrast to scientific research as a mode of existential commitment, he claims, "the artificiality and ultimate inadequacy of chess and all game analogies come most blatantly into the open" (1998, 343). The reason is that the intelligibility of science's constitutive norms itself must be yielded to something beyond the control of those who initiate scientific practice.[12] Thus, Haugeland concludes,

> the *essential* difference between science and games is that the enterprise itself is always in question. In science, unlike in any game, the repair and improvement not only of mundane but also of constitutive skills is often potentially (and often in fact) an issue. The responsibilities entailed by an empirical constitutive commitment thus necessarily extend to the very skills and concepts that are the form and substance of the domain itself. (343)

Haugeland connects this existential, empirically vulnerable responsibility for the intelligibility of the enterprise itself to a fundamental distinction between two ways in which norms come into being. Games seem to exemplify the *institution* of norms, for which the "instituted conditions [of correctness] are indeed independent of any individual's tellings, and even, on

sponsible error, in which one lets them pass, or even refuses to retract them when confronted with their unacceptability.

12. Haugeland's detailed argument (sections 13, 14, 16) for this claim draws upon the resilience of the skills needed to respond appropriately to tests of empirical hypotheses. As has been familiar in philosophy of science since the work of Popper, Kuhn, and Feyerabend, empirical science can neither afford to surrender its hypotheses too easily in the face of contrary experimental outcomes, since the conflict might be traced to mistaken auxiliary hypotheses, nor to cling to its favored hypotheses in the face of continued empirical recalcitrance. Maintaining this "precarious equilibrium" between resilient adaptation of one's familiar skills and concepts and a resolute determination to resolve incompatibilities, at whatever cost to one's verities, is what supposedly shows the constitutive dependence of scientific norms upon scientists' "existential commitment."

isolated occasions, of everyone's tellings, [though] they are not, and cannot be, independent of the 'general telling' of the community at large" (314–15). Science, however, is supposed to be a paradigmatic case of normative *constitution,* which holds itself fully accountable to an objective state of affairs, quite apart from any individual or communal responses to it. Ultimately, Haugeland's objection to Brandom is that without this more robust sense of commitment, he can never account for the constitution of objective states of affairs as authoritative over merely instituted norms.

Haugeland's attempt to distinguish empirical science from other practices and the institution of communal norms from the constitution of domains of objects fails, but its failure is instructive. I begin by noting that Haugeland is wrong about games, whose constitutive standards are indeed empirically vulnerable.[13] A provocative example is provided by Bobby Fischer's proposal for a fundamental revision in the constitutive standards of chess. Fischer argues that the extensive documentation of standard openings and responses, augmented by computer analysis, has fundamentally transformed chess such that a traditionally skillful player can be defeated by brute mnemonic force. Fischer claims that this development removes the point of chess as a test of intellectual skill and fortitude. He then proposes a fundamental revision of the constitutive standards of chess, *in order to be true to the game.*[14] Fischer may or may not be correct in his diagnosis of the present state of chess, and his prescription may or may not be an adequate or appropriate response. My point is that his claim is intelligible and debatable by appeal to empirical evidence. If the practice of chess-playing were to evolve such that the game became a test of memory rather than of analytical skill, or that the outcomes were predictable as in tic-tac-toe, then the *point* of playing the game would have been lost, and a revision or repair of the rules would be called for by the evolution of play itself.

A related oversight afflicts Haugeland's more general conception of merely instituted norms. Haugeland cites greeting gestures, folk dance performances, and conventional traffic regulations as examples of norms for

13. Haugeland (1998, 330) acknowledges that a game can go wrong empirically, but only to the limited extent that its constitutive rules make impossible demands of its equipment.

14. Fischer's proposal was that the placement of pieces on the back rank shall be determined randomly at the outset of each game, thereby defeating the analysis of standard openings and requiring thoughtful analysis from the very first move.

which the community's "general telling" could not in principle be wrong.[15] If everyone agrees to greet one another in a distinctive way (and to respond differentially to those whose greeting gestures fail), or to drive on the right rather than the left, then there may seem to be no standard according to which they could be wrong. Haugeland's analysis suppresses the community's normative accountability to what is at stake in such practices, however. In the traffic case, driving on the right or the left is entirely up to the community only because of a conceptually prior determination that this choice makes no difference to what is at stake in driving practices. If predominant human neural organization turned out to cause more accidents in countries where oncoming traffic was to drivers' left, if maintaining an unusual national standard were to be economically disastrous to automobile or tourist industries, or if converting to the dominant practice elsewhere would undermine national identity in ways that matter to its inhabitants' lives, then the community as a whole *could* be wrong about which standard to adopt.[16] Likewise, whether a greeting gesture is performed *incorrectly* rather than merely *differently* depends upon what is at stake in the performance of greeting gestures (whether that be recognition of an intention to greet, reaffirmation of community, submission to authority, or elegance of bodily presentation).

Haugeland gives normative pride of place to empirical science, because he thinks that only constituted objects can provide an independent check upon practical norms, and that science distinctively holds its constitutive norms accountable to objects. The first moral to draw from this response to Haugeland, then, is that *"objectivity" does not uniquely express an accountability of practices to something authoritative over the community participating in them.* There can be something at stake in practices to which we are accountable that is not assimilable to properties of objects.[17]

15. Haugeland appeals to the community's "general telling" to allow for the possibility that, coincidentally, each individual member of the community happens to misperform in the same way at once. On such occasions, everyone in the community would be mistaken, but only because the community itself remains in principle incorrigible.

16. I leave open for the moment how it is determined whether traffic safety, economic viability, or national identity are or should be at stake in driving practices, and whether these stakes are worthy of making an issue of the organization of those practices.

17. Haugeland might object that for him 'object' is a purely formal construct, such that whatever stakes a practice is accountable to would perforce be constituted as objects in his sense. He has himself argued that objects such as rooks in chess are not physical objects, not even as a matter of token identity. My reasons for resisting this assimilation are developed in chapters 6, 8, and 9.

Granted, it may not yet be fully evident what is at stake in a practice, the stakes may themselves be contested and may change over time, but none of these points reduces what is at stake in the "institution" of norms to what the community merely happens to agree upon.

Haugeland is also mistaken, however, in characterizing what is at stake in a commitment to scientific practices simply in terms of objectivity. Haugeland only allows for scientists' accountability to the "objective correctness" of their theories, and their experimental and instrumental manipulations. Incorrectness is neither the only nor always the primary way in which scientific work can go wrong, however. Achieving a result that is true but trivial, imprecise, uninteresting, inelegant, misleading, or confused would mark fundamental failures in scientific work. Moreover, incorrectness itself is not as simple and straightforward as Haugeland makes it appear. He recognizes the complexities introduced by the holism of hypothesis testing, such that resilient adjustments of mundane and constitutive skills are needed to maintain a "precarious equilibrium" in the face of apparently incompatible results. What he does not recognize is that the constitutive standards of the sciences are themselves rich and complex, focused by what makes the practice and its outcomes *significant.* Haugeland is not wrong to recognize the empirical vulnerability of a research program and its projected significance, but he drastically oversimplifies it. Objectivity is not one generic virtue, but a contested, historically specific field, in which scientific practitioners are accountable to what has shown itself to be at stake in their reflexive engagements with the world and their own practices, and thus to the ways in which objects matter to the practice.[18]

Bodies and Perspectives

Haugeland and Brandom seek to account for normative force as arising from a commitment undertaken by an agent who holds herself accountable to norms.[19] They take such accounts to succeed if they can show how such

18. Important recent work on the history of the concept of objectivity includes Daston 1992, 1994; Daston and Galison 1992; Galison 1998b; Porter 1995; Wise 1995; Proctor 1991; and Megill 1994.

19. On Brandom's account, one agent cannot hold herself accountable to discursive norms without mutual recognition by other agents. For Haugeland, individual existential commitments can go awry because of the failings of other persons or things, which can compel the existentially committed agent to give up on the practice to which she was committed, out of fidelity to the commitment itself. In different ways, then, they take an individual agent's fidelity to her commitments to be only a necessary, not a sufficient, condition for the fulfillment of normative responsibility.

commitments can irrevocably establish their own accountability to something (objects) independent of the commitments themselves. For Brandom, the authority of objects is instituted by the practice of tracking speakers' perspectival differences against objects that can outrun any single speaker's perspective upon them. For Haugeland, fully objective authority is constituted by commitments that render not merely one's own performances, but the very standards that govern them, empirically vulnerable. By acknowledging commitments, or by undertaking such commitments existentially, we *bind ourselves* to objective authority.

The underlying problem with both views is that a self-determined commitment could have only the semblance of authority, and hence could not genuinely constitute anything to which expressions, activities, and skills are responsible.[20] As Kierkegaard (1954, 203) trenchantly noted, the authority that a commitment could secure through self-determination, however insistent and resilient it may be, is like the authority of a monarch in a country where revolution is legitimate. If the responsibility of the practitioners to whatever is authoritative for them is dependent upon their dedication, then even the most desiring or dedicated practitioners could make no legitimate *claim* upon others (or even upon themselves), and hence could find no *authority* in objects themselves. Practitioners could likewise choose to respect others' commitments, but would not be responsible to them. Thus, even if one appeals to Haugeland's more insistent and resilient conception of an existential commitment, the problem is not avoided. What needs to be understood is not how we can commit ourselves to something, however insistently, but how something already at stake in our situation can have a hold on or over us.

Despite the difficulties I have shown in Brandom's and Haugeland's accounts, further reflection on their views can obliquely help develop such an alternative account of normative force. The place to look initially is Brandom's insistence that conceptual normativity must be perspectival. The social dimension of Brandom's account plays a crucial role here: the communicative interaction among speakers enables their performances to have articulated content precisely through their interpretive efforts to mediate their perspectival differences. But how does one come to have perspectives, or perspectival differences, in the first place?

20. Later in this chapter, and especially in chapter 9, I will discuss a related problem that arises from their attempts to confer normative authority upon otherwise anormative objects.

I have already noted that Brandom appeals to differences in bodies and desires as the basis for irreducible normative differences in perspective: specifically, people's sayings and doings are oriented toward different "goods" by differences in bodies and desires. Against this background of perspectival difference, Brandom sees semantic and epistemic normativity as instituted communicatively through the mediation of objects. I think it is a mistake to take perspectival differences for granted in this way, however.[21] By reconsidering what it means to have a body and a conceptual perspective, I hope to motivate an alternative to desires or self-binding commitments to account for the normative force sustaining discursive practices more generally, and scientific practices specifically.

The differences among interlocutors' bodies, along with different desires, was supposed to be an irreducible locus of conceptual perspective. But what is a body? Presumably it cannot just be a physical object. The "perspective" upon the world provided by the differential responsiveness of a "body" of iron to the presence or absence of oxygen in its environment is not comparable to the embodied conceptual perspectives of speakers. A useful way to understand the difference emphasizes the practical coordination of bodily activity. In chapter 6, I showed how the unity and identity of perceived objects depends (albeit perhaps mutually so) upon the ability of embodied agents to track the unity of an object through movement. Such tracking presupposes the skillful coordination of bodily movements, however. Only a body that can organize and direct its movements within, toward, and in response to its surroundings is capable of practical/perceptual intra-action.[22] A body in the relevant sense is then a practical unity,

21. The danger of taking subjects, desires, or bodies for granted has been a central theme in poststructuralist philosophy and cultural theory. For representative examples of poststructuralist criticisms of the unity of subjectivity, experience, desire, or "the body," see Foucault [1978] 1980; Butler 1989, 1993, 1995; Scott 1988, 1991; Kondo 1990; Haraway 1991, chap. 9, 1992b, 1997; Minh-Ha 1989. Rather than review this literature here, however, I would like to show how these concerns emerge from within Brandom's own account, in his own terms. This rhetorical strategy in no way minimizes my indebtedness to the overlapping poststructuralist, feminist, postcolonial, and queer-theory literatures; what follows is indebted to the works cited here.

22. The best articulated philosophical account of the practical/perceptual unification of bodies as the basis for intentionality is Samuel Todes's 1963 Harvard dissertation "The Human Body as Material Subject of the World," recently republished as Todes 2001. Although I endorse much of what Todes says about the practical unification of bodily activity, I call attention to one crucial difference between our accounts. Todes characterizes bodily self-movement as sui generis, even though it always occurs in the midst of unfamiliar and uncontrolled circumstances. I would instead emphasize that bodily movement always takes place in response to the resistances and affordances of its surroundings, including both the

a capacity for a coordinated responsiveness to what thereby become distinguished as its surroundings. Bodies would then be patterns in the world, emergent boundaries between self and surroundings, rather than already naturally bounded things.[23]

Brandom's appeal to differences in bodies as a ground for differences in conceptual perspective then could not be simply based in a de facto identity of a body as an object. First, bodies in this sense are only identifiable relationally, by their active differentiation from what thereby become their surroundings. Their integrity is not internally constituted; they are of the world rather than in it.[24] Moreover, this point cuts both ways: a body's surroundings are themselves configured as significant through their responsiveness to bodily activity.[25] A related modification in naive conceptions of a body is that the boundary between body and surroundings is then not fixed by the "natural" identity of an organism. The competent deployment of tools, for example, moves them partially across the bound-

---

earth's surface and its prevailing gravitational field (not to mention pervasive and indispensable respiratory interaction with the atmosphere and nutritional and excretory exchanges). A body, as I shall argue below, is always constitutively intra-active with its environment, such that the gravitational field and earth/atmosphere boundary within which human embodiment characteristically develops must be understood as indispensable to the *phenomenon* of human embodiment (the concept of a phenomenon and its significance will be developed more extensively in chapters 8 and 9).

23. Such patterns are thoroughly normative, since they are identifiable as coordinated capacities for skillful performance. Such normativity would be explicated in terms of their character as living bodies, in the sense in which "life" is conceptually identified with the emergence of boundaries between the unified activity of a body and its practically heteronomous "surroundings." Death would then be the irreversible disappearance of such self-maintaining patterns. Mark Okrent (forthcoming, part 1) argues that the teleological character of living bodies (the organization of their activities toward the goal of sustaining and reproducing the characteristic pattern of those activities) should be understood as the most basic form of intentionality, upon which instrumentally and practically rational intentionality is grounded. I take what follows to reinforce the core of Okrent's teleological analysis, although our subsequent treatments of instrumental and practical rationality may differ in some important respects.

24. This complication may well turn out to be an advantage. If the aim is to understand the *intra*-action of norms and causes, then the inability to identify bodies except through their active, responsive differentiation from the surrounding causally efficacious world may be crucial to fulfilling that aim.

25. This point is nicely illustrated by Brandon's distinction (1990) within evolutionary theory between the material environment of an organism (definable independently of the organism's own patterns of activity as everything in the environment that might causally impinge upon the organism) and its selective environment (definable only in relation to the organism's activity as those features of the environment that are selectively relevant to its differential reproduction). The point is that the characteristic bodily activities of the organism *configure* the environment as a selectively relevant field (even though the organism's bodily activities are themselves only characterizable as responses to their surroundings).

ary, as they become prosthetic extensions of an active bodily capacity.[26] Even when a tool has been temporarily set aside, it remains integral to one's bodily capacities and their significance—the bodily skills for riding a bicycle or driving a car, and the associated changes in one's mobility, vulnerability, and so forth, are maintained even when those capacities are not being actively exercised. Such ambiguities in the boundaries of bodies become even more striking when the effective unification of practical activity involves more than one "natural body," in dancing or conversing, for example.[27] The recognition that conversation is not merely an intersection between two distinct sets of capacities, but a joint enterprise, highlights the importance of this discussion for Brandom's conception of discursive practices. Brandom himself acknowledges that the essentially perspectival character of discursive practices means that original (nonderivative) intentionality accrues to the practitioners collectively rather than individually.[28] Discursively skillful bodies may only acquire their practical unity (and

26. See Merleau-Ponty (1945) on the ambiguity of the body/world boundary; his examples include the blind person's appropriation of a cane and a competent driver's incorporation of an automobile within the practical schema of the body. Barad (1996) shows a deep appreciation of this point (expressed in terms of agency without explicit reference to bodies as the loci of agency) when she describes scientific measurements as involving "agencies of observation" that incorporate instruments. As I argue in chapter 9, according adequate recognition to the significance of Barad's point will have far-reaching implications for philosophical naturalism. Barad's inclusion of complex material arrangements among the "agencies of observation" also undercuts any mistaken tendency to limit the "tools" that can count as bodily extensions to hand tools or other directly manipulable equipment. Heelan's discussion (1983) of the role of the orthogonal cues of our familiarly "carpentered" environments in organizing space-perception and the bodily inhabitation of space exemplifies how the agencies of observation can extend well beyond the immediate vicinity of a "natural body."

A more extensive treatment of the skillful unification of bodies would also have to deal with bodily incapacities, and perhaps with the symbiotic complexity of bodies. Disease, injury, and clumsiness display the vulnerability of the practical unity of the body as competent, without thereby shifting the loci of such recalcitrance outside the bounds of bodily practice. Bodies need not be perfectly smoothly functioning unities. The bodily functioning of limbs, organs or "systems," tumors, symbiotic organisms (such as intestinal flora), cells, or subcellular organelles (such as mitochondria) would also have to be considered in a thorough discussion of bodily identity.

27. See Taylor 1991 for a brief but cogent discussion of coordinated activity as a challenge to monological conceptions of the self.

28. Brandom specifically ascribes original intentionality to the discursive community, the open-ended collective of all those whom one ought to recognize as capable of saying "we." The word 'community' is, however, potentially misleading here, since it implicitly suggests what Brandom calls an I-We conception of the social character of discursive interaction, rather than the I-Thou conception. No common ground (such as conventions or shared beliefs) need identify participants in discursive practice other than their being held accountable to discursive norms and responding appropriately.

hence, identity) through their interactions with other bodies, recognizing (and perhaps thereby constituting) one another's bodily integrity through their active and paradoxically mutual recalcitrance.[29] The typically dialogical and materially situated character of discursive practice as Brandom describes it then becomes an especially important exemplar of a more general feature of skillful embodiment.

How might this account of discursively intra-active bodies affect Brandom's central notion of a discursive scorekeeping perspective? Think more carefully about the concept of perspective itself. 'Perspective' seems to imply differences within a unity: multiple perspectives upon a single object, for example. Within the perceptual context from which the term originates, the variations are taken to be perceptual, and the unity conceptual: we *conceive* the unity of what appears manifold, and as Kant famously insisted, both a sensuous manifold and its conceptual unification are needed to experience objects. Brandom, however, insists upon the essentially perspectival character of conceptual content. Concepts unify perceptual perspectives, but something else must play the parallel unifying role for conceptual perspectives. His own account is best understood, I would argue, as taking *mutual* participation in discursive practice—"mastering the coordinated system of scorekeeping perspectives" (1994, 590)—to be what unifies conceptual perspectives, not shared conceptual content. Since perception and action are themselves inferentially articulated, Brandom thereby takes discursive practice to be what achieves the unification of the practical/perceptual perspectives through which concepts are connected to publicly accessible objects.

I need to discuss more carefully this way of understanding essentially perspectival conceptual contents, because one could easily draw the wrong inferences from Brandom's concept of perspectival differences, or from the unifying role he ascribes to discursive practice. Familiar uses of the concept of perspective too easily introduce a split between representing subjects and objects represented, so that perspectives would be characteristics of subject positions. Metaphysically, one would thereby reintroduce the split between things as they appear from perspectives and things as they are in the "view from nowhere." Brandom tries to avoid such a split by insisting

29. The locus classicus for such an argument is chapter 4 of Hegel's *Phenomenology of Spirit* (1972). Although he does not address Hegel's treatment specifically, Todes (2001) argues that the appropriate normative grounds for such interactive bodily identities is not desire (as both Hegel and Brandom argue), but need.

upon the essentially perspectival character of objectivity: every discursive scorekeeping perspective must recognize accountability to objective norms, but no perspective can ever authoritatively embody the sovereign standpoint of objectivity. The locus of objectivity would then be the shared normativity of discursive practice governing the divergent perspectives occupied by actual participants in the practice. This rejection of epistemic sovereignty may be mistakenly undercut, however, by inferences too easily drawn from the concept of a conceptual perspective. A residual notion of perspectives as epistemically self-contained positions is in tension with Brandom's and my insistence that the appropriate use of 'know' is pragmatic and interactive. Moreover, the consequent splitting of subjects and objects may have politically dangerous consequences: on the one hand, the perspectives of those ill served by prevailing political arrangements can then be too easily translated as revealing the oppressed themselves as the problem instead of disclosing a need for political transformation; on the other hand, the identity of a perspective may be itself established by a history of injury such that its political recognition ironically sustains the injury.[30]

More careful understanding of perspective can circumvent these worries. The notion I am arguing against has been expressed by both Wittgenstein and Foucault, who each claim that the subject position from which a situation appears perspectivally cannot itself appear perspectivally, and hence that the metaphor of perspective requires an aperspectival unification.[31] The basis for their arguments, however, is that perspectival viewpoints are external to the scenes they make visible (or conceivable, in the case of Brandom's extension to conceptual perspectives). A brief reflection upon perspective in painting shows why that judgment is too simple. In a painting like *Las Meninas,* which provides the occasion for Foucault's analysis, locating a definite perspectival viewing point in front of the painting presupposes taking the lines marked out within the painting by windows, walls, and other artifacts to be orthogonal.[32] The perspectival viewing point is thus constituted by the (presumed) orthogonal configuration

30. Elliott (1994) tellingly reveals the first danger through a critical reflection upon standpoint epistemologies that draws upon a phenomenology of smooth functioning, unusability, and strangeness; Brown (1995) insightfully analyzes the connection between experiential identity and injury, expanding upon ground broken by Scott (1991).

31. Wittgenstein 1922, 5.631–6331; Foucault [1966] 1970, chaps. 1 and 9, part 2.

32. Amy Schmitter (1996) brought this point home to me forcefully in the case of *Las Meninas.*

of the space. We thus approach the painting already habituated to a "carpentered environment" of orthogonal and parallel lines, without which the painting could not work in the familiar way.[33] Perspective, then, does not indicate a standpoint outside of what it discloses, but must mark an active interplay between the exploratory capacities of a habituated body and the surroundings that solicit, accommodate, and resist its explorations.

Reflection on perspective thus leads back to active bodies as mutually configuring and configured by what thereby become their surroundings. Different bodies are not different perspectives on a scene from a distance, but different configurations of their surroundings as a field of possible activity, which in turn shape the active, responsive integrity of those bodies. Bodily capacities and possibilities are not already there waiting for the emergence of the circumstances in which they could be deployed, but instead emerge as practically significant and discursively intelligible in dynamic intra-action with a meaningfully configured world.[34] The boundaries between active body and responsive/resistant world are also thereby practically and dynamically configured. Barad's account (1996) of scientific measurements as effecting objectively reproducible but ontologically situational cuts between the agencies of observation and the phenomena observed should thus be extended to all patterns of activity.[35] What is an agent, or a component of coordinated agency, depends upon particular situated patterns of responsive intra-action with surroundings, within which a more or less coordinated "body" stands out as engaging with a surrounding environment.

What I have shown is that the bodies that engage in discursive practices

33. I take "carpentered environment" from Heelan 1983, who argues that this orthogonal reconstruction of familiar surroundings effected a shift in the predominant spatiality of visual perception, from hyperbolic to Euclidean space. The importance of carpentered cues to establishing a practical spatial orientation was brought home to me by a visit to the Santa Cruz Mystery Spot, a fenced-in hillside cabin that uses conflicting carpentered cues to produce persistent illusory orientations.

34. Here one should not imagine human beings as already having the capacity to ride bicycles or see through microscopes, needing only to develop the right equipment in order to bring out this preexisting possibility. The equipment, bodily capacities, and configurations of the world as soliciting those capacities and responsive to that equipment developed together.

35. It matters to Barad that what is disclosed are phenomena rather than objects, for the latter concept suggests an inappropriate independence of objects and properties from the patterns of intra-action in which they are manifest. The significance of this point for understanding nature as disclosed by the natural sciences will be the central theme of chapters 8–9.

are not clearly bounded objects, but emergent, shifting boundaries between coordinated activity and its surroundings. The perspectives expressed in their discursive performances are not features of bodies or subjects by themselves, but practical configurations of the surrounding world as fields of significant possible activity. The ability to produce and respond to semantically contentful performances is one of the most significant ways in which embodied agents can reconfigure their surroundings as a practical field, and thereby also articulate themselves as agents. An account of how discursive practices sustain normative force should therefore appeal to the practical configuration of the situation in which they occur rather than to the commitments or other features of subjects who engage in them.

I have emphasized the semantic and epistemic significance of these arguments, but do not forget their associated political consequences. The question of how to understand bodies, desires, and perspectives is closely connected with Brandom's endorsement of a liberal pluralism of irreducibly different perspectives upon the good that I discussed in chapter 6. The problem I attribute to this position is not the recognition of fundamental normative and conceptual differences, but only the interpretation that assigns perspectives and desires to a definite, already bounded plurality of agents/subjects. As Wendy Brown has compellingly argued, the *politically* disastrous result of such reification of the plural subjects of desire is to prevent the "replacement of a politics of difference with a politics of diversity—differences grasped from a perspective larger than simply one point in an ensemble"; the latter might constructively enable us to "argue from a vision about the common ('what I want for us') rather than from identity ('who I am')" (1995, 51). Such a politics of the *common* is what I meant in the previous chapter by the sense in which the good is rightly regarded as one. The concept 'good' expresses the normative authority for what one should say and do that results from belonging together with other agents in the same surrounding circumstances. That sense of the authoritative unity of 'good' must be complemented by a recognition of the diversity of attempts to realize the good through actual reconfiguration of those circumstances. I have not here attempted to argue for such a politics of diversity. My principal argument against Brandom's reification of a plurality of perspectives on the good as embodied subjects of desire has instead been developed in terms of his explication of intentionality. It is nevertheless crucial to keep in the foreground that in part competing conceptions of political life are at stake. With these points in mind, I now return to the question of how discursive practices acquire and sustain their normative force.

Normative Force

It will be useful to review quickly both the achievements and the failures of Brandom's appeal to implicitly acknowledged desires and Haugeland's concept of existential commitment, as accounts of how norms can be binding upon agents, and thereby confer semantic content and practical significance upon their performances. Consider first what I take to be Brandom's and Haugeland's important achievements, as a background for confronting where I think their accounts require revision. Both existential commitment and implicitly acknowledged desires, conceived as expressions of the source of binding normativity, mark an active involvement in a meaningful way of life rather than just the existence of a psychological state or a conferred social status. To some extent, Brandom and Haugeland both recognize that such involvement partly constitutes subjectivity rather than being enacted by preexisting subjects: for Brandom, being a discursive practitioner with commitments and entitlements is a normative status that only arises through its dialogical recognition; for Haugeland, existential commitment seems to constitute a "self" as a way of being responsible, which one can live up to or fall short of. Finally, both Brandom and Haugeland recognize that the normativity introduced by acknowledged desires and existential commitments extends to their material surroundings. Haugeland explicitly emphasizes that the constitutive standards of meaningful practices apply to things as well as people, and that existential commitment entails a responsibility to and for the world's conformity to norms. Brandom's tactile Fregeanism applies to desires as well as beliefs, such that one's commitments outrun one's recognition of them; ultimately, desires extend to an endorsement of the "possible worlds" in which they would be realized.[36]

Difficulties begin to emerge upon asking *who* takes on existential commitments or acknowledges desires. Brandom problematically insists upon the community of discursive practitioners as composed of a plurality of individual subjects of desires and other deontic commitments and entitlements. Haugeland explicitly notes, by contrast, that the participatory

36. 'Possibility' here is neither alethic nor nomological possibility, however. In chapter 9, I introduce an alternative modal concept, real possibility, which would explicate this normative sense of possibility.

involvement that expresses existential commitment can be plural as well as singular. Yet Haugeland does not follow up on this claim; he does not discuss how "we" might be existentially committed, or even how "my" existential commitment might connect or conflict with yours. Moreover, both Brandom's and Haugeland's accounts of how norms are binding still express a strong and problematic voluntarism. Acknowledging desires is something one *does,* and presumably might cease to do. Brandom has nothing to say about how one might be bound to have desires as well as bound by those one happens to have. On his account, we remain sovereign subjects *of* our desires even if his tactile Fregeanism denies that we are fully sovereign *over* them. For Haugeland, likewise, existential commitment is a resolute, "willful" insistence, a holding onto a way of life that is the condition for my intelligibility, but may thereby itself be rendered unintelligible. It is not clear how Haugeland could articulate a sense in which one ought to be (and remain) existentially committed.

These residual voluntarist and individualist commitments are problematic because they are in tension with Brandom's and Haugeland's own insights. Both have shown that the intelligibility of projects, and with them the identities of agents, depend upon responsive interaction with other agents and their shared surroundings. Agents are vulnerable to one another's resistance and the world's recalcitrance, which (in Brandom's terms) can remove any entitlement to intend certain actions or to be a particular (kind of) person. Not everything is possible (intelligible) for us. Sartre ([1943] 1966) famously claimed that it always remains within one's power as a free subject to determine whether the impossibility of some actions or ways of being *matters.* Brandom and Haugeland show why his claim is false. We are always *already* involved in projects and ways of being. That is why both of their accounts must be expressive rather than explanatory. They do not show how norms can be generated from something nonnormative, but instead their projects crucially presuppose having already been implicitly committed to ongoing engagement with a surrounding world.[37] What we are committed to, and how seriously and sustainedly we

37. Brandom and Haugeland both do emphasize the importance of showing how the prehistorical emergence of normativity is not mysterious. Human beings have evolved from nondiscursive, ultimately from non-normative creatures, and this evolution must be intelligible. Such genetic accounts come last in their order of exposition, however; the genetic story does not explicate what normativity *is,* but only shows how their independently articulated expression of what it is to be normative is compatible with understanding other aspects of the world, including scientific and historical practices and standards.

pursue those commitments are to some extent up for choice, but it is not up for choice whether to have any commitments at all. A desire not to have any desires cannot be fulfilled.

The problematic voluntarism of Brandom's and Haugeland's appeals to acknowledged desires or existential commitments is complemented by the equally problematic scientism of their appeal to objectivity to explicate what our commitments bind us to. They are right to see meaning and action as intelligible only as accountable to something not under the agent's own control. They are also right to locate that authority and its binding force in the material surroundings that constitute a meaningful situation or field of action. What is authoritative over and constitutive of human agency and meaning, I will argue, is not the independent objective natures of things, but the emergent configuration of a situation as having something at stake in its outcome. What is at stake is always contested and "essentially perspectival," which is just what Brandom said about the objectivity of conceptual content. The crucial difference is that while objectivity would be normatively constituted, it is not itself something normative; what is at stake is normative through and through. This difference highlights the fundamental and erroneous link between the residual voluntarism and scientism to which both Brandom and Haugeland still subscribe. In a normatively inert world of indifferent and senseless objects, significance could only come from the commitments or desires of subjects. Brandom himself highlights this connection:

> The Enlightenment disenchantment of the world and its assignment to us of responsibility for the norms, values, and significance we nonetheless find in the world are two sides of one coin. Meaningless objects and meaning-generating subjects are two aspects of one picture. (1994, 49)

But what Brandom and Haugeland have both shown is that commitments and desires could only be contentful and binding if they were accountably beholden to something outside themselves.[38] In the end, their own projects once again show how distinctions between meanings and things, norms

38. I use the phrase "accountably beholden" to make a crucial distinction. Brandom and Haugeland claim that they have shown how human beings hold themselves accountable *to* something not under their control, but what they have not shown is how human beings are or could be held accountable *for* being beholden in this way. It is not just objects as normatively authoritative over us, but the normative force that makes their authority binding upon us, that Brandom or Haugeland must account for as not subject to willful determination.

and causes, subjects and objects can become dichotomies whose components cannot be intelligibly related to one another.

So what could make these relationships intelligible, between norms and causes, subjects and objects, meanings and things? Earlier, I talked about the emergent practical boundaries between active bodies and their surroundings. Such emergent patterns of bodily activity configure their surroundings as significant by their responsiveness to a body's specific and characteristic doings and sufferings. Evolutionary biology offers an instructive model here. The evolutionary environment of an organism does not consist of everything in its surroundings, but is instead defined in response to the life patterns of the organism. The selective environment is the environment as it *significantly* impinges upon the phenotypic expression and differential reproduction of the organism (Brandon 1990). Likewise, I want to argue, the world is significantly configured in its responsiveness to the emergent projects and commitments of those beings whose activities include semantically contentful performances.

Selective environments are also a useful model for understanding how different practical and conceptual perspectives on the world affect one another. The selective environment of an organism includes other organisms *and the ways in which they configure their surroundings,* whenever those organisms have selective significance.[39] That is true both of different kinds of organisms, and of conspecifics. The selective environment of an organism thus incorporates other organisms *as actively configuring their more or less shared surroundings.* The same, however, is true in reverse. The intra-action of various organisms thus constitutes a common world, even though it is manifest in significantly different ways for each of the active bodies that help configure it. Discursively articulate bodies, of course, have richer resources for responsive intra-action to their surroundings and to one another's activities. Yet discursive and other practical performances are not perspectives on the world from outside, but practical configurations *of* the world, which can only figure meaningfully as such in the world as also partially configured by the sayings and doings of others.

These considerations make it more comprehensible how practical and

39. By analogy, one could also describe the "phenotypically expressive" environment of an organism, that is, the environment as it significantly intra-acts with the phenotypically expressive repertoire of the organism. Phenotypic expression, however, is simply the organism's bodily development and activity. Thus, the developing body of the organism both configures and is configured by its environment.

discursive commitments become normatively binding. The problem with any form of voluntarism is now clear. Normativity requires something at stake in my activity whose significance for me must nevertheless not be entirely up to me (not merely a matter of desire or chosen commitment). Such stakes can arise, however, in the intra-active shaping of a partially shared practical configuration of the world. If my activity must be responsive not only to my surroundings as significantly configured by my doings, but also to those same surroundings as configured by and configuring the activities of others, then what is at stake in my doings is both significant for me and beyond my control. What Haugeland has said about the "existential" significance of the scientific constitution of independent objects thus applies more generally. Despite Haugeland's residual talk of commitment, by substituting "what is at stake" for "objects" in his claim, I can endorse his conclusion:

> The normative authority of [what is at stake in practices], by virtue of which they can stand as . . . binding on judgments, assertions, [and other doings] devolves upon them from the commitment to the standards in accord with which they are constituted. Note that this is not to suggest that the authority of [what is at stake] is somehow "delegated" or "ceded" to them by those committed to the standards. That would be to suppose that the latter already had the relevant authority and could then give it away, neither of which makes clear sense. Rather it is to say that, in committing to a constituted domain, and thereby finding [something at stake], those who are thus committed necessarily also *find* the [stakes] *as authoritative,* and acknowledge them as such. (1998, 343)

Such stakes are essentially perspectival, in the way in which Brandom claimed that any conceptual content is essentially perspectival. Yet these different practical perspectives upon what is at stake in intra-active dealings with shared surroundings are held together by a mutual "responsibility for the terms in which the enterprise can so much as be conducted" (Haugeland 1998, 342). This responsibility is not merely for one's own behavior, but for the coordinated compliance of other people and things with the seriousness of what is at stake in their intra-action.

The concept of power has an indispensable expressive role here in understanding the simultaneously normative and causal intra-action of people and things in constituting a significant field of possible activities and ways of being. The actions of others do not merely impinge causally upon me and my activities. They help shape my surroundings as a field of possible action in which something is (or even can be) at stake in what I do, and they help determine whether those stakes can be fulfilled. Only if my actions align appropriately with the actions of others and the affordances of our shared surroundings can they amount to either the achievement of something significant or a falling short of an intelligible goal. 'Power' in this sense is neither something possessed by agents nor imposed upon them from without. It instead expresses that what agents do materially transforms the world in ways that normatively reconfigure what can be at stake in one another's actions and whether those stakes can be realized.

Reflection upon Brandom and Haugeland has thus led to a conception of action as both power-inflected and accountable. Actions (which Brandom and Haugeland understand, respectively, as implicitly acknowledging a deontic commitment or undertaking an existential commitment) also thereby strive to implement and enforce a significant configuration of the world that would situate the actions of others and what can be at stake for them. Actors thereby take on responsibility for what they do and its impositions upon others (whether or not they explicitly acknowledge that responsibility or live up to it). Of course, such actions are likewise responsive and vulnerable to the actions of others and the ways *they* configure shared surroundings. Indeed, the normative status of being an actor who takes on responsibility for action is itself constituted through power-inflected *intra*-action. Thus, subjects, bodies, or desires cannot be the given grounds for explicating action and meaning, but must instead be understood as emergent from the intra-active configuration of a meaningful world. Brandom has shown that the same must be true of speech acts as well, including assertions. Utterances can only be meaningful through their normative, inferential connections to other sayings and doings. Speaking can thus never be a fully innocent representation of an unchanged state of affairs, but instead acts upon the field of possible sayings and doings of others. Words and the sentences they compose are an integral part of the world we live in. To speak is to help configure that world as a significant field of discursive practice. There is no disengaged standpoint for speaking truth that is not also an active imposition of meaning, which is both power-inflected and accountable. To say this, however, is not to deny the possibility of speaking truthfully, but to be attentive to how the demand to speak truly could ever be normatively binding upon us.

This appropriation of Haugeland's existential conception of scientific objectivity thus points to a fundamental issue confronting the account of normativity and discursive practices that has emerged within this chapter. Haugeland's entire account was oriented toward showing how the independent, modally robust natures or causal powers of objects could be normatively binding upon scientific practices. Empirical science was Haugeland's preeminent example of a constitutive existential commitment, because it is paradigmatically in science that our commitments let objects be authoritative over what we say and do. I have challenged Haugeland's claim for the distinctiveness of the sciences in two crucial respects. First, I have argued that *all* significant practices similarly constitute something authoritative over our sayings and doings, such that the intelligibility of those practices is "empirically" vulnerable. Second, I have argued that what is authoritative over such practices is not the *modally* robust objective "natures" of things, but the *normatively* irreducible stakes that emerge in intra-active practical configurations of the world.

The latter conclusion applies to scientific practices as well. On my account, the "goal" of science cannot be simply an objectively correct representation of the real independent natures of things, but must instead belong to historically specific situations within which something is at stake between alternative ways of engaging in scientific practices. That claim emphatically does not undermine the accountability of scientific practices to experimental or observational outcomes. Such accountability is indeed indispensable to scientific practices as engaged with the world. But that engagement is significant, and its accountability is binding, only because it emerges as part of a larger, historically unfolding field of meaningful activity. One cannot grasp the significance of scientific practices or the truthfulness of their articulation of nature without recognizing how scientific practices and disclosures *matter* in specific, contestable ways.

This approach to understanding the scientific disclosure of nature runs directly counter to a dominant understanding of post-seventeenth-century science and its significance. What Copernicus, Galileo, Newton, Lavoisier, Darwin, and their heirs have traditionally been supposed to have shown us is the normative inertness of nature. The natural world is bereft of human significance, and it is modern science that triumphantly reveals to us that it is so. The brute causal unfolding of natural history is necessary but not significant. This dominant conception of nature helps account for the appeal of metaphysical naturalism. Since human beings are natural creatures, and nature is devoid of norms, all meaning and understanding must be shown to be explicable in non-normative terms.

Haugeland's alternative to such conceptions of naturalism was to show how the modally robust objectivity of nature could be constituted as significant within scientific practices. My alternative is more radical. I want to show why such supposedly "naturalist" conceptions of nature as non-normative are a mistake, and how scientific understanding and its accountability to the world are differently constituted. The sciences, I claim, are caught up within historically significant, contested configurations of the world as a field of practical activity. The final two chapters of the book will be the first installment of that interpretation of the irreducibly normative significance of nature. My aim is not to abolish familiar conceptions of the objectivity and necessity of nature altogether, but to show how objective accountability and the difference between the necessary and the contingent must emerge as *meaningful* configurations disclosed within scientific practices. Such a normative conception of nature cannot but appear initially strange in the context of contemporary philosophy and the dominant culture of modern science. What I hope to show, however, is why it emerges as the only plausible response to a long history of failure to distinguish intelligibly the normative from the natural.

EIGHT

Experimentation, Theory, and the Normativity of Natural Phenomena

I CAN NOW articulate a more adequate conception of scientific practices, and the disclosure and discovery of nature they facilitate. As I have insisted from the outset, scientific practices should hold a central place in any conception of philosophical naturalism. Scientific work, broadly construed, provides an articulated conception of us as natural beings. Scientific practices also allow nature so conceived to acquire normative authority over what we say and do. That is why I suggested in the opening chapter that scientific practices are the locus for the manifest natural necessity that would be needed to vindicate recent metaphysical naturalisms. One might say in retrospect that human beings have always been natural beings causally subject to the opportunities and constraints of belonging to nature. Naturalism, however, is a commitment to the intelligibility and thereby the *authority* of whatever necessities and possibilities nature affords. Scientific practices are where such natural modalities must show themselves as authoritatively binding.

In chapter 5, I noted that in recent science studies, the term 'practices' has often served to place renewed emphasis upon the role of experimental intervention within the sciences. My critical discussion of Brandom on perception and action reinforces such emphasis, by showing that discursive articulation only becomes meaningful through its inferential intertwining with ongoing perceptual/practical activity. In this chapter, I take up the question of how to understand the conceptual/theoretical intelligibility of nature as disclosed through experimental practice. I can thereby consider (in chapter 9) whether metaphysical naturalists are entitled to familiar conceptions of an anormative natural necessity.

Relations between Theory and Experiment

Recent philosophical work on experimentation (in concert with historical and social or anthropological studies) has commonly accorded with Ian Hacking's slogan (1983) "experimentation has a life of its own." This slogan has expressed at least four distinct claims:

1. the material practices of experimentation are not simply means to observation or observed data, but always involve performances, skills, and understanding that are of scientific significance in their own right;
2. experimental work does not simply test (or interpret and articulate) theories, but instead answers to aims, opportunities, and constraints that arise within experimental practice itself;
3. the philosophical significance of experimental practices and outcomes is not limited to or determined by their theoretical interpretation; and
4. experimentation often (perhaps typically) produces and works with novel, artificial phenomena, whose occurrence is not just the exemplification of more general natural laws.

A common theme among these four claims is the rejection of two familiar alternative conceptions of the relation between experiment and theory. Ernest Nagel's claim that "experimental *laws* have a life of their own" (1961, 87; my emphasis) and the common postempiricist insistence that all observation is theory-laden in retrospect serve as familiar counterparts to Hacking's slogan.[1] Each of my four articulations of that slogan challenges an assumption common to the apparently opposed positions of logical empiricists and postempiricist theoretical holists. These challenges generally align with Hacking's contrast of representation and intervention: their collective claim is that experimentation involves interaction with and transformation of the world, whose philosophical significance cannot be adequately expressed in terms of their contribution to experiential or theoretical representations. One consequence of accepting these criticisms would be the need for a constructive philosophical understanding of experimentation as intervention rather than representation. With such an account in place, however, a response would still be required to the question

1. Galison (1997, 7–8) also cites Nagel's passage as an empiricist analogue to Hacking's slogan.

posed by logical empiricists and postempiricists concerning the relation between theory and experimentation so conceived.

One response that should be taken seriously is that the question is confused, so that we ought to stop asking it rather than looking for better answers. Hacking, for example, began his 1983 discussion of experimentation with the reminder that there have been many different relations between theory and experiment, and he noted as well that the term 'theory' itself needs to be disambiguated. Such a general question about how experiment relates to theory is also ahistorical, and abstracts from differences across scientific fields. Why think that there is one concept, 'theory,' which univocally applies in physics, molecular or cell biology, evolutionary and developmental biology, and geology, or throughout the history of any of these fields? Similarly, why think there is one goal or one general description for the extraordinary range of activities that fall under the catchall term 'experimentation'? Perhaps the philosophical fantasy of a general account of *the* relation between theory and experiment should be replaced by detailed, field-specific, historical narratives such as in Peter Galison's projected historical trilogy on experimentation, material culture, and theory in high-energy physics in the twentieth century. The need for such narratives will likely become all the more clear as more of them are written. Certainly Galison's story (1987, 1997) does not readily map onto biology or geology, where there is nothing like an autonomous tradition of theory distinct from experimentation or fieldwork except perhaps in severely limited domains (population genetics and a few areas of mathematical biology). Refusing to ask the question might also accord with Fine's "Natural Ontological Attitude" (in Fine 1986), which would forgo the very idea of a global philosophical interpretation of science, in favor of more localized work within specific scientific contexts that provide their own interpretive frames.

Despite considerable sympathy for this response, I nevertheless resist it. As one consideration, note that Hacking and Galison themselves in the end also resist the urge to give up this kind of generalization altogether. Hacking proposes an account of the "self-vindication" of the laboratory sciences, and the experimental realism that it leads to. Instead of rejecting the issue of theory/experiment relations, his account significantly increases the number of elements in question (not two, experiment and theory, but fifteen, with correspondingly complicated patterns of interrelation and mutual adjustment). Galison explicitly proposes to replace logical empiricist and postempiricist historiographies with an account of how "the language and practice of experimentation, instrumentation, and theory are distinct, but linked in interesting ways" (1997, 13). In his conception, the "trading zones" in which such links develop involve the "practical coordination of action and belief" rather than the translational equivalence of linguistic representations.

But I resist the dismissive response for my own reasons as well. One may well doubt that there can be a philosophical theory of *the* relation of experiment to theory (let alone of theory to observation), and yet still recognize a need for better ways to talk about how material practice is connected to conceptualization and explanation in the historical development of the sciences. The authors of historically situated, field-specific narratives still need terms in which to write them and some understanding of what the use of those terms commits them to. Furthermore, this issue in the philosophy of science responds to the more general philosophical perplexities expressed in the dualism of nature and normativity. The previous two chapters should heighten recognition of this point. Recall that Haugeland's central claim against Brandom's account of discursive practices (paralleling McDowell's 1994 criticisms of Davidson) was that Brandom cannot adequately account for how discursive practices are accountable to objects.[2] Haugeland in particular argues that Brandom and Davidson cannot adequately make sense of the objective normativity of scientific research, and must settle for the "mere coherence" of theoretical discourse.

If Haugeland's and McDowell's criticisms are on the mark, their arguments would not be the first occasion when important work in the philosophy of language has been prompted in part by critical reflection upon the relation between scientific theory and experimental practice. The causal accounts of reference associated with the work of Hilary Putnam, Saul Kripke, and Keith Donnellan were prompted in significant part by reflection upon the difficulties confronting postempiricist theoretical holism. The work of Kripke and Putnam has contributed in turn to the ongoing revival of philosophical naturalism, for which a central question has been

2. McDowell's critique of Davidson instead appeals to the requisite authority of "experience" rather than objects, but in both cases, the objection is that Brandom's and Davidson's accounts of the normativity of discursive practice floats free from any adequate objective constraint, and hence cannot actually explicate the normativity of language use. I discuss McDowell's argument more extensively at the end of chapter 2 and the beginning of chapter 6.

how to understand the relation between the causal interactions disclosed by the sciences and semantic and epistemic norms, including the meaning and truth of scientific theories. Hence, appropriate reconsideration of how to pose and respond to questions about the relation between experimentation and theoretical conceptualization in science must be central to my reconsideration of philosophical naturalism.

Although substantial work must be done to reformulate these questions, I need not begin altogether anew. I therefore note at the outset several important points from recent discussions of the autonomous life of experimentation that I simply accept. First, one should think of experimental practices, phenomena, or systems rather than observation as one pole of what is in question. My criticism in chapter 6 of Brandom's retention of a vestigial epistemological role for perception apart from action should reinforce this point, although reinforcement is hardly needed. The philosophical history of empiricism shows clearly enough that the turn from public interaction with things to the supposedly inner realm of experience is a dead end. Second, conceptions of theory as a formalizable structure (whether as an axiomatized system of sentences or a structured group of models formalizable in terms of set-theory or state-spaces) should be replaced by looser amalgamations of models (which can be mathematical, verbal, physical, pictorial, or some hybrid combination).[3] Again, recent work in this direction by philosophers of science is reinforced both by the history of failed efforts to provide adequate formalized reconstructions of scientific theories and by the implications of the turn to discursive practices and the priority Brandom rightly ascribes to material over formal inferential relations. A third important advance, already foreshadowed in chapter 5, is a shift from static to dynamic conceptions of experimentation and theorizing. On the one hand, that involves a shift from thinking about *an* experiment, or even *an* experimentally created phenomenon, toward the ongoing development of experimental systems (Rheinberger 1997) or experimental traditions (in the sense in which Galison distinguishes image and logic traditions in physics). On the other hand, a more dynamic conception involves a shift from talk about theoretical models, to talk about modeling, the use of models, in part for the further development and articulation of the models themselves.

3. For extended discussions of such a conception, see Cartwright 1983, 1999; Giere 1988, 1999; Rouse 1996b, chap. 8.

These constructive developments are augmented by other lessons taken both from the failures of empiricist and theoretical-holist accounts and from my own discussion of Brandom on discursive practices. Guiding what I say in the remainder of this chapter will be two such conclusions. From the failure of empiricism, and also in a different way from the difficulties of scientific realism and the associated causal theories of reference, I draw the moral that relations between theory and experiment should not be conceived as relations between linguistic representations and something nonlinguistic, whether those nonlinguistic relata are conceived as experiential observations or as causal interactions. From the failure of theoretical holisms,[4] I draw the moral that theory and experiment should also not be conceived as entirely intralinguistic, whether in the classically Carnapian mold of logical relations between theories and observation sentences or post-Quinean variations that take all of the relevant sentences to be intratheoretical. At first glance these two conclusions may seem to be at odds: surely experimental work must be conceived as either intralinguistic or extralinguistic. Additional assumptions are needed for these to be exhaustive alternatives, however. The rejection of the underlying assumptions common to the failures of empiricism and theoretical holism has been perhaps most strikingly formulated in Davidson's claim (1986) that "there never was a language." The point, which could also have been learned from my discussion of Heidegger and Neurath in chapter 1, is to avoid treatments of how sentences are meaningful that take them as elements in a larger system (such as a theory, a language, a conceptual scheme, or a culture). Empiricists and scientific realists have each tried to connect these larger intralinguistic systems to something extralinguistic; theoretical holists have tried to incorporate experience and reference to objects as intralinguistic or intratheoretical. A better alternative to both approaches is to think of linguistic performances as inextricable from larger patterns of practice that are not themselves understood on the model of a language or a theory. Such an approach is thus the inverse of theoretical holism: instead of assimilating experimentation and observation within networks of implicitly presupposed theoretical sentences, it treats discursive articula-

4. I refer to holisms both in philosophy of science (Feyerabend, familiar interpretations of Kuhn, and their followers) and in philosophy of language (exemplified by Rortyesque readings of Davidson in terms of conversation or vocabularies). I have elsewhere argued (Rouse 1987, chap. 2; 1998; 2002b) that the common reading of Kuhn as a theoretical holist is mistaken.

tions as extensions of other practical capacities and the norms to which they are accountable.

Having thus moved away from conceptions of scientific theories as linguistic entities, and even as formalizable semantic structures, however, one might then question the need to think about language at all in this context. Why not conclude that one lesson to draw from past failures is that problems in the philosophy of science cannot be solved by a philosophy of language? Put that way, I am inclined to agree. Yet even if theories are not themselves linguistic entities, an adequate understanding of scientific theorizing cannot avoid addressing how the development and stabilization of new phenomena within experimental practices is connected to the articulation of new ways to talk about them. A better philosophy of language will not solve all the problems, but an inadequate understanding of scientific talk has long been a source of philosophical confusion. Hence my commitment to starting from discursive practices rather than linguistic representations.

My criticism of Brandom's account of perception and action is especially important in this context, however. Recall that instead of beginning with the autonomous institution of intralinguistic norms, which would then be connected inferentially to objects through perception and action, my approach asks how the normative interrelations among some practical interactions with the world constitute what can then be understood as semantic relations. "Content" is not a property of particular performances, or of whole sets of them, but instead accrues throughout discursive practice as a whole. The material-inferential relations between patterns of talk and particular practical interactions (including experimental practices) both articulate the meaning or content of what is said and express what is going on in the practical interactions. In the case of experimental science, I will argue, this expressive role of scientific discourse is not something external to the phenomena investigated, but is a constitutive component of the phenomenon itself. Thus, the common presumption, that the natural phenomena studied by the sciences are what they are entirely independent of language and culture, turns out to be mistaken. In particular, I will argue that the boundaries of causally interactive systems only become determinate through their incorporation of discursive patterns and norms. On the other hand, it would also be a serious mistake to suggest that the meaning of linguistic expressions (as supposedly established independently through intratheoretical or intralinguistic relations) somehow determines what is happening in these material interactions. The mistake in both cases is to think that material phenomena have definite bounds, or discursive practices express definite meanings, apart from their mutually constitutive interrelations.

This claim should become less perplexing upon considering how it arises in thinking about specific material phenomena. My strategy in what follows is to begin by considering experimental practices in the sciences as themselves causal interactions. Experimental understanding is developed through causally efficacious intervention. By thinking through what it means to characterize experimentation in this way, I will show how the conceptual/discursive dimension of experimental practices emerges as an indispensable and inseparable aspect. It is not merely the deliberately arranged phenomena established in scientific laboratories that illustrate this line of argument, however. Part of the point of initially conceiving experimental practices as *causal* interactions is to insist that it is not through the intentions of scientists or their deliberate performances in constructing an experiment that discursive practices and causal interactions are mutually implicated. Most philosophical discussions of causality take the boundaries of causally interacting systems (objects or events) to be already determinate, without asking how such determination occurs. The point of my argument is to show that this presumption is illegitimate.

This line of argument has an obvious further significance in the context of naturalism. Metaphysical naturalists nowadays often aspire to explain intentionality in terms of the causal-functional roles of mental representations or linguistic expressions within an overall economy of human behavior. Such an account presupposes that sense can be made of causal relations or causal roles independent of intentional or semantic determinations. If my line of argument can be vindicated, then this presumption is mistaken. Thus, mine is emphatically not a reductionist strategy of showing how intentionality and the conceptual content of scientific talk can be reduced to or explained in terms of causal relations. It is similarly not an emergentist strategy of showing the genesis of intentional content from a simpler set of "preintentional" interactions. Rather, my strategy, like Brandom's, is expressivist. I show how correct application of the concept of causal interaction already involves conceptual/discursive normativity all the way down. The result is not then to reduce causal relations to intentional ones, but to show the inseparability of material and discursive interactions with the world.

Causal Interactions

Scientists aim to understand the world in part by interacting with it in revealing ways, and such interactions are always in part causal interactions. What is it, then, for one system to interact causally with another? To pose the question in this way, of course, is already to side with anti-Humeans, for whom material or causal interactions take precedence over whatever regularities or laws they exhibit. I shall have considerably more to say about why one ought to take this approach, but first I need to offer a clearer sense of what it amounts to. Wesley Salmon's distinction between causal processes and causal interactions is a useful starting point. Salmon (1998) begins with processes like an electron or photon traveling through space and asks when their intersection amounts to a causal interaction. He concludes that, "very roughly, when two processes intersect, and both are modified in such ways that the changes in one are correlated with changes in the other, . . . we have a causal interaction" (1998, 135).

At least two problems confront this initial suggestion, however. The first is indicated by a counterexample posed by Nancy Cartwright. Under the right circumstances, a pseudoprocess like the movement of a luminous spot cast by a rotating beacon satisfies Salmon's criteria for interacting causally: on Salmon's original account, the intersection of the moving spot with a piece of red cellophane would count as a causal interaction if "someone were to install a red lens in the beacon [a few nanoseconds before the intersection]. In this case the luminous spot would become red (because of the cellophane at the wall) and remain red (because of the red lens on the beacon) after going past the red cellophane. It would seem that the luminous spot had transmitted a mark" (Salmon 1998, 18). What the counterexample shows is that the initial situation was underdescribed. Salmon omitted the surrounding circumstances of the moving spot and the red cellophane. He mentioned the beacon and the intervening lens, of course, but they are irrelevant to his analysis, strictly speaking; they only characterize the situation that presented a problem for his analysis. Salmon described the situation *as if* it were shielded from other ways of marking the process in question, but omitted the shielding from his description.[5]

5. Salmon himself acknowledged the force of Cartwright's counterexample, but responded in a different way than the one I present below. His initial response was a retreat: "With great philosophical regret I realized that it was necessary to add a counterfactual condition, that is, that the color of the luminous spot would not have changed if the mark had not been introduced into the process" (1998, 18). Later, he championed an alternative

What is needed to block Cartwright's counterexample is inclusion of the actual circumstances that shield the interaction from other potentially relevant processes. Causal interactions do not occur in a vacuum, so to speak, unless the vacuum is itself part of the relevant setting.[6]

The second problem with Salmon's suggestion involves a more subtle incompleteness in his description. Characterizing physical interactions as mutual interactions through which both systems are modified tacitly presupposes a standpoint outside the interactions themselves, from which both modifications are equivalent. Understanding nature as disclosed through experimental interaction, however, requires explication of causal interaction from within. For this aspect of scientific interaction with the world, I supplant Salmon's account of causal interactions with Karen Barad's reflections on Niels Bohr's conception of measurement interactions. A measurement is a material interaction that effects a cut between the "agencies of observation" and the "object" system, between what measures and the event or process it measures.[7] Like Salmon, Bohr and Barad take such interactions to be objectively manifest "in the sense that 'permanent marks [are] left on bodies which define the experimental conditions' " (Barad 1996, quoting Bohr). The difference is that Barad and Bohr seek to describe the overall configuration of an interaction such that the marks on one component of this configuration disclose or measure the other in some definite respect.

My question now comes into better focus. The aim is to understand the natural world as disclosed through scientific interaction. The question now has two components. First, what is a measurement interaction, such that there is a distinction between a measuring system and what it measures? Second, what specifies the proper boundaries of the measuring sys-

---

analysis proposed by Phil Dowe (2000): "Dowe took the crucial step. He pointed out that causal processes transmit conserved quantities; and by virtue of this fact, they are causal" (Salmon 1998, 253). I am dubious that the conservation laws can be adequately understood without counterfactual import, and even more dubious that an adequate account of causation can be confined to transformations expressible as the transmission of conserved physical quantities. My principal response to Salmon and Dowe, however, is to stand on the merits of the alternative approach developed below and extended in chapter 9.

6. I only know of this exchange from Salmon's description. From Cartwright's subsequent work (1999) on the role of "nomological machines" in generating regularities or laws, however, I suspect that her point in raising the counterexample was similar to mine.

7. It is crucial for my purposes that measurement interactions need not be limited to those processes that are actually interpreted by someone as measurements. Rather, I am trying to consider what could appropriately be taken as measurements of what; in Brandom's terms, measurement is a normative status rather than a normative attitude.

tem and the system measured?[8] This latter question is clarified by reflection upon some familiar but mistaken answers. Consider first the bounds of the measuring system. Bas van Fraassen proposed to add a naturalistic element to his constructive empiricism by insisting that science itself should determine what is observable: "what is observable . . . is a function of facts about us *qua* organisms in the world" (1980, 57–58). Yet what van Fraassen mistakenly assumed prior to any scientific considerations is that the relevant boundaries for a scientific measuring system must be individual human organisms. Why presume in advance that the bounds of the human organism are ultimately the boundaries of the scientific measuring system (the systems relevantly affected by the causal capacities of what is measured), rather than insisting that those boundaries should be specified from within scientific practices or measurement interactions themselves?

Equally familiar is the assumption that objects and their properties (where events can also be "objects") are what is measured. What the object of a physical measurement is, however, should also not be determined by philosophical preconceptions. Haugeland made a similar point in criticizing Fred Dretske's specification of the proper object of a perception from among its causal antecedents:

> Why, despite being *via* (proximal) stimuli, [is] perception *of* distal objects? . . . It is crucial not to suppose at the outset that *objects* are an unproblematic fixed point, and [that] the only issue is how perception gets to be *of them*. . . . Rather, the "objecthood" of perceptual objects and the "of-ness" of perception go hand in hand, something like the interdependence of target and aim, or puzzle and solution. (1998, 246)

Similarly, the object of a measurement interaction is intelligible only in relation to the measuring system and vice versa. As Barad has noted, such mutual ontological dependence makes it appropriate to refer to measurements as intra- rather than interactions.

I therefore begin with the intra-action as a whole, the *phenomenon,* rather than with its component systems separately.[9] Something is disclosed

8. The question of how and where the boundaries of measuring and measured systems are constituted has also been a central concern in Barad's work (1996, 1998, forthcoming), and my discussion below bears important affinities to hers.

9. This starting point also characterizes Karen Barad's development of agential realism as an alternative ontological and epistemological position for feminist science studies (1996, 1998, 1999, forthcoming). The priority of phenomena (and the correlated reconception of objectivity) are also at the core of her interpretation of Niels Bohr's philosophy/physics (rather than his more widely discussed notion of complementarity, or his insistence on the

as a natural object only as a component of specific physical phenomena. It is easy to forget that objects only exist within phenomena. Nancy Cartwright offers the telling reminder that, however much one talks about, or even models theoretically, the behavior of "hypothetically isolated objects" such as hydrogen atoms, "real hydrogen atoms appear in an environment, in a very cold tank, for example, or on a benzene molecule, and the effects of the environment must be reflected in [our models and other discursive articulation]" (1983, 137). My fidelity to what is at stake in the naturalist tradition requires taking this recurrent feature of the world seriously, by refusing to take objects for granted without recognition of their inextricable involvement in such phenomena. An object shows up as a definitely bounded system with determinate capacities *within* such phenomena, through the modification of some other aspect of the phenomenon.[10] The marks of a material intra-action on one of its components are thereby a measure of the material capacities of another, and these capacities are initially properties of the phenomenon as a whole rather than of its components.[11] As I noted earlier, the failure to recognize this distinction clearly, and the willingness to assign to objects what properly belongs to phenomena, is a significant deficiency in John Haugeland's illuminating discussion of the normativity of scientific understanding.

The point of beginning an interpretation of *causality* with intra-active phenomena rather than given objects or events has an illuminating analogue in the familiar post-Quinean recognition of the role of an unanalyzed "home language" in the determination of *meaning*.[12] Without a magic language of self-interpreting signs, Quineans have argued, the meaning of an

---

ineliminability of classical concepts from adequate descriptions of measurement interactions in atomic physics (on the latter point, see Howard 1994 and Barad, forthcoming). There are many close affinities between Barad's agential realism and the position I develop in the last two chapters of this book, and these chapters could not have been developed in this way apart from extensive conversation with Barad.

10. As I indicate later in this chapter, and discuss in chapter 9, such phenomenally situated objects and object-kinds are reidentifiable as the same objects (or kinds of object) in other phenomena through their practical-inferential role in intra-actions among these phenomena.

11. The point is not to deny that causal capacities can be properly attributed to objects or object-kinds, but only to note that these attributions and the objects to which they are attributed are derivative from the primary attribution of capacities to phenomena.

12. Recall from chapter 5 that this point was pivotal in distinguishing normative from regularist conceptions of practices. The inability to characterize a practice except in relation to another practice was central to Turner's objections to practice talk (1994), which I reinterpreted as a failure to recognize the difference between the two conceptions.

expression-in-use can only be fixed against the background of a home language. Nothing is privileged about the language used for this purpose, since its expressions can in turn be interpreted against some other background language. What stops a regress of languages is the practical situatedness of interpreters. Interpretation always occurs at a location in "discursive space," as an actual encounter between an expression-in-use and the already assimilated capacities of an interpreter to use and respond to other expressions.

Material intra-actions are also situated in this way.[13] Just as an expression has meaning only through encounters with a "body" of interpretive capacities,[14] so objects only display definite boundaries and properties through intra-action with a body of physical capacities. Their capacities and bounds are defined by the marks they can make upon another component of a phenomenon. Like a home language, however, a measuring system does not specify an absolute anchor, but only a location relative to other material systems. Thus, juxtaposition to the annual rotation of the earth around the sun marks the duration of a material process; the periodicity of that rotation can be measured in turn, but only by another process whose periodicity is unmeasured in that context.

How, then, is the distinction between a measuring system and an object measured established in a particular interaction? I begin with Barad's example:

> Consider an experiment in which light is scattered from a particle. The scattered light may be directed towards a photographic plate rigidly

13. As I show in the third section below, this connection is more than an analogy: the discursive interpretations that Quine characterized are, and must be understood as, causal intra-actions, but causal intra-action turns out only to have definite structure (components distinguishable as causes and effects) through its intra-action with discursive practices. It is no objection to this connection to contrast the apparent abstraction of discursive space with the concreteness of physical spacetime. Discursive space is not a free-floating field of possibilities independent of particular materialized patterns of utterance and response; spacetime is likewise also not a metric independent of the physical systems within it.

14. Expressions themselves, of course, are not already given as meaningless noises or marks in need of interpretation. The boundaries of semantically significant expressions are parasitic upon interpretative norms rather than being a given datum (think of how learning a new language involves learning to discriminate aurally its semantically significant sounds, and not just attaching meanings to already discrete noises; consider likewise what is involved in learning to recognize written characters in a novel script). The same is true of the capacities of interpreters whose responsiveness allows expressions to be meaningful: in Brandom's terms, being a discursive deontic scorekeeper (one who tracks and responds appropriately to changes in speakers' commitments and entitlements) is a normative status rather than an objective property.

> fixed in the laboratory and therefore used to record the position, or the light may be directed towards a piece of equipment with movable parts used to record the momentum of the scattered light. The first case essentially describes the process of taking a picture of the particle with a flash camera. In that case, the light is part of the measuring apparatus. In the latter case, the light's momentum is being measured and hence it is part of the object in question. (1996, 171)

Nothing about the interaction between the light and the particle alone fixes what property is defined by their intra-action, or even the boundary between the measuring and the measured components of the phenomenon (between effect and cause). Does that mean that what is measured depends upon what someone chooses to take as a measurement of what? No. What effects the cut is the larger configuration of apparatus that constitutes a particular material *arrangement*. Different material arrangements constitute different determinations, with different boundaries between the measuring system and the object measured. My point is not that human practices and conceptualizations have no role here. The problem is how to understand that role. The cut between the component systems in a causal intra-action cannot be instituted magically, by fiat. If human concepts and practices have a role, it must be as part of a larger material arrangement; that is another way to express the descriptive incompleteness of Salmon's account of causal interactions. Discussion of "our" role in experimental arrangements, however, must be postponed for the moment.

First, I must explore some further characteristics of measurement intra-actions. Barad's example illustrates the joint indefiniteness of position and momentum familiar from quantum mechanics: for position to be definite requires a physical arrangement of the measuring apparatus incompatible with the arrangement needed for a precise definition of momentum. Yet the example is not peculiar to microphysics; similar incompatibilities show up in straightforwardly macroscopic phenomena.[15] An example from experimental evolutionary biology shows the parallel clearly.

Consider the intra-actions between plants and their environments sche-

15. Barad herself also very clearly avoids the mistake of treating such relations as distinctive to microphysics. She takes the incompatibility of the cuts constituted by different arrangements of measurement apparatus to be a perfectly general characteristic of measurement interactions. The quantum indeterminacies are simply an especially revealing manifestation of this more general feature of measurement, rather than a localizable peculiarity of microphenomena.

matized in figure 1.[16] In experiment 1, a collection of plants grown under similar environmental conditions produces a range of phenotypic expression along some set of parameters. The variation in the resulting phenotypes might seem to measure the differing adaptive response of each organism to these environmental variations along these parameters. Now consider experiment 2, a standard norm of reaction experiment in which clones of the same organism are grown under differing environmental conditions, and measured along the same phenotypic parameters. Here the outcome measures the relevance of these environmental differences for the phenotypic expression of this organism, since the selective "environment" for an organism is not its entire physical surroundings, but only those aspects differentially relevant to its life pattern (Brandon 1990, chap. 2). In experiment 1, the varied environments are an apparatus for determining the capacities of these organisms; in experiment 2, the organisms are instead part of an apparatus to measure aspects of their own selective environment.

Now consider experiment 3, which might seem initially to conjoin the first two: each organism grown in environmental conditions E1 in the first experiment is cloned and grown in different environments as well. Where is the boundary between measuring system and object in this third experimental configuration? It might seem attractive to interpret the combined experiment as a mutual determination of organism and selective environment along the measured phenotypic parameters: the varied environments fix more precisely which elements of the original environment are selectively significant, which then permits a more definitive measure of the adaptive response of the various organisms to that environment. That interpretation is untenable, however. The second part of the experiment that varies the environmental conditions is needed precisely because the surrounding conditions in abstraction from the life patterns of an organism are not the selectively relevant environment. So what the "primary" part of the experiment reveals is not the relative adaptation of the measured organisms to a single measuring environment, but to different selective environments.

Even the simplest configuration of the third experimental arrangement, which shields individual organisms from intra-action with their conspecifics and competitors or symbionts, would still not measure the adaptive

16. These examples and the difficulties of interpreting them were initially brought to my attention by conversations with Sonia Sultan concerning her work on phenotypic plasticity. See Sultan 1995 for an overview.

Figure 1

EXPERIMENT 1: Various organisms O_n are grown under similar environmental conditions E_1, and measured along some set of parameters $P_{a,b,c}$ (e.g., biomass distribution among their component parts, CO_2 uptake rate, total biomass, etc.).

WHAT IS MEASURED: the *adaptive response* of these organisms to this environment along these parameters.

	O_1	O_2	O_3	O_4	O_5
E_1	$P_{a,b,c}$	$P_{a,b,c}$	$P_{a,b,c}$	$P_{a,b,c}$	$P_{a,b,c}$

EXPERIMENT 2: a single organism is cloned and grown under a variety of environmental conditions, and then measured along the same parameters from experiment 1.

WHAT IS MEASURED: These "norm of reaction" experiments measure the *biological significance* of these environmental variations along these parameters for this genotype.

	E_1	E_2	E_3	E_4	E_5
O_1	$P_{a,b,c}$	$P_{a,b,c}$	$P_{a,b,c}$	$P_{a,b,c}$	$P_{a,b,c}$

EXPERIMENT 3: For *each* organism from experiment 1, we perform experiment 2, by cloning the organism and growing it under the same set of environmental variants used in experiment 2.

WHAT IS MEASURED: the phenotypic plasticity of the genotypes along these parameters within this range of environmental variation.

	E_1	E_2	E_3	E_4	E_5
O_1	$P_{a,b,c}$	$P_{a,b,c}$	$P_{a,b,c}$	$P_{a,b,c}$	$P_{a,b,c}$
O_2	$P_{a,b,c}$	$P_{a,b,c}$	$P_{a,b,c}$	$P_{a,b,c}$	$P_{a,b,c}$
O_3	$P_{a,b,c}$	$P_{a,b,c}$	$P_{a,b,c}$	$P_{a,b,c}$	$P_{a,b,c}$
O_4	$P_{a,b,c}$	$P_{a,b,c}$	$P_{a,b,c}$	$P_{a,b,c}$	$P_{a,b,c}$
O_5	$P_{a,b,c}$	$P_{a,b,c}$	$P_{a,b,c}$	$P_{a,b,c}$	$P_{a,b,c}$

EXPERIMENT 3a: A variant of experiment 3 in which some or all of the organisms being measured are also part of the surrounding "environment" for the other organisms measured.

WHAT IS MEASURED: The selective significance of these environmental variations for the *complex* of organisms O_1–O_n along the parameters $P_{a,b,c}$ of O_n.

response of the various experimental organisms along specific phenotypic parameters, for the different experimental arrangements effect subtle shifts in the "organism" being measured. Experiment 1 measured the response of organisms (incorporating their developmental history) to a specific selective environment. Experiment 3, having cloned the organisms from experiment 1, divides the situation differently. It measures not the adaptive response of a developing organism, but the phenotypic plasticity of its genotype, since the development of the organism in various environments has shifted from what is measured to part of the apparatus measuring the genotype as phenotypically expressed.[17] The shift from single to multiple environments for the organisms measured in experiments 1 and 3 parallels (in reverse) the shift in Barad's example from a detector with fixed parts to one with movable parts. The scattered photon is shifted from part of the measuring apparatus to what is measured; in the biological examples, the developing organism is shifted from what is measured by the environment to what measures the phenotypic plasticity of the genotype. The biological case has no obvious quantitative parallel to $h/4\pi$ (h-bar/2) as the inverse limit to the determinacy of position and momentum, but the joint incompatibility of the experimental arrangements that would determine an organism and its selective environment is nevertheless an analogue to the material incompatibility of position and momentum measurements.

Now consider a crucial feature of measurement interactions, their repeatability. Barad does not identify a specific particle, photon, and detector, but sketches an arrangement that could be realized by many actual particles and detectors. Likewise with these botanical examples (or, to note parallel conceptions, Ian Hacking's "chance setups," and Nancy Cartwright's "nomological machines").[18] The marks of its intra-action with one system can determine an aspect of another only because these intra-actions exhibit repeatable *patterns*.

17. Just to complicate life, I note that one could also construct a fourth experimental arrangement, which incorporates multiple variations upon the third, by growing sets of cloned organisms in groups of variant environments, thereby measuring the adaptive response of the various phenotypically plastic genotypes to particular ranges of environmental variation. Sultan (1995) presents the results of one version of such an experimental arrangement, which measures the adaptive significance of phenotypic plasticity itself.

18. It might seem that Hacking's (1965) and Cartwright's (1999) discussions can refer to specific physical objects, such as a particular highway or coin-tossing device, and a particular planetary orbit or mechanical contrivance. Yet even when the device is specified in this way, the trials that determine a long-run frequency are iterable, as are the motions that instantiate a regularity in the nomological machine.

This claim might seem to betray my anti-Humean allegiance to the primacy of physical intra-actions over any regularities they display, but not so. The pattern is in the intra-actions themselves, not in the recognition or representation of a regularity.[19] This point highlights a familiar feature of experimental practice, which nevertheless confounds inductivist and nomological interpretations of experiments, namely, the importance of a singular experimental phenomenon. As Cartwright has noted:

> The trick of the outstanding experimenter is to set the arrangements just right so that the observed outcome means just what it is intended to mean; and that takes repeated efforts, usually over months and sometimes over years. But once the genuine effect is achieved, that is enough. The physicist need not go on running the experiment again and again to lay bare a regularity before our eyes. A single case, if it is the right case, will do. (1989, 92)

Such cases show that no actual regularity ever need be involved, for what matters is not repetition but repeatability. Thus, Humeans rightly see a conceptual relation between causal interactions and repeatable patterns, but they misunderstand it.[20] The Humean identifies causality with a regularity, an actual constancy of conjunction. The repeatable pattern of a physical phenomenon is, however, normative rather than simply regular.[21] That a particular intra-action is causal indicates that under the right circumstances its pattern would recur, but there need be no actual regularity

19. Sometimes, of course, a phenomenon does incorporate the recognition of a regularity, but in that case, the capacity for recognition is itself part of the apparatus that belongs to the phenomenon, and the phenomenon is then not the regularity itself, but the larger configuration of the world that enables the regularity to be manifest. That is why Cartwright sees regularities as properties *of* nomological machines and Hacking ascribes long-run frequencies to chance setups.

20. Thanks to Sanford Shieh for pointing out this way of understanding the connection to Humean accounts of causation.

21. A different variety of Humeanism might seem to capture my contrast between repetition and repeatability by construing causal relations as involving regularities across possible worlds, and not merely the actual world. Thus, a unique event might be said to exemplify a transworld regularity. For such accounts to mark a genuinely Humean alterative to the normative account of causal intra-action being developed here, however, there must be some account of epistemic access to possible worlds that is independent of the normative proprieties that express the reproducibility of the phenomenon. Otherwise, the transworld regularity would not be the antecedent basis for distinguishing which phenomena are genuinely causal configurations, but would be only a subsequent way of expressing the propriety of drawing the distinction in one place rather than another.

that it instantiates.[22] That was Bohr's point in redefining the objectivity of phenomena in terms of the reproducibility of "permanent marks . . . left on bodies which define the experimental conditions" and the unambiguous communicability of "all relevant features of the experimental arrangement" marked in this way (Barad 1996, 171, 172, quoting Bohr). That was also my inference from Cartwright's objection to Salmon: a causal intra-action has not been correctly specified without including the shielding that effects its bounds.[23] To repeat the crucial point: causal intra-actions must be in principle *reproducible* and unambiguously *specifiable,* but their actual reproduction or descriptive specification is neither part of nor necessary for their occurrence.

I can now return to how the proper bounds of intra-acting systems become determinate. Consider Salmon's interpretation (1998, 71) of an example parallel to Barad's (ironically, he discusses the same case). A photon collides with an electron, and Salmon takes the scattering of the photon to mark a (probabilistic) causal capacity *of the electron.* That is not so, however. By itself, the scattered photon cannot "measure" something so specific as the electron; without appropriate shielding from more complicated intra-action, what is measured by the scattering of the photon can only be a less well defined configuration of matter and energy. As materially appropriate apparatus is incorporated within the measuring system, its correlated object system becomes more definite. Under the right circumstances, the change in momentum of the photon can indeed be a mark of the electron's causal capacities alone, but that increases the complexity and structure of the measuring system.

This point is quite familiar in different idioms. My description exemplifies what Carnap or Neurath would have called the "material mode." In the "formal mode," the point is widely acknowledged: a precisely described event is typically a test implication not of a single hypothesis, but only of its conjunction with auxiliary hypotheses. If the event did not

22. It would be a mistake to conclude from this point that causal intra-actions always implicitly obey strict laws, since there need be no description of the "right circumstances" that can be adequately specified without employing ceteris paribus clauses. Indeed, Cartwright employs the same normative language to discuss the role of idealizations in connecting disparate manifestations of the same causal capacities: "the 'ideal' circumstances are the 'right' ones—right for inferring what the nature of the behavior is" (1999, 84).

23. Note well that the shielding is not external to the causal intra-action, but is part of one of the intra-acting components, namely, the one so marked by the intra-action as to measure the effects of the other (Barad calls this component of a phenomenon the "agencies of observation"). See my discussion below of J. L. Mackie 1974 on causality.

occur, the falsity of the consequent would be distributed disjunctively across the entire antecedent conjunction of hypotheses. If the antecedent were less articulated (for example, if imprecise values were assigned to the initial conditions, or if certain relevant conditions were unspecified), then the resulting ambiguity would be exported into the consequent, which would either have less precise values or would itself be conditional.

The point is also familiar in well-known discussions of causality. Mackie (1974), for example, distinguished the total cause of a phenomenon from its more specific components that "we" pick out as significant by holding a causal background constant experimentally or hypothetically. For Mackie, Salmon's discussion would be an anthropocentric specification of part of the total cause of the modification involved, because Salmon tacitly eliminated the causal background. But it is only when the event in question is sufficiently "simple" and undifferentiated that its occurrence indeed marks only its intra-action with a system as extensive and unarticulated as Mackie's postulated total cause. A more complexly structured event can indicate a more sharply defined object. Ceteris paribus, the more relevant interrelationships incorporated within the measuring system, the more specifically bounded is the object system.

From my standpoint, the problem with Carnap's and Neurath's preference for the formal mode is that they help themselves to semantic and inferential norms without an adequate account of how those norms become contentful and binding in practice. Mackie's problem is different. First, Mackie arbitrarily identifies genuine causal interactions with only one of the cuts that a local configuration of the universe might realize.[24] Mackie thereby recognizes as full-fledged causal phenomena in nature only those in which isolated events or "effects" impose minimal structure upon what they measure. He takes any more highly organized system to be anthropocentric, a reflection of the human interests that construct or posit a "normal" background against which localized partial causes stand out. This distinction of natural causality from anthropocentric normativity is doubly mistaken, however. First, it is not just human interests that localize intra-action. Consider planetary orbits: the trajectory of Mars *in the relative isolation of the solar system* does indeed mark Mars' intra-action with the gravitating masses of the sun and the other inner planets. A Mackiean total cause is unnecessary if the relative spatial isolation of these bodies is

24. Mackie then interprets ordinary causal descriptions as anthropocentrically chosen sets of INUS conditions.

part of the phenomenon. More important, Mackie does not recognize that the activities that he dismisses as anthropocentric are as much a part of a physical configuration as planetary momenta or electromagnetic fields. Human experimental and conceptual practices must be understandable as components of a physical system, not an external imposition upon the natural world. They might, of course, be understood to be components of a *different* physical system, namely, the causal relations between an independently specifiable causal intra-action and the discursive performances it evokes from scientists. Such a conception would only be viable if the proper bounds of the original causal intra-action are identifiable apart from the discursive practices in the first place; that is precisely what my arguments will eventually deny.

The cut between measuring and measured systems is an actual material configuration that composes a phenomenon. The "configuration" incorporates not everything there, however, but only everything *relevantly* interrelated. Organisms and their selective environments are especially clear illustrations of this point: the selective effect of an environment upon a biological population is not the influence of everything within its habitat region, but only of those things that intra-act with the capacities of the organisms themselves, or indirectly, with the capacities of other things that do intra-act with their capacities. The organisms, their adaptive traits, and the boundaries of the relevant population are likewise not identifiable apart from their characteristic patterns of intra-action with an environment (as is especially striking in cases of symbiosis). The organisms, that is, are not independently identifiable entities, but only sustain themselves as organisms (and as a population) through ongoing intra-action with their surroundings.[25] Recall my discussion of this point from a different perspective in considering human bodies in chapter 7.

25. Okrent (forthcoming) offers a provocative philosophical conception of intentionality that starts from the teleological character of this self-maintaining organismic intra-action with changing environmental surroundings. Okrent's account thus seems to contrast sharply with Brandom's in one crucial respect: for Brandom, the linguistic intentionality characteristic of humans who engage in discursive practices is "original intentionality," from which the intentionality of other organisms is derivative, whereas Okrent begins with the intentional directedness implicit in organismic teleology, and shows how rational and discursive intentionality can develop on that basis. Although an argument for this claim will have to await the appearance of Okrent's book, I think that my inversion of Brandom's account of perception, action, and desire makes possible a reconciliation of these conceptions, indeed, one that vindicates the core orientation of this book toward a naturalism that incorporates discursive normativity and scientific understanding within the natural world.

Lest one mistakenly think that such considerations are confined to organisms, consider the example of the motions of the sun and the inner planets. The very large numbers of solar neutrinos and photons that continually intersect the trajectories of these planets do not belong to their gravitational and mechanical environment. In different terminology (Cartwright 1999), they are not components of the "nomological machine" that produces the regularities of planetary motion. What materially affects these motions are the capacities realized and disclosed through a specific pattern of intra-action, to which photons and neutrinos do not contribute. Hence my earlier insistence on the primacy of intra-active phenomena over objects and their properties. Objects and properties only acquire determinacy in the phenomena that mark their capacities; even their boundaries become determinate through the intra-action, and not the reverse.[26]

The Normativity of Nature

My discussion so far runs counter to familiar intuitions about the anormativity of nature. Skeptical readers may suspect that normativity has already been surreptitiously smuggled into the discussion through my insistence upon the repeatability of a material configuration. I have two primary responses to these concerns. The first has already been mentioned, but bears repeating. Without the "directedness" of a repeatable cut between the measuring system and the object system, one needs to presume a "God's-eye" standpoint outside the interaction that could register the marks of their mutual determination without further physical interaction with them.[27] I reject that presumption, on broadly naturalistic grounds. That is why the

26. In chapter 9, I discuss how to understand objects and object-kinds as material-inferentially dependent upon multiple phenomena, such that it makes sense to speak of objects as independent of any particular phenomenon of which they are intra-active components, without thereby implying their independence of or priority to phenomena.

27. It might initially seem that the standpoint tacitly omitted from Salmon's description of causal interactions is that of an observer (a measuring apparatus) that is not part of the causal interaction being studied, but otherwise part of the natural world. Yet for the description of this interaction to be complete, the marks that one causally intra-active system constitutively leaves upon another must be manifest as determinate marks without the system being further marked by any causal interaction with the observing system that registers those marks. That requirement places the observing system outside the physical domain altogether, able to acquire information about physical processes without physical interaction with them. That is why I describe it as a "God's-eye" point of view. For Salmon, the marks of a causal interaction of two physical systems must be manifest to "God's eye," which could see nonphysically.

two basic differences between Salmon's and my conceptions of causal interactions are interestingly connected: the "measuring system" (effect) is distinct from the "object system" (cause) because the initiating and shielding conditions of the intra-action specify from within the intra-action itself what object system is actually determined by the measuring system, in what respects.

My second response turns the tables against those who might worry that this account of causal intra-action tacitly and illegitimately presupposes an unexplicated role for intentionality. Intentional interpretation itself is an example of a material intra-action on my account (just as any good naturalist would expect!). Causal determinacy and intentional significance are mutually implicated rather than either being the independently intelligible ground for the explication of the other. To see why this is so, notice that intentional interpretation (exemplified by Davidsonian radical interpretation or Dennett's intentional stance) can only be undertaken toward some physical system: it must begin with "marks on bodies," whether these are movements, vocalizations, or inscriptions.[28] Now, how should one characterize the interpretive system (the "agencies of observation") that explicitly attributes beliefs and desires or commitments and entitlements to an object system incorporating those marks? The post-Quinean tradition assigns intentional interpretation to systems that have command of a background language, and that also ascribe to themselves (at least implicitly) intentional states and acting for reasons. But that is not sufficient. One cannot interpret marks on bodies as intentionally contentful except in relation to the circumstances that occasion them. Thus, any system that attributes intentional content must incorporate those circumstances as part of its own measuring apparatus to determine those contents. Only a being intra-actively situated in the world can engage in intentional interpretation.[29] Ongoing intra-action between the interpreter and her surroundings is part of the "apparatus" that "measures" the intentional con-

28. This point already emerged in chapter 6 in my criticism of Brandom's attempt to assert the independence of discursive norms from perception and action: there could not be genuinely discursive norms unless they applied to publicly accessible marks on bodies.

29. The underlying point is widely acknowledged, for example, by Davidson when he recognizes that his account has "erased the boundary between knowing a language and knowing our way around the world generally" (1986, 445–46). But he does not recognize that this affects how to characterize properly what it is to be a speaker engaging in intentional interpretation.

tent of the marks the object system makes within those same surroundings.[30]

But now consider that the ascription of intentional content is not the only possible measuring configuration involving some putative intentional system's response to its surroundings. Post-Quinean intentional interpretation incorporates those surroundings within the agencies of observation, and thereby interprets the marks made by that system in the register of a language antecedently understood. Yet this measurement configuration presents an ambiguity exactly parallel to the interactions of a scattered photon with a detector and an organism with its environment. The particle could either be measuring or measured, depending upon the configuration of the parts of the detector as relatively fixed or movable; the organism likewise could be either measuring or measured, depending upon its interaction with a fixed or variable environment. The same is true of utterances and actions. While they can be "measured" (assigned an interpretation) through the interpreter's intra-action with her surroundings, they can also be a "measuring device." Given a specific interpretation in the home language and appropriate background assumptions, those marks are a measure of the environment (roughly, do those circumstances confirm or falsify *this* expression?). The marks then express part of the *semantic* significance of those circumstances, namely, their correct description in these words. The meaning of an expression and the correlative truth about the world are thus not simultaneously determinate, for they involve mutually incompatible practical configurations of the world.

When intentional interpretation is thus understood as a measurement intra-action, concepts become part of the material configuration of the world. 'Position' is what a detecting apparatus with relatively fixed components can measure, whereas 'momentum' is defined by an apparatus whose components can move appropriately in relation to one another.[31] Such ma-

30. The utterances and actions of the "object-system" are thus the measurements elicited by its intra-action with the measuring circumstances that constitute the "intentional configuration."

31. Howard (1994) offers a speculative but highly plausible reconstruction of Bohr's claim that even quantum physical measurements must always employ classical concepts to characterize the interaction between experimental apparatus and the object system. Howard concludes that, while Bohr's claim makes sense on this interpretation, it should not be accepted, because it "concedes too much to the worldview of classical physics" (226). If 'classical concepts' are not the concepts of classical physics, however, but simply the concepts articulated by an experimental configuration, then the force of Howard's objection is less clear. Barad (forthcoming, chap. 7) argues that 'classical concepts' should not be identified

terially embedded concepts are ineliminably normative, of course: an apparatus with internally fixed components does *not* measure momentum, but only because it fails to indicate distinct momenta by distinct marks.[32] Concepts in this sense are the normativity of physical arrangements *as reproducible.*[33] A concept is then *expressed* by the term 'position' when the use of that term is accountable to the kinds of apparatus and marks that can define position. Such implicitly definite concepts thus cannot be expressed unless the world contains the needed apparatus, namely, systems that hold others (and themselves) to account for appropriate responses to things, and for correct uses of marks that thereby become linguistic expressions. These systems, however, are neither individual minds nor social collectives, for the circumstances to which concept-use is materially and discursively accountable are part of what establishes their definiteness as concepts. Material intra-actions become causally structured measurements, such that their object systems take on definite boundaries and properties as causes of these measuring effects, when they are appropriately situated within a larger pattern of practical engagement with something at stake in the outcome.

The appropriate use of such concepts extends beyond the specific circumstances in which they are actually employed. That follows from the

---

with Newtonian physics, but only with the concepts specified by a particular configuration of apparatus.

32. The sense in which a detector with internally fixed components fails to measure momentum might initially seem to be interestingly different from the sense in which it fails to measure specific heat or electrical charge. Collision with an internally rigid detector might be thought of as a "bad" measure of momentum (the collision intra-acts with the momentum of the measured system, but does not distinguish among possible momenta, except at the single point at which the measured momentum is sufficient to break the rigid arrangement of the detector), whereas such a collision neither instantiates nor exemplifies charge or specific heat at all (on the role of experiments as exemplifying some of the properties they instantiate, see Elgin 1991). Yet the case of the breaking of the rigidly fixed components of a detector (or the failure to break them) is not yet a momentum measurement, even a "bad" one, without further apparatus and practical performances that measure a definite breaking point for a detector so constituted. Hence, one would have to conclude that the concept of momentum has no definite application within that material configuration, which is just what one is also inclined to say about specific heat or charge (which, of course, is not to say that those concepts aren't appropriately applicable to these components under different circumstances). Thanks to Karen Barad for convincing me of this point.

33. It is important to recognize that reproducibility is an inherently normative notion, since the same phenomenon can be produced in various ways. Whether the same phenomenon has recurred or a different phenomenon is occurring depends upon what is at issue and at stake in the claim. The importance of this point for understanding experimental practice will be introduced at the very end of this chapter, and discussed more extensively in chapter 9.

recognition that a phenomenon is not just a particular event, but an open-ended pattern of actual and possible events that the particular event embodies. This feature of concepts can be explicitly articulated through the use of counterfactual conditionals. From what happens under particular experimental conditions, one infers what *would* happen under other conditions. Such inferences, however, are only explicit articulations of commitments already undertaken implicitly in using the concept at all.[34] Of course, the full extent of the counterfactual commitments undertaken in employing a concept is not already definitely determined in advance. Which counterfactual inferences are appropriate depends not just upon what actually does happen in subsequent experimental practice, but also upon what collateral concepts and commitments are undertaken or required in the context of subsequent work, and what is at stake in undertaking or rejecting such commitments. As I indicate in chapter 9, these considerations are indispensable to the normative force as well as the normative authority of ongoing scientific work.

Such patterns of practical engagement, through which prosthetically embodied agents are responsible for (and to) their engagement with the world,[35] *allow* the world to show itself in particular ways (including, upon

34. This expressivist conception of counterfactuals inverts the aspiration to understand causal connection through a proper analysis of the logic of counterfactual conditional statements. The logical form of counterfactual conditionals and their inferential relations does not determine where the boundary belongs between correct and incorrect specifications of causal intra-actions, but only expresses the consequences of drawing it in one place rather than another. Which of those consequences should be accepted and which rejected is then a practical scientific project rather than one conducive to formal logical or semantic analysis. The idea that counterfactual conditionals are explicit expressions of material-inferential commitments implicit in concept use goes back at least to Sellars's classic paper "Concepts as Involving Laws, and Inconceivable without Them" (1980), although my account of the normative authority governing such material inferences is not the same as Sellars's. Sellars locates the counterfactual authority of the use of scientific concepts in laws, already fully determinate structures of causal relations among definite objects. I find that authority in what is at stake in ongoing practices, which certainly outruns present uses, but not in ways that are presently fully determinate. See my discussion of laws and models in chapter 9 below.

35. The significance of the ambiguous boundaries between agents and their surroundings first emerged in chapter 7, in a discussion of Brandom's perspectival account of conceptual content. The above discussion of measurement makes this point central to scientific practices, as instruments, laboratories, and theoretical models become prosthetic components of a bodily engagement with one's surroundings. The world is thereby *configured* differently as a field of practical engagement and understanding. As Kuhn once famously said of Lavoisier, "after discovering oxygen, [he] *worked* in a different world" (1970, 118). Discovery, however, is then not a psychological phenomenon, but instead expresses a complex assimilation of instruments, concepts, and their skillful deployment within the practical-discursive competences that constitute agency.

occasion, showing the incoherence or inefficacy of those forms of engagement). I thus reiterate the importance of Haugeland's distinction (1998) of the normative constitution of objects-in-phenomena, "letting things be," from both the silly idea that norms *create* objects and the serious but mistaken idea that they merely make it possible for already constituted objects or events to "count as" something else. One might say that practices constitute objects-in-phenomena by configuring a setting within which the boundaries between one object and another make a difference, and are answerable to the differences they make. Yet here, grammar is misleading. Constitution involves the emergence and maintenance of normative accountability and stakes to which objects and practitioners are accountable. Practices are not something *else* that accounts for or accomplishes the formation and location of such patterns and boundaries within the world. They *are* the establishment and maintenance of boundaries and the accountability that sustains them. Practices do not preexist the objects, boundaries, and stakes that they constitute.

I can now show more clearly why my conception of causality does not illegitimately presuppose a prior conception of intentional content or normative accountability. On my account, causality and normativity (including the normativity of intentional content) are internally related to one another. This claim is illuminated by comparison to Brandom's and Davidson's treatments of the relations between thought and talk. Regarding conceptions of intentional content as essentially linguistic, Brandom (1994, 150–53) has drawn a distinction between "analogical" and "relational" accounts. Linguistic conceptions of either sort draw upon the contentfulness of sentences as the basis for understanding the content of intentional states such as beliefs and desires. Such projects might seem to require an account of sentential content that is prior to and independent of any attributions of intentional content to beliefs and desires (similarly, a naturalistic account of intentional normativity seems to require a conception of causality or natural law that is prior to and independent of any conceptions of normativity). Brandom notes that this requirement does indeed apply to analogically linguistic conceptions of intentionality (of which I take Jerry Fodor's theory as exemplary). Fodor (1975, 1981) treats the content of a belief by analogy to the content of a sentence (roughly, a belief is a sentence in the language of thought, and the content of the belief is the content of the postulated sentence). He then hopes to provide an independent construal of the intentional content of such sentences through a causal/functional theory of their representational content. But it is also possible to have a relational linguistic account such as Davidson's (or Brandom's own). Relational theorists argue that the contents of beliefs and desires are essentially, and not merely accidentally, expressible by sentences in a public language. Nevertheless, such sentences are only contentful through a practice of interpretation/attribution in which content is always attributed simultaneously to expressed utterances or actions and unexpressed beliefs and desires. Brandom quotes Davidson on the key point: "Neither language nor thinking can be fully explained in terms of the other, and neither has conceptual priority. The two are, indeed, linked, in the sense that each requires the other in order to be understood, but the linkage is not so complete that either suffices, even when reasonably reinforced, to explicate the other" (Davidson 1984, 156; cited in Brandom 1994, 151). Davidson and Brandom do not, then, require a conception of linguistic content independent of the attribution of content to beliefs and desires in order to understand the latter as essentially expressible linguistically.

I am advancing a similarly relational conception of causal structure and intentional normativity, and the difference between my account of causality and that of more traditional metaphysical naturalists parallels the difference between analogical and relational conceptions of how intentionality is linguistic. I am arguing that intentional attribution and acknowledgment *are* causally intra-active phenomena (just as Davidson and Brandom would argue that beliefs and desires *are* linguistic). Traditional metaphysical naturalists also want to construe beliefs and desires as causal phenomena, but do so analogically. They therefore need an independent, anormative conception of causality, which could then be used as the basis for their subsequent account of intentional content and its associated proprieties. I do not need in this way to show that the intelligibility of causal structure is independent of the appropriate ascription of intentional-discursive proprieties, because my account of the causal structure of "ordinary" causal intra-actions and that of intentional attribution or acknowledgment is relational. Neither is intelligible apart from the other—there is no causal *structure,* no determinacy to the components of a causally intra-active system, apart from its involvement with discursive normativity—and yet intentional attributions are themselves causal intra-actions. Note, of course, that my relational conception of causality and intentional normativity then *incorporates* something like Brandom's or Davidson's relational conception of talk and thought, but to a very different philosophical end, namely, as part of a nondualistic philosophical naturalism.[36]

Brandom's expressivist account of discursive articulation may also help avoid other possible misunderstandings here. Brandom argues that all explicit articulations of norms rest upon and are accountable to norms implicit in practice; similarly, I claim that causal intra-actions are implicit determinations of the object-system measured (of its identifying boundaries as well as the properties and capacities ascribable within those bounds). There is an air of apparent paradox to claims that the implicit is more fundamental than the explicit. That a norm or a boundary is implicit within an intra-action can only be vindicated by making it explicit. Indeed, *what* was implicit can only be characterized as a now-explicit articulation of it; its having-been-implicit only emerges through explication. The same issue characterizes the constitution of causal intra-actions: that the ebb and flow of tides marks a gravitational intra-action with the moon (rather than with the local force-field in the Firth of Forth, or the Mackiean total cause of all that some set of water molecules does and does not do in some time interval *t*) can now be seen to depend upon the emergence of discursive norms. Such normativity expresses the "agency" of the practically constituted cut between agencies of observation and object systems. Yet what those norms provide is a *particular* setting within which the capacities of things become manifest through the marks they leave upon a discursive practice. The gravitational capacity of a body is only manifest in the trajectory of another body, against a background of other influences; the implicit capacity of those marks to make a *semantic* difference is in turn only manifest in an ongoing pattern of talk and implicitly attributed beliefs. Yet that pattern only continuously constitutes itself as *talk* (rather than as noise or evocative cries) through its responsiveness to what is at stake in its articulation, and what it thereby comes to be "about." Without a pattern of utterances responsive to one another and their circumstances in semantically significant ways, the world would indeed have no semantic structure, and hence no boundaries or properties.[37] Yet being

36. Thus, I share Davidson's aim to show how his ineliminably normative relational conception of thought and talk is compatible with claiming that intentional states are also causal, but I am claiming that a relational conception of normative and causal conceptions succeeds where Davidson's anomalous monism failed.

37. Dennett (1991, 379–80) distinguishes two ways in which correct attribution of a predicate can depend upon human capacities and practices. Something is "lovely" if there are or have been entities with the perceptual capacities and aesthetic responsiveness needed

responsive to "things themselves" is constitutive of that pattern itself as discursive. Neither sentences nor things can be what they are without the other.[38]

Understanding discursive practices as causally intra-active in this way does not provide an alternative metaphysics of nature, one that would substitute a metaphysics of phenomena for a more familiar naturalist metaphysics of nature as composed of material objects or events, or as determined by laws. Such an alternative metaphysics would treat phenomena as simply a more complex kind of object, and claim that the world (nature) as a whole is composed of them rather than of some other kind of object or process. As in other supposedly naturalistic metaphysics (such as physicalism), that claim would still tacitly depend upon the possibility of representing the world in a language that is not itself constitutively part of the world represented. Understanding discursive practices as causally intra-active instead denies that one can intelligibly do metaphysics in the sense of saying what the world as a whole is like. Frank Jackson has recently recognized that this rejection of any traditional metaphysics of nature

to recognize its loveliness, whether or not such an entity has ever actually been in a position to make that judgment. Someone is a "suspect," however, only if actually suspected by someone; the counterfactual judgment that one would have been suspected under different circumstances is not sufficient. Many philosophers have implicitly taken the dependence of truthful disclosure upon discursive practices to be a "lovely" dependence: the structured possibility of semantic significance is independent of any actual instantiation of a discursive practice. One way of reading my argument is that semantic significance depends upon there being actual discursive practices and norms in much the way that being a suspect depends upon actually being suspected. This point, of course, also bears affinity to Quine's insistence that linguistic meaning (and hence the possibility of truth) is not specifiable apart from translation into a language *antecedently* understood.

38. The causally intra-active dependence of language use has often been overlooked because of the salient phenomena of false utterances and failures of reference. These phenomena provided a principal motivation for Fregean and Husserlian accounts of language use as expressing an ideal sense, whose prior determinacy would allow the determination of whether the world accords to what was said. I have instead been arguing that language use is a form of practical, bodily activity, which like any other practical comportment acquires its coherence from mostly successful intra-action with surrounding circumstances. Similar insistence that language is a worldly phenomenon are expressed in Quine's and Davidson's appeals to the principle of charity, Brandom's and McDowell's recognition that contentful utterances must be objectively accountable, and Heidegger's claim that "making statements is being toward the existing thing itself" ([1953, 218] 1996, 201). What sustains the illusion that the possibility of error shows that language use can be meaningful independent of causal intra-action with its actual surroundings is the wordiness of the world. Language is so pervasive, so familiar, and consequently so inferentially intertwined with other aspects of language, that it may seem as if intralinguistic relations float free of other causal intra-action. For further discussion of this point, see Rouse 2000.

follows from Quine's naturalist commitment to the indeterminacy of meaning:

> It is insufficiently realized just how radical the Quinean position is. . . . Suppose [as Quine does] that it is impossible to effect a partition among the possibilities independently of how things actually are. No mental state, no linguistic item, no diagram, no system of semaphore, divides the possibilities, except *relative* to how things are. Then we can never say, diagram, depict, semaphore, think, . . . how things are. (1998, 53)

My position accords with Quine's in just this respect. To say, as I do, that causally intra-active systems are always components of phenomena is not to say how things are in this sense. My claim instead expresses that discursively articulated scientific understanding always intra-actively *belongs* to the world ("how things are") and cannot magically represent the world from outside. The world in turn is not the totality of things (or of phenomena), but is instead the unbounded setting within which anything is intelligibly said or done. To talk about phenomena is simply to make explicit that all meaningful boundaries are intra-actively established within the world.[39]

Much more needs to be said, obviously, to work out my underlying conception of scientific practices and the disclosure of nature. The point, however, should be clear. Many discussions of whether semantic, epistemic, or even moral norms can be "naturalized" presupposes a conception of nature ("how things are") that is disconnected from the scientific disclosure of the natural world through discursively articulated experimental practice. Since metaphysical naturalism acquires its rationale from an acceptance of that scientific disclosure, these conceptions of naturalization are misguided. My point is analogous to McDowell's criticism of Davidson for illegitimately helping himself to the concept of a body of beliefs. Metaphysical naturalizers and many of their critics alike help themselves illegitimately to the concepts of (already-determinate) objects, causes, or laws in ways that cannot be reconciled with their own commitments to naturalism. Before they could be entitled to such concepts, they must adequately attend to the material and conceptual apparatus through which scientific practices articulate and become accountable to objects, events, causes, or laws.

39. The mistake of trying to understand the world as a totality (as if one were outside the world) is closely related to the conception of a language as a representational system. Indeed, this criticism can be seen as an analogue, in the "material mode," to Neurath's "formal mode" rejection of Carnap's conception of a language. See chapter 1.

Experiment and Theory Revisited

I now return to the question that both logical empiricists and postempiricist holists had asked about the relationship between theory and observation, which has been transposed by more recent work into the relationship between theory and experimentation. My discussion in this chapter helps show how and why that conceptualization as a relation between two distinct components of scientific work is a mistake.

At issue in this question are scientific concepts and claims, and how their meaning and truth value are accountable to the world (or to experience of it). The logical empiricists could conceive this issue as they did through a prior commitment to the implicit role for the formal structure of a language in securing meaning. Empirical laws could have a life of their own, on the presumption that one could in principle articulate a formally closed and consistent observational language in which such laws could be expressed and justified independently of their inferential relations to "theoretical sentences." Postempiricist holists denied the independence of theoretical and observational languages, while nevertheless retaining the underlying commitment that concepts are only meaningful through their place in the systematic structure of a language or Quinean total theory.[40]

Hacking rightly objected that to understand utterances as meaningful only through a possibly implicit commitment to an underlying theory requires an expansion of "the word 'theory' to denote all sorts of inchoate, implicit, or imputed beliefs," which then makes omnibus claims to theory-dependence "trifling" (1983, 175–76). Yet accepting the force of this objection reopens the question of how to understand the conceptualization of experimental practices. The most common response to this issue (Hacking's included) has been to appeal to causal relations between canonical uses of natural kind terms and the contextually relevant objects and properties. My account also invokes causal intra-action as indispensable to conceptual articulation, but it differs from the most familiar causal-naturalist approaches in at least four crucial respects.

40. The correlative shift from "language" to "theory" as the systematic context that conferred meaning upon particular concepts was mediated by a shift toward defining that context in terms of a substantially contentful, unrevisable core (a research program, paradigm, ideal of natural order, or "hard core") and whatever collateral commitments that were then articulated in relation to it, rather than by a deductively closed formal structure. Of course, the critical norms for the distinct identification of a theory were then semantic rather than syntactic. The semantic significance of the core commitments was thus taken

1. My account ascribes ontological and analytical priority to phenomena over entities and their properties or relations, which are only defined (definitely bounded as reidentifiable entities, properties, or relations) by their contribution to the repeatable patterns of phenomena.[41]

2. The inferential/discursive relations among the terms that denote distinct concepts are an ineliminable part of the phenomena themselves. The practices through which the normativity of natural phenomena is discursively expressed constitute the boundaries that articulate the components (cause/effect, measured/measuring) of causal intra-actions themselves. To repeat: they "constitute" them in the sense of allowing them to be significant and relevant in a definite pattern, not in the sense of "making them up."

3. The recognition that differentially repeatable phenomena and the causal patterns they articulate are ineliminably normative shows that the causal-functional roles of terms cannot be the basis for a naturalistic expla-

for granted, instead of being instituted through the stipulative interpretation of a formal syntax.

41. It is no objection to my account that the *entities* investigated in particular scientific fields may well remain relatively stable over time, while the specific *phenomena* through which they are most prominently manifest may be continually shifting. Scientific intelligibility is not to be found in phenomena in isolation, but in their inferential interconnectedness. Objects and object-kinds are constituted by their inferential role in the more complicated phenomena through which even materially incompatible phenomena are related to one another. For these purposes, the anaphoric and indexical apparatus of discursive practices are indispensable in constituting the normativity of object-identity.

Brandom (1994, chaps. 6–8) offers a richly articulated account of such a response, albeit in the "formal mode." He thus argues for the analytical priority of propositions over singular terms, which are articulated anaphorically through their inferential roles in propositions. My emendation turns Brandom's account inside out by taking the material mode to be primary, emphasizing the priority of phenomena over objects. This approach has the virtue of incorporating intralinguistic performances within the broader practical normativity of agential intra-actions with the world. The practically intra-active ability to distinguish words within sentences, and to reidentify them in other sentential contexts through the inferential relations between appropriate performance in those different contexts, is thus itself a crucial case of the priority of phenomena over entities. Note that words similarly provide relatively stable links among the more ephemeral sentences spoken and heard, and yet their status as words arises only through their contributions to sentences.

Returning to the original objection, it is then no surprise (and no objection to my approach) if objects or object-kinds provide the stable links among the changing phenomena of interest within scientific practices; such stabilization and inferential interlinking are precisely the practical-inferential role through which objects and object-kinds acquire definition within the phenomena of scientific practice. They are nevertheless only defined through this inferential role, which presumes the priority of the phenomena generally (albeit not of any particular phenomenon). This point will be discussed further in chapter 9, when I consider the relation between causally intra-active phenomena and causal laws.

nation of semantic or epistemic norms. Displaying the characteristic causal roles of scientific concepts is expressive rather than explanatory.

4. This incorporation of larger patterns of inferential relations within the "apparatus" through which causal relations are articulated does not surreptitiously bring back the problems of a theoretical holism, which must posit "theoretical" commitments underlying any semantically contentful utterance or action, and which cannot readily account for the critical assessment of these core commitments. Although this point requires more extensive development than I can provide here, the crucial difference arises because I take the normativity of inferential connections to be established not by the systematic self-enclosure of a "theory," "research program," or "conceptual scheme," but by an open-ended *responsibility* for further inferential articulation.[42] The commitments undertaken through concept-use always outrun the actual material and discursive circumstances of their use, such that they can turn out to have been mistaken.

However, this account of conceptual articulation as part of experimental phenomena themselves may seem to pose an opposing difficulty. It is perhaps now understandable how experimentation can have a life of its own, but at the expense of making the relatively autonomous development of discursive, theoretical understanding unintelligible. If scientific concepts are always practically and inferentially embedded within particular material phenomena, what could it mean for *theory* to have a life of its own? Why doesn't my approach, in which conceptual articulation is always intimately a part of causal intra-action, lead to an experimentalist analogue to the old discredited positivism that tried to reduce theoretical conceptualization to its manifestation in specific empirical phenomena? Have I been thrown back in the position of Hempel's theoretician (1965), whose dilemma expressed how logical positivism made theoretical concepts empirically meaningful only by rendering them scientifically superfluous?

Even if my account did have the consequence that all meaningful theo-

42. Brandom characterizes this antisystematic way of understanding the holism of conceptual articulation as "tactile Fregeanism": "Frege's own favorite metaphor for our cognitive relation to [conceptual content] is that of *grasping* rather than *seeing*. One can grasp an anaphoric chain as one grasps a stick; direct contact is achieved only with one end of it, and there may be much about what is beyond that direct contact of which one is unaware. But direct contact with one end gives genuine if indirect contact with what is attached to the other end. . . . A *tactile* Fregean semantic theory . . . incorporates, as two sides of one coin, both the possibility of ignorance of and error about our own concepts and the possibility of genuine aboutness of those concepts and genuine knowledge of the [phenomena]" (1994, 583).

retical conceptualization was very tightly linked to experimental practice, the problem would not be as acute as the one articulated by Hempel, for two reasons. First, the "empirical basis" it provides for the interpretation of theoretical-conceptual work is much less impoverished. Theoretical meaning would be inferentially linked not to an artificially restricted "observational vocabulary," but to the causal nexus of experimental and observational practices. Moreover, experimental practice is itself conceived as incorporating inferential norms for the application of concepts far beyond immediate practical contexts. Any theoretical articulation that floated relatively free from its use within such an inclusive conception of experimental practice might indeed plausibly be open to criticism as "just mathematics" or "empty speculation" rather than theoretical understanding. Second, it is worth remembering that in most scientific disciplines, theory does not have a life of its own. Theoretical physics and mathematical population genetics do not typify the life of scientific theorizing, except perhaps in the imagination of some philosophers who identify with those scientific practices that most resemble their own.

Yet I want to conclude by briefly suggesting how my approach actually does accommodate the relatively autonomous development of some theoretical practices, in ways that also further display the ineliminable discursive character of experimentation itself. Recall that theorizing is to be understood here as practices of *modeling*. A diagram, an equation, a computer program, or a physical entity is a model if it is *used so as to be held accountable* to the outcomes of operations performed on it, to their inferential relations to one another and to what is to be modeled, and ultimately to what is supposed to be at stake in these performances.[43] One important role of such models is to extend the inferential articulation of phenomena, especially through the substitution-inferential patterns that constitutively discriminate objects, properties, and relations. The differences between constants, parameters, and operators in mathematical models, for example, discriminate different patterns of appropriate substitutions or other transformations, and whatever significance is thereby ascribable to the outcomes of those inferential operations. Functional diagrams in turn identify the intelligibility of functionally equivalent substitutions, whose trackable effects can thereby constitute functional roles and

43. For a more extensive discussion of theorizing as modeling practices (articulated, however, in a Davidsonian truth-theoretic idiom rather than the inferentialist idiom that I now find more felicitous), see Rouse 1996b, 224–36.

the components that answer to them. Software programs for calculating the intersections of known constraints upon the structural assembly of large molecules (or their earlier analog analogues in machined physical models of three-dimensional structures) discriminate possible from impossible spatial configurations of components. And so forth.

Such models, whatever their form, are often strongly dependent for their significance upon their inferential relations to particular physical systems.[44] The operations performed in a computer program written to model protein folding, for example, may be senseless apart from the accountability of their inputs and outcomes to an antecedent concern with specific aspects of protein chemistry. Theoretical practices can come to have a life of their own, however, when the operations internal to such theoretical modeling can themselves be modeled in illuminating ways.[45] The software program may involve subroutines whose structure makes them suitable for deployment in other programs for quite different purposes; or the software may itself exemplify higher-level features such as efficiency, parallel processing, or finite computability. If these other deployments are sufficiently complex and consequential, then they can spin off as a relatively autonomous practice.[46] Whether these operations and whatever discriminable fea-

44. Remember also that such theoretical modeling is always itself materially realized in particular ways, which affect the subsequent development of the practice. This point is often neglected, because philosophers take for granted the institutional and material practices involved in, for example, training a generation of scientists to be able to use and evaluate new mathematical or other theoretical tools. This point can be seen especially clearly, however, when the material realization of theory involves more than just the inculcation of new skills in scientists. Consider, for example, the different material/practical manifestations of the periodicities that show up through x-ray crystallography of large biological molecules. These patterns can be manipulated physically (recall Watson's and Crick's machine-shop models of helical configurations of DNA components), analytically (in those cases in which the relevant Fourier series are tractable), or computationally (where the calculative capacities of computer systems enable different kinds of operations to be performed on the same measurements). Such manifestations may be equivalent in one sense (they can themselves be analyzed as mathematically interchangeable), but they are not equivalent in practice, since they cannot be effectively applied in all the same circumstances. Even looking at scientific theory in its most "autonomous" forms always involves dealing with simultaneously causal/material and inferential/discursive dimensions of scientific practice.

45. Latour (1987, chap. 6) discusses such metamodeling as an iterable process of quasi-bureaucratic formalization. He regards such theoretical work as primarily representational in its function—the $n^{\text{th}} + 1$ level of forms stands in for the nth level. While he acknowledges that formalization "changes the scale" of scientific work to allow new operations to be performed, he accords less recognition to how it can change what is at stake in the practice and thereby allow theory to have more of a "life of its own."

46. Galison (1996, chap. 8) provides an extended discussion of one example of such autonomous modeling practices in the development of Monte Carlo simulations, which

tures they disclose come to have a theoretical significance apart from their deployment within protein modeling will depend upon the articulation of inferential relations to other practices. More fundamentally, it will depend upon whether the differential outcomes of various operations disclose something else at stake in those inferential relations. Thus, for example, the symmetries disclosed through operations on theoretical models, as well as the explanatory role that emerged for symmetry and symmetry breaking, have contributed to the relative autonomy of "fundamental" theory from phenomenology in twentieth-century physics.

Such shifts in what is at stake in a theoretical model, and hence in what the model is "about," also highlight the discursive, inferential role of experimentation. Understanding the specific phenomena created in scientific laboratories, after all, is typically only the proximate goal of experimentation. Experimental phenomena and experimental systems are themselves typically deployed to model some other aspect of the world. Morgan's group developed the *Drosophila* system, for example, as a model of relations between chromosomal structures and phenotypic traits generally, not of any species-particularity of *D. melanogaster*. As Hans-Jörg Rheinberger has clearly shown, the inferential relations that constitute experimental systems as models can undergo substantial shifts. Thus, for example, Albert Claude's use of the ultracentrifuge on the chicken sarcoma system he was using to model the "viral" transmission of cancer transformed the sarcoma system itself into a model of the "normal" functional significance of subcellular organelles. Such transformations of what is at issue and at stake in experimental practice are among the most provocative illustrations of how experimentation has a life of its own.

Such shifts nevertheless point toward a crucial limitation in my discussion of scientific practices in this chapter. Throughout the discussion, I have emphasized the constitutive repeatability of causally intra-active phenomena, including those phenomena realized in experimental practice. But experimental science is not primarily in the business of repetition. Rheinberger showed this point ironically by invoking Alfred Hershey's celebrated conception of scientific happiness, which was "to have one experiment that works, and keep doing it all the time" (Rheinberger 1997, 24–25). The point is ironic, because such repetition of a phenomenon is pre-

were originally introduced as stochastic models of particle interactions in nuclear fission, but quickly spun off into a distinct focus of specialized scientific work, applicable to an open-ended variety of problems.

cisely what experimental science never does. Once an experimental phenomenon has been stabilized as repeatable, experimental practice outruns it; to the extent that the phenomenon continues to be reproduced, it is only as a component of related phenomena, which often subtly shift what is at stake in the work. As I shall show in chapter 9, one of the most important reasons for talking about the normativity of natural phenomena, and about the discursive significance of experimental practice, is to understand better this aspect of scientific research. Scientific practices are not just directed toward, or accountable to, the phenomena that they currently work to stabilize and articulate. Scientific research is always directed beyond itself, toward something at stake that is not yet adequately realized or understood. Giving central place to this futural dimension is indispensable to an adequate understanding of scientific practices and their accountability to nature and natural phenomena.

NINE

Natural Necessity and the Normativity of Scientific Practices

Naturalism and the Representation of Natural Necessity

Having reflected upon scientific practices and discursive practices more generally, I can now return to the questions about naturalism with which I began this book. Philosophers today sometimes ask whether epistemology, ethics, or the intentionality of mind and language can be "naturalized." To naturalize a domain in the intended sense would be to show how its normative dimension (or perhaps its appearance of normativity) can be reconciled with an acceptance of the natural sciences and the world as they disclose it.[1] The underlying presumption, however, is that one would not be *entitled* to accept at face value the apparent normativity of some phenomenon unless philosophical explication could assimilate it to a world of scientifically understandable facts, causes, or laws.[2] That commitment in turn presumes that normativity should be regarded, at least prima facie, as "unnatural": the facts, causes, and laws that the sciences disclose are not themselves irreducibly normative.

1. This conception of the "naturalization" of semantic, epistemic, or moral/political norms should be read expansively. The opposition between naturalists and antinaturalists has often been defined in terms of whether norms can be naturalized in a stringent sense (of which elimination, reduction, or supervenience are the most salient examples). Yet many philosophers who oppose these strong conceptions of the naturalization of semantics, epistemology, or ethics do so only because they believe that a less stringent naturalization is sufficient. Those who think that philosophical accounts of normativity need only be consistent with the scientific disclosure of nature, or capable of showing how a non-natural source of normative authority could be realized by natural beings, still accept that normativity must be "naturalized" in these less stringent ways. Thus, the passages I quoted from Christine Korsgaard and John Haugeland in the introduction suggest that even they still accept this broad conception of the naturalization of normativity.

2. This commitment to demonstrate the naturalistic respectability of any acknowledgment of norms is commonly fulfilled with promissory notes, which enjoy wide philosophical circulation. The question of when such promises have to be redeemed, and how much argumentative collateral they require, has not often been explicitly addressed. All too commonly,

Those who endorse demands to naturalize semantic, epistemic, or moral/political norms differ vociferously, of course, about what a naturalistic explication would require. Proposals have included such diverse possibilities as the natural scientific elimination, reduction, supervenience, determination, predictive utility, or mere compatibility of these norms.[3] With considerably less fanfare, however, naturalists have differed almost as widely in their conceptions of nature as explicans. For the purposes of naturalistic explication, the natural world has been presumed to consist variously of material objects, natural kinds, causal powers, natural laws, or accessible possible worlds. The differences among these alternative naturalizations are not decisive, however, with respect to the concerns that have been at the forefront of discussion in this book. If I am right, the most significant difficulties confronting metaphysical naturalisms apply to assumptions common to all of these alternatives.

At the outset of the book, I proposed that two commitments jointly articulate what is at stake in aspirations to a naturalistic philosophy. A naturalist can countenance neither arbitrary impositions upon scientific practice nor appeals to the mysterious or supernatural. I focused upon these two commitments because together they capture what has been most philosophically compelling about naturalism. The Nietzschean naturalist commitment is not to countenance philosophical appeal to any "supernatural" principles, structures, relations, or grounds. This commitment expresses a condition of philosophical understanding; such appeals to the supernatural mark a willingness to settle for what now seems to be only the semblance of understanding.[4] It also reflects a recognition that eliminating residual supernaturalism may be difficult. Contemporary naturalism, after all, arose from criticism of the great foundational programs of twentieth-century philosophy, logical analysis and phenomenology, for their willingness to settle unwittingly for a supernaturalist "magic language" concep-

rival proposals are dismissed for their failure to meet demands that one's favored alternative also cannot meet without considerable handwaving.

3. Not everyone who has construed normativity in these ways would claim to be naturalists. Some advocates of the (mere) compatibility of norms with natural causality, or even of the predictive value of normative attribution or the supervenience of norms upon nature, identify naturalism with reductionist or eliminativist proposals, and thereby take their own claims to be antinaturalist. Since I am primarily concerned with how nature is being conceived in these accounts, I treat all such positions as variations upon metaphysical naturalism.

4. Nietzsche diagnosed here a failure of more than just understanding, of course, but I will not address this aspect of his criticisms.

tion of meaning and the understanding of meaning.[5] Nietzsche explicitly noted a vestige of theism in such semantic piety: "I am afraid we are not rid of God because we still have faith in grammar" (1954, 483). The Quinean naturalist commitment reflects both the historical role of the sciences in articulating human self-understanding as natural beings and the recognition that any attempt to assess that self-understanding from a standpoint outside of ongoing inquiry would itself violate the Nietzschean commitment.

The Nietzschean commitment requires that a philosophically adequate explication of normativity in any domain must eventually show how the relevant norms could be accessible and applicable to such natural beings, and it must show how norms acquire and sustain whatever binding force they seem to have for such beings.[6] The Quinean commitment imposes three constraints that hang together uneasily.[7] First, naturalists are committed to the philosophical primacy of the natural sciences as they have actually developed historically.[8] Naturalists can take any particular feature of contemporary science to be defeasible, but a naturalistic explication of

5. I take the perspicuous phrase "magic language" from Wheeler (2000), who both applies it to the classical semantic theories of the early twentieth century and also in passing cites Nietzsche as an honored ancestor of those who undertake a commitment not to invoke a magical conception of language use.

6. This second constraint is close to what Christine Korsgaard has dubbed "the normative question": "When we seek a philosophical foundation for morality we are not looking merely for an explanation of moral practices; we are asking what *justifies* the claims that morality makes on us" (1996, 9–10). Naturalism, of course, leaves open the possibility of a negative answer to this question for any putative normative domain, although one must then surely account for how and why the initial appearance of normative authority and force arose.

7. It should be noted that I hereafter use "Quinean naturalism" not to refer to the actual positions taken by Quine himself or his commentators and followers, but to a looser and more widespread identification of what there is in nature with what the sciences can disclose.

8. Some contemporary naturalists are also comfortable appealing to a definite conception of the *future* course of the sciences. Thus, John Searle expresses a commitment to the prospect that intentionality will turn out in the course of further research to be a straightforwardly biological property of certain organisms, whereas both Paul and Patricia Churchland foresee a quite different resolution of various philosophical issues by the anticipated continuation of some current research programs in neuroscience. In both cases, however, it matters that the future science to which they appeal be methodologically and perhaps ontologically continuous with contemporary science. Searle would not take his view to be vindicated, for example, if the disciplinary successors of contemporary biologists were to appeal to shared mystical intuitions that certain biochemical pathways or neural transmission patterns were intrinsically intentional, nor would he do so if the biological character of intentionality were taken to follow from the allegedly biological character of a divine presence within bodies.

normativity cannot depend upon the results of a science too discontinuous with contemporary science. A naturalist cannot, that is, identify 'science' through criteria of adequacy substantially unmet by any current science, or by imagining the science of beings with cognitive capacities that substantially exceed those of humans, or a science that would only emerge after some massive, revolutionary reconstitution of the extant sciences. Second, what matters for naturalism is science as responsive to and accountable for something beyond its own claimings. Science must be criterial for the "nature" of naturalism not because of de facto deference to its authority, but to the (defeasible) authority itself. Thus, naturalists regularly refer to "mature" sciences, successful sciences, or sciences at their best.[9] Third, that authority must itself be open to naturalistic explication. It is not uncommon for those who would naturalize ethics or intentionality to take for granted the authority of the natural sciences, but a thoroughgoing philosophical naturalism must surely account for the achievement of scientific understanding from human beings' interactions with their surroundings. Naturalism in the philosophy of science thus seems to occupy a pivotal place within any comprehensive philosophical naturalism.

Care must nevertheless be taken in identifying the appropriate task for a naturalistic philosophy of science. Alexander Rosenberg's insightful survey (1996) of naturalistic philosophies of science also identifies, as a central issue for naturalists, the need to understand the authority of contemporary science in ways consistent with commitment to naturalism. Rosenberg takes for granted, however, that this question is epistemological: "How can we be confident that science—whether interpreted realistically or otherwise—is objective knowledge of significant truths, even loosely interpreted?" (1996, 18).[10] I take the more crucial question for naturalists in-

9. Larry Laudan's aspiration (1987) to a "normative naturalism" clearly expresses this dimension of recent naturalistic philosophy, but it is implicit in Boyd's invocation of "mature" sciences, and generally in the opposition of most philosophical naturalists to any hint of epistemological relativism.

10. Rosenberg's discussion of recent prominent naturalist philosophies of science (those of Larry Laudan, Ronald Giere, and Philip Kitcher, among others) is cogent and incisive. His epistemological orientation has another unfortunate consequence, however. Rosenberg frames his discussion as a confrontation between naturalistic philosophies of science and relativist-constructivist sociology and history of science. That is an unfortunate choice for two related reasons. First, constructivists in science studies today often conceive themselves as naturalists (and indeed do so in opposition to many of those whom Rosenberg considers naturalists, precisely because the latter are allegedly too willing to take familiar expressions of scientific norms at face value). Second, this framing conflates two questions that need to be kept separate. The question that naturalists should ask, I think, is an analogue within

stead to be ontological. How does a naturalistic conception of nature incorporate scientific understanding as a natural phenomenon? The answer to my question may well have epistemological consequences; only some ways of understanding science and nature may turn out to be naturalistically justifiable (or reliable). Yet the initial aim is to understand the world, not standards of belief.

This ontological question has been obscured by a presumption that its answer is obvious. Nature is the world as *represented* (correctly) by the natural sciences. Many naturalists straightforwardly express a commitment to representationalism in their very conception of naturalism: the representation of nature in terms of regularities, laws, or causal powers provides the baseline for their attempted naturalizations of semantic, epistemic, or moral norms. Yet naturalists who begin with the philosophy of science have also typically conceived the place of natural science within nature representationally. The demand that philosophical reflection upon the sciences be conceived as itself part of the scientific enterprise has most often been realized by choosing some theory or theories as the preferred scientific location of epistemology. Quine's naturalized epistemology, for example, is often misconstrued representationally, as taking some conjoining of physics, neurophysiology, and behaviorist psychology to be the privileged setting for a scientific theory of knowledge. Those who would conceive of knowledge psychologically nowadays more often appeal to cognitive science than to Quine's behaviorism, but they do explicitly characterize the resulting theory of knowledge as itself continuous with scientific theories. Rosenberg (1996) has similarly argued for the primacy of Darwinism as the site for a naturalistic philosophy of science, while the Edinburgh school's Strong Programme was conceived as a naturalistic theory of knowledge drawing upon a sociology of interests rather than any natural scientific theories. Even David Stump's defense of pluralistic naturalism (1992), which recognizes and challenges other naturalists' commitments to any single preferred discipline, nevertheless also seems to accept

philosophy of science to what Korsgaard (1996) has called "the normative question" in ethics. That question is not the same as how to defeat a skeptic (or relativist). Rosenberg refers to the relativist-constructivist challenge as "a form of sophistry" (1996, 26). Yet the figure to whom the normative question is posed is more akin to Glaucon than either Thrasymachus or Protagoras. The Glauconian naturalist does not need to be convinced that humans are natural beings or that the sciences broadly construed defeasibly provide the best setting for understanding ourselves and the world. Rather, she needs to articulate an account of what is "natural" sufficient to sustain these prior commitments upon reflection.

the primacy of scientific representations (theories and their associated "methods") as the appropriate setting for a naturalized philosophy of science.

Two fundamental problems have always confronted any attempt to articulate a naturalistic conception of nature on the basis of a representationalist account of how the sciences engage the world. The first problem, less often recognized, concerns the place of explicit representation within the overall economy of scientific practices. The widespread and enduring successes of the natural sciences provide the principal motivation and ground for Quinean naturalism. Yet further argument is needed to show that that success should be construed as *representational* success, and hence that the authority rightly accruing to scientific understanding extends to any particular interpretation of the theoretical claims or experimental methods that yield those successes. As Thomas Kuhn rightly noted:

> Scientists can agree [on] an apparently permanent solution to a group of outstanding problems and still disagree, sometimes without being aware of it, about the particular abstract characteristics that make those solutions permanent, . . . that is, . . . without agreeing on, or even attempting to produce, a full *interpretation* or *rationalization* of it. (1970, 44)

Such scientific disagreement (or more accurately, such implicitly divergent interpretations among scientifically competent practitioners) often becomes most intense and far-reaching concerning precisely those theoretical commitments that have informed philosophical naturalism. The centrality of Darwinism, quantum field theory, or cognitive science to any adequate understanding of nature is nowadays almost unquestionable. Yet no single explication of what Darwinism, quantum field theory, or cognitive science tells us about the world holds comparable authority, and naturalists who would appeal to particular interpretations of those theoretical traditions as their philosophical baseline cannot take for granted the authority carried by that tradition as a whole. This argumentative gap is all the more acute for any naturalistic metaphysics that floats relatively free from particular scientific theories. Naturalistic accounts of causality, natural law, or natural kinds and their properties cannot without further argument inherit the authority of the scientific tradition. Nor can one rescue the representationalist presumption by acknowledging the contestability of *which* interpretation of scientific theories or methods should be the basis for naturalism while insisting that *some* such interpretation should be used. One cannot presume in advance that successful science needs to resolve these interpretive issues.[11] A naturalist might instead follow the lead of Arthur Fine's proposed "natural ontological attitude," in "tolerating all the differences of opinion and all the varieties of doubt and skepticism that science tolerates" (1986, 150). Such a naturalism could accept that a fully adequate scientific understanding of the natural world (where 'understanding' denotes a practical competence in theoretical modeling, experimental manipulation, and their reflexive development in research) could be consistent with divergent interpretations of what the acceptance and use of that understanding says representationally.

The second challenge confronting representationalist interpretations of naturalism is the more widely discussed issue of how to construe representation naturalistically. The standard approaches identify representation with a causal/functional role supposedly accruing to tokens of syntactically identifiable types, whether these tokens are mental states or utterances. The difficulties of these standard approaches are well known, notably the need to appeal to an unanalyzed conception of causality or reference to pick out the correct referents of 'cause' or 'referent'; the consequent inability to specify a unique reference relation against a background of "deviant" mappings of sentences onto the world that are consistent with all the evidence or all the truths; and the difficulties in accounting simultaneously for both successful reference and misrepresentation, since it is hard to see how the causal role of mental or linguistic tokens could map them onto the misrepresentations that nevertheless obviously do occur.[12] A successful

11. Hilary Putnam (1990, chap. 5) notes that Quine can avoid having to specify which interpretation of a theory provides the basis for a naturalistic philosophy because his version of a disquotational theory of truth and reference cuts across such interpretive niceties: "If the theory that microstructures *are* what dispositions turn out to be is a 'good' theory (scientifically speaking), then it becomes true—or as true as anything is, in a Quinian way of thinking—as soon as we adopt it. The work of actually *reducing* dispositions one by one to (disjunctions of) microstructures is unnecessary. We have all the advantages of 'theft over honest toil'" (93). But that is just to say that Quine's naturalism is, in the end, not representationalist at all. A representationalist naturalism requires that *science itself* specify the proper interpretive relation between scientific theories and what there is, and that means that it must be able to account for how one interpretation really is the correct one.

12. Among the most important critiques of naturalistic theories of representation are Putnam (1981, chaps. 1–2; 1990, chap. 5), Brandom (1994, esp. 26–42, 206–12), and Wheeler (2000). For an especially detailed analysis of how naturalistic accounts of the causal role of mental tokens turn on crucial ambiguities in the concept of a symbol, see Horst 1996.

response to such arguments is indispensable for any thoroughgoing naturalism that identifies the natural either with what fulfills the content of successful theoretical representations or with the regularities or causal powers that establish the representational connections between scientific theories and the world. I do not believe that such responses will succeed, although I shall not here rehearse the well-known arguments against naturalistic theories of representation or the various attempts to circumvent them. The consequences of failure would be especially ironic, however, since as Putnam and Wheeler each pointedly remark, the standard appeals to representation seem to invoke a "magical" conception of language in support of what is supposed to be a thoroughly naturalistic metaphysics.[13] A second irony is that commitment to a representationalist approach would locate the central issue for philosophical naturalism not in the domains of the best-established and most widely applicable scientific theories and practices, but in the programmatic commitments of controversial research programs in cognitive or evolutionary psychology.

The difficulties with metaphysical naturalism in general, and representationalist conceptions in particular, are well known. Despite these difficulties, however, the apparent lack of a viable alternative has been a powerful consideration in favor of a naturalism that starts from representations of natural, causal necessity. If a scientific understanding of the natural world is not construed as correct representation of causal or nomic necessity,[14] how could it be construed? My response so far has been twofold. My critical response has shown why taking representations of natural necessity as the basis for understanding normativity might render impossible the joint fulfillment of the Quinean and Nietzschean commitments. The problem of manifest necessity shows that an adequate philosophical naturalism must accommodate within scientific practices the normative authority and force of any appeals to necessity. The dualism of nature and normativity suggests that familiar conceptions of nature in terms of anor-

13. One reason for invoking the name of Nietzsche in characterizing one of the two dimensions of naturalism is that he was perhaps the first to see clearly the theological residue in supposedly natural accounts of language. Recall my earlier quotation from *Twilight of the Idols:* "I am afraid we have not got rid of God because we still have faith in grammar."

14. Note that "correct representation" need not be construed realistically here. An empiricist might construe correctness in terms of empirical adequacy, and a constructivist might do so in terms of conformity to relevant social norms. What interests me is the possibility of interpreting scientific understanding and the nature that it understands in terms other than those of correct representation.

mative causes or laws might block reconciliation of the Quinean and Nietzschean dimensions of naturalism. My reflections upon Quine's indeterminacy thesis challenge any attempt to localize semantic and epistemic normativity within a particular objective domain of facts or necessities (even, as semantic eliminativists would have it, an empty domain). My discussion of sociological constructivism and Turner's criticisms of practices-as-social-regularities undermines attempts to shift from natural to social regularities or necessities to understand semantic and epistemic normativity.

The constructive part of my response has been to accept the challenge to articulate an alternative conception of scientific understanding, which neither starts from the representation of natural necessity nor succumbs to its characteristic dualism in other ways. This conception places scientific practices rather than theoretical representations at the center of a naturalist philosophy of science. Several crucial reconceptions of the notion of a practice have been necessary to accommodate such a project. First, practices have been reconceived in terms of their normative accountability rather than their performative regularity. Second, the semantic significance of practices has been reconstrued in terms of discursive practice rather than linguistic representation. Furthermore, consonant with my reconsideration of Quinean indeterminacy, I have argued that discursive practice is not a distinct domain of semantic facts or norms, but an aspect of the practical disclosure of anything and everything. All phenomena are semantically significant (note well, however, that semantics has been construed inferentially and that material and practical inferences are incorporated among the inferences that collectively constitute semantic significance). Third, my criticism of Brandom has shown that this unboundedness of semantic-discursive significance amounts not to a kind of linguistic idealism, but the converse: the distinctive linguistic components of discursive practices are semantically significant because of their place within patterns of material intra-action. These larger patterns are discursive not because everything is part of language, but because language is involved with everything. Fourth, I argued that scientific practices disclose natural phenomena rather than objects, in a sense in which scientific practices are themselves understandable as natural phenomena. Intentionality does not have to be naturalized, because intentional interpretation is itself a natural phenomenon in just the sense in which scientific experimentation discloses (as well as consists of) natural phenomena. Finally, this account enables understanding of the primacy of experimental practice in science. The sciences allow the natural world to be intelligible through their dealings with it. Experimental practice involves material intervention that transforms the world and is also simultaneously a conceptual articulation. The problem of how to understand the relation between experimental practice and discursive articulation dissolves, because the supposed separation between them is an artifact of mistaken philosophical assumptions.

Two crucial tasks nevertheless remain before this conception of scientific practices as the naturalistic disclosure of nature can be regarded as a defensible philosophical program. The first task is to show how to understand the *modal* character of scientific understanding in terms of the normativity of scientific practices. After all, philosophical emphasis upon causal or nomic necessity has not taken prominence merely because philosophers have lacked the imagination to conceive alternatives. The ways in which scientific understanding outruns mere description of actual events to explicate what is possible and impossible are sufficiently robust that any adequate account of science must account for them. The familiarity and prominence of appeals to laws within scientific discourse likewise suggest that a conception of nomic necessity is ineliminable from the scientific disclosure of nature, even if it cannot take central place in accounting for the normativity of science or nature.

The second task is to show how such a conception of scientific practices can indeed understand normative accountability in the sciences. I began the book with Heidegger's and Neurath's reasons for concluding that appeals to logical, transcendental, or nomic necessity fail to express the normative force of the necessities supposedly disclosed within scientific practices. These arguments ring hollow unless I can account for the normativity of the sciences more adequately in other terms. Such concerns are also heightened by recent philosophical preoccupation with the specter of epistemological relativism, which has arisen in part as a response to some philosophical movements toward naturalism.[15] I have argued elsewhere

15. Concerns with relativism seem to have arisen in two waves in the latter half of the twentieth century. The shift away from logical empiricist conceptions of the accountability to logical norms was thought by some philosophers to render scientific authority merely "subjective" (Scheffler 1967). In Hacking's succinct summary, philosophers' recognition in science of "the remnants of an historical process of becoming and discovering created for themselves a crisis of rationality" (1983, 1). The emergence of detailed sociological attention to that process renewed philosophical fears that naturalists might wittingly or unwittingly turn out to be epistemological relativists. As I noted above, Rosenberg (1996) places that fear at the center of his discussion of naturalized philosophy of science. Unlike many

(1996b) and suggested here that the mistake animating this specter is not the inclination toward relativism (nor toward history or sociology of science as supposed wellsprings of relativism), but the underlying commitment to epistemology. What is needed here, however, is not yet another deconstruction of this specter, but a constructive articulation of how scientific practices are normatively accountable in ways that have binding authority over what people say and do. As it turns out, these two tasks are intimately connected. Understanding how to express the modal significance of scientific understanding and how to articulate its normative authority and force turn out to be mutually implicated.

Causal or Nomic Necessity

My discussion of scientific practices in chapter 8 started from a conception of experimentation as causal intra-action with the surrounding world. That in turn stemmed from my earlier inversion of Brandom's account of discursive practices; I argued that discursive practices should be understood as powerful extensions of practical/perceptual intra-action rather than the reverse.[16] This commitment to the philosophical primacy of causality is central to conceiving my project as a more adequate philosophical naturalism. I have characterized scientific practices as themselves causal intra-actions that disclose nature as consisting of causally intra-active phenomena. Any understanding (scientific or naturalist-philosophical) of the natural world must be situated in the midst of such causal intra-action, as itself causally intra-active. Extending Brandom's expressivist conception of philosophy, I present naturalist philosophy as making partially explicit (and thereby extending) implicit competences in coping with the causal circumstances in which we find ourselves.

---

philosophical critics of the sociology of scientific knowledge, Rosenberg recognizes that his criticisms might apply all too well to his fellow philosophical critics.

16. Putting the point this way should not be read as positing a prediscursive "stratum" of practical/perceptual intra-action, to which discursive practices can then be added, or from which they emerge. Once discursive articulation is in place, it is pervasive, such that all human intra-actions have discursive significance. The pervasiveness of discursive normativity can be shown phenomenologically, or as a consequence of my discussion of Quinean indeterminacy in chapter 3. In either case, the primacy I here attribute to causal intra-action arises as an articulated expression of what we are *already* doing when we intra-act with the world materially-and-discursively; it cannot provide a genetic account of the "origin" of discursive normativity. Later in the chapter, I shall consider the temporality of discursive normativity, including what to say about the origins of normativity (and what preceded it), and how to locate evolutionary accounts of human beings within the kind of naturalistic philosophy I am advocating.

The paradigm cases of such philosophical explication are Brandom's treatment of logical vocabulary such as conditionals, singular terms, or 'true' as a prosentential operator. He displays the expressive role of logical locutions by asking what their use would add to discursive practices that lacked them. This strategy may seem problematic, however, when applied to causal vocabulary. Causal talk is pervasive, for it encompasses not merely the use of 'cause' and associated terms such as 'effect,' but also the inferential patterns expressed by counterfactual conditionals, active verbs, dispositional adjectives, and referential expressions that invoke or depend upon capacities.[17] Practices lacking all such locutions might be too impoverished to sustain ascriptions of semantic content at all, so I cannot similarly ask what they would add to an antecedently intelligible descriptive repertoire. Indeed, the pervasiveness of causal talk reinforces my insistence upon the primacy of practical/perceptual intra-action in explicating discursive practices. To make explicit the expressive role of causal locutions is not so far from making explicit the expressive role of semantically contentful expressions generally. Indeed, the ineliminability of causal locutions from even a minimally adequate descriptive vocabulary could be seen as the preeminent lesson to be learned from the failures of the deductive-nomological theory of explanation.[18]

Philosophers have nevertheless been reluctant to acknowledge this lesson. My commitment to the primacy of causal intra-action runs counter to the still predominant philosophical trend. Hume and his many followers deny the intelligibility of causality in any robust sense, and seek to eliminate it in favor of some distinctive kind of regular occurrence. Whether these are construed as simple regularities, regularities that support counterfactuals, INUS conditions, or regularities (essential properties) across possible worlds, regularists substitute *exhibited* rules for causal intra-actions themselves.[19] Regulists differ from Humeans only by substituting

17. I do not claim that all uses of these classes of expressions express causal relations, but only that their causal uses are extensive, and that the causal uses take priority in the sense that an understanding of any noncausal uses (e.g., of counterfactual conditionals) must at least implicitly grasp how they differ from causal contexts, but not vice versa.

18. That presumes, of course, that the lesson had not already been learned from Kant!

19. The primary sense of exhibited rules would be actual regularities. Possible worlds semantics, however, has encouraged the treatment of possibilities as if they were actual. I therefore extend the category of exhibited rules to incorporate such conceptions. One is offering a regularist conception of causality, whether one takes the relevant rule to be exhibited in actual sequences of events or in sequences extended to encompass counterfactual events as "actual," in the sense of being identifiable across possible worlds.

instead a *governing* rule or law, whose necessity is merely exemplified by any actual sequences.[20] Neither account, I suggest, takes causal intra-action seriously.

My discussion in chapter 8 suggests a novel diagnosis of why philosophers have been so strongly inclined to identify causality with rules or regularities, however. Hume's perplexity about causality was focused upon the supposed *connection* that causal claims posit between discrete (disconnected!) objects or events. Yet this perplexity only arises once one has established or presumed the initial disconnection to be overcome. The root error underlying the familiar difficulties plaguing the Humean tradition is the presumption that the world somehow already comes naturally composed of discrete objects. My argument in the preceding chapter suggests that this presumption might be especially illegitimate for naturalists, if we grant that scientific understanding only arises through experimental intra-action. My line of argument thus moves in the opposite direction from that taken by the fashionable presumption that causality can be understood in terms of alethic modalities attached to natural kinds. The latter approach responds to the difficulties with traditional conceptions of scientific understanding by construing the boundaries of natural objects and object-kinds as *more* robust and rigid than earlier regularists and regulists did. Genuine objects and object-kinds are taken to be rigid designators, trackable across possible worlds on the basis of their essential properties.[21] My suggested alternative is to think of objects only as practically constituted components of repeatable phenomena.

20. It is tempting to treat Kant as the classical exemplar of causal regulism, parallel to Hume's more widely noted regularism. Whether Kant in the end is a regulist about causality, however, depends on how one interprets the *necessity* with which he thinks phenomena are related by the concept (rule) of causality, which in turn involves the relation between theoretical and practical reasoning in the critiques. Brandom (1994) persuasively reads Kantian necessity as normative rather than alethic, such that this "regulism" might be seen not as an elimination of causal intra-action in favor of causal laws, but as instead a (mistaken) commitment to the possibility of the complete explication of causal intra-action. Yet even Brandom, as noted, does not endorse the implication of this hypothetical Kantian position that causal intra-actions should themselves be understood to be constitutively normative, but only takes them to be non-normative events or patterns to which normative significance accrues when they are appropriately caught up in discursive practices.

21. Roughly speaking, such approaches would be modal regularisms if they gave primacy to whatever properties or structures *turn out* to be invariant across possible worlds (David Lewis's counterfactual analysis of causation might exemplify such a conception). More commonly, they would be modal regulists if natural kinds (and their natural instances) were determined by the laws (structural invariances across possible worlds) that govern actual events.

My approach also contrasts, however, to conceptions of causality that start from a supposed prior understanding of agency or manipulability (von Wright 1971, 1974). First, such accounts are usually non-naturalistic. Von Wright, for example, hoped to explicate causality in an intentional idiom, in which the logical relations involved in an agent's acting for reasons provided the explicans. By contrast, I take the intra-active phenomena that constitute causal relations as the basis for understanding intentionality. As agents, we are already situated in the midst of ongoing patterns of causal intra-action with our surroundings, which provide the basis for explicating agency. Moreover, von Wright's account took for granted the identity of the agents who paradigmatically "bring about" effects. I instead take the boundaries of all causal agencies to be constituted by the material/practical configuration of phenomena. "We" are not agents apart from the phenomena in which we participate (including, for example, the complexes of apparatus and circumstances through which effects are brought about), any more than are the objects whose causal capacities or behavioral regularities are invoked by anormative naturalists. Finally, my approach highlights an important deficiency in any account that grounds causal talk solely in capacities to bring about effects; such capacities are no more basic than our understanding of ourselves as causally affected, and hence as vulnerable to being marked by intra-action with other causal agencies. Agency accounts of causality hence do not take seriously causal *intra*-action.

When starting with causally intra-active phenomena, however, I must then show how to understand objects and agents as having causal capacities that are not circumscribed by the particular phenomena of which they are components. To insist upon the priority of phenomena is not to deny the reality of objects, but only to recognize their embeddedness in patterns of intra-action. Consider, for example, water, whose paradigmatic role as a natural kind definable by essential properties is exemplified by Putnam's original Twin Earth thought experiment. Whether various performances and capacities belong to water (rather than to some other superficially similar substance) was supposed to be determined by the a posteriori necessities resulting from the chemical structure of water as an essential characteristic of the natural kind. I start instead with various phenomena: phase transitions, aqueous solubility, electrolysis, combustion of hydrogen or methane, and so forth, not to mention the manifold phenomena of biological metabolism or meteorology. Water has the normative status of an object-kind through the practical-inferential relations among these various phenomena, notably but not exclusively substitution relations. What is condensed in an atmospheric dehumidifier can normally replace glacial runoff in variously configured phenomena without altering the characteristic resultant patterns. Thus, the ontological priority of phenomena over objects and object-kinds is the ground for the priority Brandom rightly ascribed to propositions over singular terms in understanding discursive practices. Indeed, the latter priority is both an expression of the former and an instance of it, since the capacity to discriminate and appropriately interchange situated materializations of inscriptions and utterances such as 'water' or 'wot -ər' itself exemplifies a complex, intra-active phenomenon).

Note the immediately obvious advantages of starting from proprieties of practical, material inference rather than from object-kinds and relations. I can readily cope with the pull in both directions from boundary questions about identity. Is salt water water? Is polluted water, strikingly exemplified by the 1975 Cuyahoga River fire? Water hydrogen-bonded to a DNA-molecule? Is blood water, despite its higher viscosity? What about heavy water, whether pure deuterium oxide or a mixture with the more plentiful isotope? The crucial point is that such cases are sometimes intersubstitutable and sometimes not, depending upon which phenomena are in question and what is at stake in the identification. Indeed, the case of heavy water illustrates the importance of something only loosely attended to in Putnam's thought experiment (1975a) about "water" on Twin Earth (the hypothetical Twin Earth supposedly contains a substance phenomenologically indistinguishable from our ordinary water, but with a complex chemical composition abbreviated as XYZ).[22] Whether the hypothetical XYZ is water surely depends crucially upon how it intra-acts with H_2O or the other substances that characteristically intra-act with H_2O. Is a mixture of XYZ and H_2O also phenomenologically indistinguishable from either material in isolation? In the absence of further *phenomena* in which the characteristic intra-actions of these substances can be interconnected, it is unclear what one ought to say about XYZ, because it is insufficiently clear what it can do and what can be done with it.

The introduction of a kind-term, 'water,' explicitly articulates practical-

22. Putnam's thought experiment has other crucial omissions from a naturalistic standpoint, of course, since the phenomenological features of water cannot be so readily disconnected from the relative simplicity of its molecular structure. (Does XYZ expand when it freezes? Does it show up inside a closed container in which a candle is burned, as it undoubtedly did in some of Boyle's pneumatic experiments?)

inferential norms connecting various phenomena. As with all concepts, its content exceeds the circumstances of its past and present use, so that understanding water involves recognizing its open-ended inferential accountability to further involvements in phenomena, whether those involve phenomenological differences (such as phase transitions), composition (hydrogen bonding in aqueous complexes), decomposition (electrolysis), incorporation (within bodies and their cells), or transformations (as in reaction with metallic sodium). The significance of such a term can change quite fundamentally with changes in the range of practical inferential involvements it can be used to express. Thus, for example, my approach offers a natural interpretation to the phenomena that Ian Hacking (1983) takes to be decisive for his commitment to "entity realism." Hacking advocates realism about those scientifically characterized entities (or kinds of entity) that can be reliably manipulated experimentally. Such manipulability effects a significant transformation in their normative status as objects, because the phenomena in question do not simply mark the interconnected capacities of an "object" (the object-component of various phenomena), but instead cross the boundaries between what we merely encounter and what is incorporated within the agential capacities for which we are partially responsible. I will say more about that significance later in the chapter, in considering what is at stake scientifically in such transitions.[23]

Even though the pervasiveness of causal talk casts suspicion upon the strategy of asking how it augments an antecedently intelligible acausal discourse, the expressive role of causal locutions can still be addressed in terms of the inferential roles such locutions fulfill. The standard construal of such roles is as a function from the circumstances in which the locution is appropriately used to the consequences of its correct application. On any account of causality, many of the *consequences* attendant upon appropriate attribution of causal relations are clearly normative: the *understanding* achieved through the articulation of causal patterns, or the *responsibilities* and *liabilities* that follow from them, for example.[24] The philosophical

23. Hans-Jörg Rheinberger (1997) usefully discusses this difference in what is at stake in scientific work as the transition of an entity from an epistemic to a technical thing. Although Rheinberger does not put it this way, what changes is its normative status as an entity.

24. Many philosophers would cite prediction and whatever capacities for intervention and control of one's surroundings it confers as the semantically decisive consequences of correct causal attributions, because they take predictive success to be articulable in terms of non-normative regularities. The arguments in chapter 8 strongly suggest that that hope is forlorn, but I will consider those arguments primarily in their application to the circumstances rather than the consequences of the appropriate use of causal locutions.

tradition has, however, insisted that at least the *circumstances* under which the various causal locutions are employed must be anormative.[25] If this were correct, then causal talk would function at least in part as inference tickets from anormative circumstances to normative consequences. My arguments in chapter 8 suggested that causal intra-actions themselves do not constitute such an anormative basis for the introduction of normatively consequential causal locutions. The boundaries of causally intra-active systems are only definable within the constitutively normative patterns of phenomena. But these arguments might not seem compelling to many defenders of causal anormativity, since causal laws rather than singular causal phenomena have more often been taken as the locus of anormative natural causality. The boundaries of the causally intra-active systems in any particular case might then simply be among the normative consequences of anormative causal laws. My preceding arguments would then merely reinforce Humean suspicions about the intelligibility of causal connections apart from their subsumption under causal laws. Hence, I need to consider how to accommodate *nomic* necessity within the broadly practical-normative approach that I am advocating, and whether causal laws should indeed be understood to articulate nature as anormative.

Attempts to develop a naturalistic conception of nature in terms of laws has always confronted the difficulty that much of scientific understanding cannot easily be couched in those terms. The biological and geological sciences in particular do not usually formulate their theoretical claims as laws, and in fact have yielded few if any candidate natural laws even with the aid of philosophical reconstruction. Perhaps this simply reflects the subordinate place of these disciplines, or a deficiency in their current state of development.[26] Such a stance, however, may abandon the naturalistic

25. Most philosophers have claimed or presumed that the interactions, regularities, or laws that provide the circumstances for the appropriate use of causal locutions can themselves be described in a non-normative vocabulary (of which scientific vocabulary is a subset, possibly a proper subset). Davidson and Rorty, who deny that there can be any sharp distinction between normative and non-normative vocabularies, nevertheless insist that irreducibly normative vocabularies must nevertheless be occasioned by non-normative causes. These causes are then semantically inert; they insist that a belief can only be interpreted or justified by another belief.

26. Strictly speaking, this need not be regarded as a deficiency. Alexander Rosenberg (1994), for example, argues that the complexity of biological systems, the computational limits of human cognition, and the human interests at stake in biological understanding dictate that biology ought to be a relatively more instrumentalist science than the physical sciences.

commitment to avoid arbitrary impositions upon scientific practice. Why should the absence of lawlike generalizations count as a failure or subordination of biological understanding rather than as an inappropriate standard to apply in this setting? The point is not that naturalists must accept whatever regularities of practice prevail within a science as setting the standards for its success, but only that the norms to which a science is accountable must remain in appropriate contact with current practices. There must, that is, be reasons explicable from within the practice for why its goals as a science ought to incorporate laws, or to offer epistemic obeisance to those sciences that do.

Yet the familiar arguments against nomocentric conceptions of the special sciences are not what concern me in rejecting a regulist or regularist conception of causality, for two reasons. One reason is that such arguments do not go far enough. Despite the absence of distinctive laws in many of these sciences, their explanatory and predictive capacities might still ultimately depend upon some form of nomological necessity, derived from their supervenience or other dependence upon physical and chemical laws, or from natural selection as a very high-level and abstract biological law whose particular consequences cannot be usefully characterized nomologically.[27] The more fundamental reason, however, is that what is expressed as nomological necessity, even in the most plausible cases from the physical sciences, is better understood in terms of the ineliminable normativity of causal intra-action. Moreover, once it is clear how such a construal would go, it is not hard to see why the allegedly special character of sciences other than physics is a misconstrual; they are normative in the same way that physical theories are.

Although laws need not be expressible mathematically, mathematical expressions have provided the most prominent and plausible exemplars of natural laws. Newton's laws, Boyle's law, the Wiedemann-Franz law, Maxwell's equations, or the laws of thermodynamics at first glance seem to illustrate both the centrality and the pervasiveness of nomological necessity within scientific understanding. The question I ultimately want to raise, already suggested in chapter 1 by my extension of Neurath's and Heidegger's arguments against the normative force of logical or transcendental necessity, is how to construe the normative force of such supposed nomological necessities. I approach that question, however, by first taking up

27. Rosenberg (1994) takes the latter position.

Nancy Cartwright's recent discussion (1999) of the scope of such expressions. Philosophers have rarely considered the scope of lawlike expressions to be problematic; they have been taken either as universally quantified sentential expressions or as abstract semantic structures that admit of no scope limitations on their admissible models. The relevant question has instead seemed to be whether to construe their unquestioned generality as claims to truth, or only to some form of instrumental reliability. Cartwright argues, however, that the underlying "fundamentalist" presumption of the unlimited scope of mathematical or other putative nomological expressions is a mistake.

Cartwright's argument starts close to my own emphasis upon phenomena. She maintains that, except in very special circumstances (often, although not always, contrived), the behavior of objects is not accurately described by laws, because of the multiplicity of causal factors or governing laws involved in their behavior. Maxwell's equations do not, for example, adequately describe the movement of a charged body in an electromagnetic field if the body also has a mass affected by a coincident gravitational field. Gravitational and electromagnetic influences combine vectorially, of course, but other forms of causal intra-action cannot be so neatly accommodated. But surely, the response would go, the laws of classical mechanics or quantum mechanics have universal scope, such that $F = ma$ or the Schrödinger equation govern all motions. That is what Cartwright denies:

> Think again about how we construct a theoretical treatment of a real situation. Before we can apply the abstract concepts of basic theory—assign a quantum field, a tensor, a Hamiltonian, or in the case of our discussion, write down a force function—we must first produce a model of the situation in terms the theory can handle. . . . I want to focus on the multitude of 'bad' cases, where the models, if available at all, provide a very poor image of the situation. These are not cases that disconfirm the theory. You cannot show that the predictions of a theory for a given situation are false until you have managed to describe the situation in the language of the theory. When the models are too bad a fit, a theory is not disconfirmed; it is just inapplicable. (1999, 26)

The situation may be more tractable in other terms; for Cartwright's example (adapted from Neurath) of a $1,000 bill in a windy plaza, fluid dynamics may provide a more adequate partial model of the bill's motion, but without thereby providing any way of construing a total force acting on it.

Phenomena like the irregular trajectory of a paper bill in a swirling wind illustrate the tradeoffs between truth and meaning discussed in chapter 8. To the extent one can give a definite interpretation to $F = ma$ in these circumstances (in which gravitation is the only force for which there is a well-defined force function), the equation so interpreted is falsified, but it makes much more sense to say that it lacks any definite interpretation here. There is no determinate application of the concept of force in these circumstances. Fundamentalists will presumably understand the failing to be epistemic; one simply does not know how to express the force function that would represent the force exerted by the swirling wind. And in some cases, such a response is appropriate; the range of models that successfully articulate the concept of force has, after all, been substantially extended over time. That history does not, however, support its extension to this particular case, let alone its universalization. No one doubts that the bill's trajectory results in substantial part from its causal intra-action with the wind, but that is not sufficient to show its conceptual tractability within the mathematized regimen of classical mechanics. 'Force,' after all, is an abstract concept that assimilates a wide variety of situations. What unites them as forces is precisely their articulability within the mathematized practice of classical mechanics. And in practice, no scientist would account for this situation by attempting to model it as a vectorial composition of forces.

Yet surely, you may retort, classical mechanics is not irrelevant to the motion of the bill. The gravitational force on the bill partially explains why it is more likely to be found within a short distance of the ground rather than higher (only partially, because there are some possible aerodynamic configurations of the bill which would make that substantially less likely). So it would indeed be a mistake to limit the concept of force strictly and solely to where there exist well-defined force functions that provide models for its mathematically precise application. This point shows that Cartwright's objection to nomological fundamentalism is a special case of a more general failing, however. Consider the problems with earlier attempts to define theoretical concepts (whether one by one or collectively through testable "partial definitions") as disjunctions of their presently observable or measurable applications. Such attempts failed in two crucial respects (apart from a misguided attachment to the epistemological primacy of sense experience). Because they fixed the scope of a concept in advance by its currently articulable applications, they failed to allow for the possible subsequent extension of its scope through the development of new or extended models. Because they took these applications to confer a definite meaning upon its applications, they also failed to allow for the appropriateness of vague applications such as my construal of the significance of the gravitational force on the likely height of the swirling bill. But similar failures characterize the universalizing fundamentalist interpretation. Universalist-nomological readings also fix the scope of the relevant concepts in advance (as unlimited!), and hence fail to allow for the possible intractability of some situations in those terms.[28] Fundamentalists would likewise confer definite meaning on all applications of a concept, in construing them as nomological.[29] The problem with nomological fundamentalism is its underlying commitment to semantic realism.

How then should one construe the scope of lawlike expressions? Cartwright argues that, to the extent that they are regarded as laws, such expressions can only apply ceteris paribus. But that cannot be the whole story. Ceteris paribus laws say nothing about what happens when other things are not equal, but classical mechanics can say quite a bit about situations for which there is no adequate model for articulating a force function. Cartwright instead treats ceteris paribus laws as merely partial expressions of causal capacities and their consequent tendencies. Aspirins have the capacity to relieve headaches and, under the right circumstances, tend to do so. Similarly, the capacity of the earth's gravitation to attract other masses in its proximity tends to move them toward its center (absent other countervailing causes). But where do such capacities reside?

Cartwright still wants to say that causal capacities belong to objects (as instances of relevant kinds such as acetylsalicylic acid or gravitational mass), because she wants to distinguish relatively robust capacities embodied in the natures of things from apparently more coincidental "capacities." Fiber optics illustrates the latter case. The reliable transmission of discrete light pulses is not a robust capacity of fiber cables, for the pulses

28. Remember that such intractability in no way shows the theory to be false in any part of its domain, but only that its domain is limited. The fact that many motions cannot be informatively modeled in classical mechanics does not show anything problematic about the theory's application to cases that can be so modeled.

29. One might reply that where the application of the law cannot determine a definite trajectory, it could still apply probabilistically. That response, however, merely exports the difficulty from 'force' to 'probability.' In the case of the windblown bill, determining the long-run frequency with which it is to be found within some definite distance of the ground under these conditions requires a model (a "chance set-up," as Hacking 1965 characterized it) that both relevantly delimits "these conditions," and does so in a way that allows the probability to have a definite value.

normally disperse, creating mutually destructive interference among successive pulses. A nonlinear chirp in the pulse, however, results from a shift in the fiber's index of refraction through its causal intra-action with the pulse. Cartwright concludes:

> The effects of the chirp combine with those of dispersion to produce the soliton. Thus stable pulse shapes are not a general phenomenon of low-loss optical fibres. They are instead a consequence of two different, oppositely directed processes that cancel each other out. . . . Solitons are indeed a stable phenomenon. They are a feature of nature, but of nature under very special circumstances. (1999, 30–31)

The problem with this distinction is that the capacities Cartwright attributes to object-kinds are circumstantial in the same way. Aspirins only relieve (the right kind of) headaches under those conditions in which digestion introduces their active constituent into the blood in sufficient concentrations, and in the absence of other substances that interfere with their intra-action with inflammation. Forces likewise only induce characteristic trajectories under the right circumstances; as Cartwright herself notes, for example, under some circumstances, a repulsive force between two bodies can actually move them closer together (1999, 59–61). So what Cartwright should have concluded is that capacities of *both* sorts are features of natural phenomena, rather than of object-natures.[30] The difference she rightly recognizes between the two cases can then be understood as a difference in the substitution-inferential normativity of the phenomena.

The scope of causal laws, then, like the boundaries of causally intra-active systems, is constitutively normative. Cartwright does not put it this way, insisting only upon the "open-endedness" of the capacities that generate lawlike regularities where they do occur. But Cartwright's account implicitly accords with my conclusion, in two ways. First, we can note the characteristically normative vocabulary needed to articulate the relations between circumstances and outcomes. Whether one talks about the "nor-

30. Cartwright does insist that behavioral regularities only result from such capacities within "nomological machines, fixed (enough) arrangements of components, or factors, with stable (enough) capacities that in the right sort of stable (enough) environment will, with repeated operation, give rise to the kind of regular behavior that we represent in our scientific laws" (1999, 50). I am arguing first, that the notion of *nomological* machine should be extended to my conception of phenomena (not all phenomena exhibit observable regularities), and second, that the attribution of capacities to object-kinds is derivative from their appropriate attribution to phenomena, via practical, material substitution inferences.

mal" expression of a capacity, what happens under the "right" circumstances, or the "appropriate" configuration of the phenomenon, the modalities needed to express causal capacities as Cartwright understands them are normative rather than alethic (or better put, the use of nomological or alethic-modal locutions expresses the normative force of causal intra-action). More significantly, however, this point is implicit in Cartwright's own deepest philosophical commitments. Cartwright's empiricism is an *antinaturalistic* epistemological stance, in which epistemic norms are governed by practices of empirical measurement. Her empiricism then provides the basis for determining the scope of lawlike expressions. How far should one construe the scope of classical mechanics, for example? Cartwright's prior commitment to empiricism determines her answer on empiricist grounds:

> Mechanics is true, literally true we may grant, for all those motions all of whose causes can be adequately represented by the familiar models which get assigned force functions in mechanics. For these motions, mechanics is a powerful and precise tool for predictions. (1999, 27)

What then determines the representational adequacy of the models?

> When we have a *good-fitting* molecular model for the wind, and we have in our theory . . . systematic rules that assign force functions to the models, and the force functions assigned *predict* exactly the right motions, then we will have good scientific reason to maintain that the wind operates via a force. (1999, 28; my emphasis)

Cartwright's conception of empirical adequacy thus contains vestiges of an empiricist theory of meaning; unlike van Fraassen (1980), for whom empirical adequacy is circumscribed by a fundamentalist semantics for scientific theories, Cartwright circumscribes the domain of empirical assessment of theories by the availability of tractable models.[31]

Although Cartwright's own work has been oriented toward under-

31. Cartwright's conception of the tradeoffs between scope restrictions and falsification presumably avoids the problem of gerrymandering theoretical domains to ensure empirical success by requiring that scope restrictions are only appropriate where one has at least a qualitative understanding of other causal influences that could account for the inapplicability of a theory. Hence, the importance of being able to model *all* of the significant causes of a motion or other effect for it to fall within the evidential domain of a theory. The windblown bill does not falsify classical mechanics because of the recognizable *causal* efficacy of the wind, despite the inability to model it mechanically.

standing the putative mathematical laws articulated in physics and economics, the lessons of her account extend more generally.[32] Composition and decomposition provide especially instructive cases, given the roles assigned to objects and properties in more traditionally conceived metaphysical naturalisms. Consider claims to explain features of the world manifest at one level of organization in terms of constituent entities or processes. The traditional reductionist hierarchy suggesting that something like string theory, if successful, would provide a theory of everything implicitly assumes the universal scope of such compositionalist claims. Yet Cartwright's arguments suggest that it is only appropriate to understand hadrons as composed of quarks and quarks of vibrating strings to the extent that there are or could be models that construe (with sufficient empirical adequacy) the properties and contextual performances of hadrons in terms of what is manifest in quark-phenomena, and these in turn on the basis of phenomena whose articulated expression invokes vibrational modes of strings. In other contexts, the supposed constituent entities cannot be understood as distinguishable components of the phenomena, any more than salt water is composed of the sodium hydroxide and hydrochloric acid that might have been combined to make it, or in Cartwright's example, than the causal intra-actions modeled in fluid dynamics are compositions of forces. Claims to ontological supervenience ring hollow in the absence of systematic connections that would give some practical (intra-active) *sense* to the properties or states hypothetically attributed to the supposed constituents. Similar points can be made about other articulative systems, such as the *Drosophila* cytogenetic work that partially embodied the classical understanding of genetics, or the treatments of prokaryotic systems through which molecular genetics was initially articulated.[33] Regardless of whether models take the form of mathematical expressions, their scope extends

32. Philip Kitcher's recent critical response (1999) to Cartwright mistakenly attributes to her the claim that she is only discussing those sciences whose theories are expressed mathematically. The considerations discussed below, as well as Cartwright's willingness to take genetics to exemplify fundamentalism (1999), show that his reading is mistaken.

33. Kitcher (1984) rightly claims that such model systems *in use* are "illustrations" of what the theory amounts to in classical genetics, in the sense that the content or scope of the theory is hardly limited to what it says about the model system, and yet the theory cannot be given an adequate abstract formulation apart from some such illustrations. I take that to be an analogue to Cartwright's point about the inadequacy of generalizations like $F = ma$ in the absence of its model force functions. As noted in chapter 5, I endorse Kitcher's claim for the philosophical primacy of the practice of genetics, while rejecting his conception of practices as regularities in the performances or beliefs of a scientific community.

only as far as one could establish robust connections between the models and the phenomena they are supposed to illuminate.

Not surprisingly, as a naturalist I reject Cartwright's prior commitment to empiricism as an unwarranted imposition upon scientific practice. Her conception still works remarkably well from my perspective, however, because the strong practical-normative bent of her empiricism removes the most problematic elements from the empiricist tradition. She construes the empirical basis of science as measurement intra-action rather than sense experience, takes causal understanding to be both irreducible and epistemically more basic than regularities, and recognizes the empirical "meaning" of theoretical expressions to be open-ended yet not unlimited in scope. In all three respects, Cartwright's empiricism parallels my naturalistic conception of scientific practices as expressive extensions of causal intra-action with the world. Nevertheless, there are several ways in which I take Cartwright's empiricism to provide an inadequate account of the normativity of scientific understanding. Reflection upon these difficulties will point toward the second topic of this chapter, concerning the normative authority and force that sustain scientific practices.

Consider Cartwright's empiricist understanding of the scope of a theoretical conceptualization (in my terms, the scope of its normative authority). A theory such as classical mechanics is only applicable, she argues, where there are (or could be) models providing a good enough predictive account of the measurable phenomena. Elsewhere, it is simply not clear what the theory would say about those circumstances. But what is the appropriate basis for assessing a predictive account as "good enough" to secure a phenomenon within the domain of the theory? Astronomical models are sometimes doing well to accord with available evidence within an order of magnitude, yet in other circumstances, rather small discrepancies can indicate a serious failure. Such evaluations are further complicated by possible challenges to the adequacy of available measurements, and to the understanding of other causal factors that lend confidence to a decision whether or not to approach a phenomenon with alternative theoretical and methodological tools. Cartwright's basic strategy for answering such questions is to appeal to the *purposes* for which one seeks (or should seek) an understanding of the phenomenon. A good enough fit between theoretical model and data is one that makes the model suitable for understanding and coping with a phenomenon in the ways that (should) matter.

This, however, displays why Cartwright's antinaturalist commitment to empiricism is a contentious response. For why assume that the *empirical* adequacy of theoretical models should always trump other considerations? The debates over fundamentalism and the unity of science provide a striking illustration. Cartwright herself has long insisted upon the opposing pull of empirical adequacy and explanatory power as scientific desiderata, and the need to make appropriate tradeoffs between them. She once noted that

> at heart the [quantum] theory works by piecing together in original ways a small number of familiar principles, adding corrections where necessary. This is how it should work. The aim is to cover a wide variety of different phenomena with a small number of principles. . . . It is no theory that needs a new Hamiltonian for each new physical circumstance. (1983, 139)

Presumably this commitment to a modicum of theoretical unification would hold even if such tailor-made Hamiltonians turned out to be more empirically adequate in each case than are the standard models with appropriate phenomenological corrections appended. For in being tailored to particular situations, each "theory" would purchase its empirical success at the price of understanding, in the form of the ability to connect these successes inferentially and thereby to allow their extension to other circumstances. So empirical deficiencies need not always override the quest for unity. Indeed, Cartwright acknowledges that the quest for unity and understanding places a more overriding constraint upon prediction and control within empirically circumscribed domains in the demand for consistency among various locally adequate models, for "we do not want colour patches to appear in regions from which the laws of physics have carried away all matter and energy" (1999, 33). But why not? *Because that is impossible,* you may be inclined to say. But why should an empiricist like Cartwright say what is possible or impossible beyond those situations she can model with some empirical reliability? Indeed, how could she even say what is improbable absent an appropriate chance setup? Empiricism, even Cartwright's very sensible empiricism, is ill suited to grasp *how* the world can make claims beyond those actually captured in its empirically adequate models. Must we then resurrect conceptions of nomological *necessity* to capture the force of these claims?

Consider Cartwright's reasons for imposing empirical adequacy as a constraint on the scope of theoretical claims. Her reasons are simultaneously metaphysical and moral-political. Metaphysically, she sees in scientific fundamentalism an unfortunate variant of theology; the quest for a unificationist accommodation of the world's unruliness uncomfortably parallels the theological casuistry required to accommodate evil within the greater goodness of God. Although "we yearn for a better, cleaner, more orderly world than the one that, to all appearances, we inhabit . . . it will not do to base our methods on our wishes" (1999, 12–13). The reason it will not do is that the practical inferences drawn from such metaphysical commitments matter, both to the science we practice and increasingly, to the way we live. Cartwright is worried that a fundamentalist faith leads to empirically unwarranted decisions, with baleful consequences, both in specific cases and in the general direction of scientific methodology:

> I care about our ill-supported beliefs that nature is governed by some universal theories because I am afraid that women are dying of breast cancer when they need not do so because other programmes with good empirical support for their proposals are ignored or underfunded [in favor of the programme that] is our best shot right now at a theory of everything. . . . The great challenge that now faces philosophy of science [is] to develop methodologies, not for life in the laboratory where conditions can be set as one likes, but methodologies for life in the messy world that we inevitably inhabit. (18)

A nonreductive naturalism responds to these concerns more adequately than does empiricism, however, because it speaks to them directly. The problem with the fundamentalist aspiration to an all-encompassing nomological theory is not that it goes beyond the empirical evidence it can accommodate, or even beyond any empirical meaning assignable to its terms. The problem is that it goes beyond the world, losing contact with the causally intra-active circumstances to which moral-political concerns are bound, with many of the same baleful consequences that concern Cartwright. These circumstances have normative import that extends beyond our present understanding, as the setting for our possibilities and the issues and stakes that emerge in their realization.

From Nomological Necessity to Practical Normativity and Real Possibility

My consideration of causal intra-action and nomic necessity has led back to where my discussion of normativity left off at the end of chapter 7. There I argued that Brandom's and Haugeland's otherwise promising strategies for understanding practical normativity still shared two reciprocal mistakes. They construed the normative authority of practices (including science) as conferred upon *objects,* while locating the normative force of that authority in the (deontic or existential) commitments undertaken by speaking and acting human *subjects.* I suggested that these conceptions could not avoid the dualism of nature and normativity that has blocked any adequate understanding of how such authority could be binding. Proceeding further along those lines, however, requires grasping the world disclosed by the sciences as *normative.* The preceding discussion of natural phenomena, causality, and the scope of scientific laws has now cleared the way to proceed.

The connection becomes clearer through a brief comparison of Cartwright's empiricist account of nomological expressions with Brandom's and Haugeland's implicit projects of construing causal and alethic modalities normatively. I can then develop the implications of this comparison for a more adequate understanding of what is at issue and at stake in scientific practices, and how these stakes have normative authority and force. This conception of the normativity of scientific practices and scientific understanding brings to completion my reconception of philosophical naturalism.

Cartwright and I give explicative priority to causal intra-actions over nomological necessity, in opposition to regulist or regularist explications of causality. She has concluded that natural laws, even the most apparently fundamental laws, only hold ceteris paribus. The "ceteris" in question, however, are primarily the causal capacities at work in particular settings. Lawlike regularities only show up within "nomological machines," phenomena whose arrangements enable particular causal capacities to manifest themselves clearly by shielding them from intra-action with other causes.[34] Such capacities, I have argued, belong to objects-in-phenomena rather than to objects in abstraction; references to objects express the practical-inferential norms that connect the capacities displayed in one

34. Several initial responses to Cartwright (1999), notably Kitcher 1999 and Winsberg et al. 2000, express the worry that the concept of a nomological machine is vacuous: "Her thesis that laws only hold relative to the operation of a nomological machine appears to reduce to the claim that laws hold where they do" (Winsberg et al. 2000, 404). Yet this conclusion overlooks the role of causal capacities in Cartwright's account of nomological machines. What gives content to the concept is the description of a nomological machine in terms of the causal capacities utilized or shielded against in its construction.

phenomenon to those manifest in others. Not surprisingly, then, causal laws have an ineliminable normative dimension as well: the scope of causal laws or regularities is open-ended, and how expansively one ought to construe their authority depends (for Cartwright) upon empiricist epistemic norms, as well as upon the moral-political consequences revealing what is at stake in particular scientific practices.

Cartwright's empiricism, I have argued, is misleading both because it subtly misconstrues the appropriate scope of nomological expressions and because, in any case, it would settle by epistemological fiat what ought to be open to discussion from within scientific practices themselves. But how, then, should one construe the normative authority that governs scientific practices and scientific understanding? How, moreover, do Cartwright's other moral-political concerns have bearing on *scientific* understanding? Here it will be useful to revisit my discussion of normativity in chapter 7. The normativity of nomological necessity and causal intra-action are only implicit in Cartwright's account. Brandom and Haugeland, however, more explicitly propose to understand the legitimate uses of alethic and causal modalities in the irreducibly normative terms of deontic or existential commitments. With recognition that Cartwright also construes supposed necessities in normative terms, the comparison will show how better to understand the expressive significance of causal or modal locutions and their practical-inferential counterparts, and how to understand the normative authority and force that govern scientific practices and the understanding of natural phenomena.

For Brandom, claims of nomological necessity express material-inferential incompatibilities among the commitments and entitlements taken on through the use of theoretical-descriptive concepts. For Haugeland, doing science seriously requires an existential commitment to sustain an excluded zone of conceivable but impossible phenomena. Two points can be highlighted through their comparison to Cartwright's discussion of ceteris paribus laws and nomological machines. First, as I already noted, Brandom and Haugeland ultimately locate the source of the normative authority governing scientific practices in human desires, interests, or commitments. Cartwright conceives the issue in similar terms, but her treatment may also usefully point in a different direction. Second, Brandom and Haugeland take impossibility as the crucial modality to be explicated. Cartwright does not address this issue explicitly, but her treatment points toward the centrality for understanding science and nature of an alternative modal concept ('real possibility').[35] Grasping the importance of this alternative sense of possibility, I argue, will vindicate the strategy for understanding normativity foreshadowed in my discussion of Heidegger and Neurath in chapter 1.

I first need to locate Cartwright's conception of the normativity of science in relation to Brandom's and Haugeland's conceptions. Recall that Brandom takes the normative force of commitments to be expressed by the attribution of desires, where 'desire' itself expresses a normative status caught up in networks of commitment and entitlement rather than a de facto psychological state. Although Cartwright does not explicitly discuss these issues, her conception initially seems close to Brandom's. She desires that the sciences ameliorate people's conditions of life, on the basis of the most reliable attainable understanding of those conditions. Recognizing that others start from different conceptions of the scientific good, for example, a fundamentalist desire to articulate a unified theoretical understanding, Cartwright implicitly treats all such conceptions as within the space of reasons. She argues first that the attainable satisfactions of this latter desire do not really yield what is sought (in the absence of at least approximately adequate models, a fundamentalist faith yields only the semblance of understanding); moreover, seeking to satisfy it also sometimes blocks the achievement of practical goals of ameliorating human life that everyone ought to desire. She concludes that everyone should recognize the limited but open-ended scope of theoretical articulations, and apply whatever theory empirically best satisfies human needs and desires in a particular situation without demanding a unified theoretical regimentation.

Where Cartwright and Brandom would adjudicate such competing desires in a common space of reasons, Haugeland argues that that strategy does not take science sufficiently seriously. Science, he argues, establishes an autonomous domain, and its normative authority within that domain is unqualified. The commitment not to allow conceivable but impossible phenomena cannot just be one desideratum to be weighed among others, but is constitutive of scientific understanding. What makes science both

35. 'Real possibility' should not be read as a straightforward analogue to alethic and nomological possibility. Each of these latter modalities are interdefinable with a parallel conception of alethic or nomological necessity, respectively (alethic possibility $\Diamond$ is equivalent to $\sim \Box \sim$). The relation between real possibilities and other modalities will be discussed below.

difficult and worth doing is the need to sustain a robust excluded zone of impossibilities that will not be tolerated. To limit the scope of theoretical claims because one cannot clearly articulate what phenomena are excluded in certain situations, as Cartwright would, Haugeland thinks would be tantamount to giving up *doing science*. He would undoubtedly acknowledge that many persons participate in science without maintaining such overarching existential commitment to scientific understanding, whether because their primary commitment is to Cartwright's moral-practical concern for human welfare or because science is instrumental for their career and personal goals.[36] Their scientific work, however, would be parasitic upon the uncompromising existential commitments of others to sustain the kind of understanding that ultimately allows science to be suitable for these other goals.

I argued in chapter 7 that the normative considerations available to Brandom and Haugeland are not sufficient to cope with such disagreements. My preceding discussion of causality and causal laws provides the basis for a more adequate conception of the normativity of scientific understanding, which does not reproduce the dualism that still places authority in objects independent of us, while grounding its normative force in our commitments to do so. Science discloses not objects or laws independent of us and our concerns, but phenomena that we are a part of. That recognition is crucial to the normative force of scientific understanding. I see Cartwright's allegiance to "life in the messy world that we *inevitably* inhabit" (my emphasis) as a partial move in the same direction. Moreover, the modality that expresses its authority is not nomological necessity (or its normative analogues in incompatible commitments and exclusions), but what I have been calling real possibility. Here, too, Cartwright can be read as moving in similar directions—she objects to fundamentalism for foreclosing some otherwise revealing scientific possibilities, by only acknowledging those that can be regimented within a unifying scheme of necessities. Such a conception of real possibilities can only be adequately developed, however, in the context of an understanding of scientific practices as normative rather than as governed by regularities of belief or performance.

36. Haugeland does not discuss the extent to which human beings may live without any sustained existential commitments. Whether he takes such commitment to be rare but indispensable to the possibility of a fully human life even for those without such commitment, or whether it is widespread even though in most cases only partially sustained, is unclear from his recent writings.

On my account, the normative force of scientific understanding comes from causal intra-action. Recognition of this connection between causality and normativity has long been blocked by a dualistic conception of causality. The normative dimension of causal intra-action has been made inconspicuous by building it into taken-for-granted boundaries of objects. Ironically, the result has been to limit the normative significance of causal intra-action, by treating it as *contingent.* Recall from the introduction and chapter 1 that necessity has often served as a stand-in for normativity; in that context, causal circumstances have often been thought insufficient to explicate normativity because they might have been different. But the conception of causal relations as contingent draws upon a mistaken conception of the agents responsive to norms. Such agents are not one distinct and discrete object (a body), contingently surrounded by some objects impinging upon them rather than others. Bodies, as I argued in chapter 7, are not objects in this sense, but rather only become unified as active bodies through ongoing (causal) intra-action with their surroundings. Moreover, the boundary between body and surrounding world is not fixed in these intra-actions. Who "we" are depends upon how our surroundings are configured. That, in turn, is a normative question: what are the phenomena (as repeatable patterns) to which our activities belong?

This interdependence of bodies and their surroundings should be especially clear in modern, technologically intensive societies. Human beings have assimilated a substantial technical apparatus on the side of our own capacities for practical activity; in the absence of buildings, vehicles, clothing, the clearing of farms, yards, and roads, a pervasive discursive apparatus, and a plethora of other familiar equipment, human beings would not be the agents we are. We are prosthetically embodied beings, for without this apparatus, our bodily lives would indeed be nasty, brutish, and short. Human bodies have not always had this particular array of prosthetic extensions, of course. In the past, human beings were very differently embodied, in the sense of having differently developed capacities and patterns of activity, and different possibilities and stakes. Human agency thus belongs to a complex array of causally intra-active phenomena. Indeed, not all of our prosthetically extended capacities as agents can be identified with what is deliberately fabricated or employed as equipment, for symbiotic intra-action with microbial organisms, metabolic incorporation and expulsion of material from the surrounding atmosphere, waters, and soils, and the diurnal and annual patterns of solar light and heat are partially incorporated within the dynamic patterns constitutive of embodied human agency.[37]

These considerations give a new complexity to Cartwright's seemingly commonplace reference to "the messy world we inevitably inhabit." Its inevitability is primarily deictic: we (as understanding agents) belong to *this* world in all its spatial and temporal particularity, and its causally intra-active capacities.[38] But what the inevitabilities of this world are is less immediately clear. The relevant obscurity is not primarily ignorance of laws, but a partially open field of possibilities. Cartwright, as we have seen, has expressed concern about the foreclosure of therapeutic possibilities that might prolong the active, meaningful lives of many women who will otherwise die relatively young from breast cancer. Although she does not discuss this point, the relevant sense of possibility is neither logical nor nomological; the range of possibilities in those senses is far too extensive and remote to matter to us. Cartwright instead has to be referring to real possibilities. Her claim is that the desired outcome could actually be achieved in *these* circumstances, appropriately and feasibly rearranged (including revisions in what people say and do). It would not merely be a philosophical mistake to construe the uncertainty about these possibilities as nomological ignorance, for Cartwright's claim is that to do so is also a scientific and moral failure. When scientific fundamentalists confer priority upon research aiming toward a systematic nomological grasp of the phenomena manifest in breast cancer, she argues, they effectively foreclose some of the most important real possibilities for reconfiguring the available preventative and therapeutic responses and thereby do harm.

My point in emphasizing the modality of real possibilities goes significantly further than Cartwright's concerns, however. She follows Neurath

37. Todes (2001) gives an especially prominent place to vertical orientation perpendicular to the earth's surface in establishing fundamental characteristics of human embodiment: only when balanced within the vertical field can we effectively orient ourselves asymmetrically forward toward objects dispersed horizontally in our surroundings. Todes, however, describes these aspects of human embodiment merely as phenomenological invariants of human experience rather than as patterns of causal intra-action with the earth's gravitation. A very different kind of intra-active configuration of human agency is highlighted in work such as McNeill 1976, Crosby 1986, or Diamond 1997. These studies argue that the political, cultural, and biological expansion of urban societies in general, and Euroasian societies in particular, has been made possible by symbiotic intra-action with microbes and domesticated animals and plants.

38. I am indebted to Brian Cantwell Smith (1996) for his much more extensive insistence upon the importance of deixis for understanding intentionality and normative accountability.

in advocating an interdisciplinary, encyclopedic science that would bring multiple theoretical and experimental resources to bear on socially salient concerns. I regard real possibility as the most important modality for understanding scientific practices more generally; indeed, it can show when Cartwright's encyclopedic empiricism ought to yield to a more "fundamental" theoretical articulation. Haugeland's attempt to articulate a normative conception of nomological impossibility as the primary modality for understanding science will be the foil for my argument, in significant part because his account is the best developed alternative that actually tries to link scientific practice and the modality of scientific understanding. This discussion, however, will show the indispensable convergence for philosophical naturalism of the issues I have been discussing in the past several chapters: the priority of causal intra-action over regulist or regularist appeals to laws; of phenomena over objects as the locus of causal capacities; and of a normative conception of what is at stake in science over more standard appeals to the authority of objects and objective necessity sustained by scientists' and other agents' subjective commitments.

Haugeland's existential conception of the constitution of objectivity is a useful foil for my argument because he gets so much right about the normative authority that governs a scientific understanding of nature. He, too, looks to scientific practice as the primary locus for the normative accountability of science. His account of practices gives priority to scientific understanding, and he recognizes that understanding is a kind of practical competence. His articulation of the precarious equilibrium established in scientific understanding, and the indispensable resilience of the skills that convey that understanding, is a valuable articulation of what such competence involves. Moreover, Haugeland rightly emphasizes not only that scientific understanding requires a kind of participatory involvement in what is at stake in scientific practices, but that these stakes must show how practices are accountable to something not determined by "us." The place where he goes wrong is in his conception of the modal expression of these concerns in terms of objective impossibilities sustained by the existential commitments of those who take science seriously.

For Haugeland, what is at stake in scientific practice is the objectivity of understanding. He takes this to be important, because only by appropriately holding ourselves accountable to objects can we distinguish scientific understanding from the merely social-normative recognition of appropriate performance. Haugeland insists upon a distinctive kind of error that the sciences highlight:

> If all but one in a group of [scientific] results turn out to be compatible with one another, but incompatible with that one, then the outlier stands to be rejected as erroneous. That such a result is an error is not simply a matter of consensus of experts; indeed, they might all agree that each performance was properly executed. Rather, the error is a matter of a *structure* [emphasis mine] of relations among a number of actual [scientific] findings. . . . What is authoritative for this new sort of error is not any consensus or collective judgment but rather the locus of incompatibility among those related exercises—which is to say, the constituted *object* [emphasis mine]. It is at that locus, that point of intersection, at which the many results concur, and the one stands apart. But the kind of potential error for which the object itself is authoritative—normatively criterial—is nothing other than *objective correctness.* (1998, 337)

Haugeland's account thus concurs with familiar nomological conceptions in locating the modality of scientific understanding in a structure of alethic necessities (impossibilities) accruing to kinds of objects; he differs primarily in taking the normative force through which those necessities are manifest and authoritative to be an existential commitment rather than an inexorable necessity.[39]

I have already explained why it is a mistake to take objects separate from us, rather than phenomena in which we participate, as the relevant focus of scientific understanding. I now want to show how this mistake is related to the also mistaken focus upon necessity (impossibility) and objectivity as the correlative modal and normative construals of how phenomena are authoritative. Haugeland is obviously not wrong to see scientific understanding as accountable to worldly phenomena, but he is mistaken in the narrowness of his construal of the stakes in such understanding, with important consequences. Catherine Elgin cogently captures what is initially wrong about Haugeland's characterization of the stakes in scientific understanding solely in terms of truth and objective correctness:

39. In effect, Haugeland is saying that for most of human history, natural necessity has not been authoritative over human discursive and other practical performances, because the agents involved have not recognized and acknowledged what we understand as the relevant necessities. The distinctive contribution of science has been to hold human performances accountable to objective necessities, and thereby to understand our situation in a

> If our overarching epistemic objective is the acquisition of new truths, we would be ill advised to turn to . . . science or philosophy. . . . Still, science and philosophy are plainly cognitive enterprises. . . . Cognitive excellences and deficiencies are many. We value good questions, apt remarks, illuminating experiments, as well as potentially fruitful hypotheses, insightful studies, significant discoveries. We disparage irrelevant or obvious statements, tortuous, tenuous, or tendentious arguments, unimaginative hypotheses, ad hoc explanations. . . . By broadening our [epistemological] vista to include understanding of all sorts, we will be better equipped to deal with the issue [of what sort of knowledge is worth having and seeking]. Not being restricted to facts, understanding is more comprehensive than knowledge ever hoped to be. (Elgin 1991, 196–97)[40]

But if the stakes in scientific understanding are not just objective correctness in Haugeland's sense, how else can the normative accountability of scientific understanding make sense?

One important consideration is that objectivity and objective correctness are themselves not as simple and straightforward as Haugeland makes them appear. He does recognize the practical difficulties introduced by the holism of hypothesis testing, such that resilient adjustments of scientists' conceptual and practical skills are needed to maintain a "precarious equilibrium" in the face of apparently incompatible results. What escapes his notice is that the constitutive standards of the sciences are themselves rich and complex, focused by what makes these practices and their outcomes *significant.* This point emerges especially clearly in recent work on the history of scientific objectivity: what is at issue in the quest for objective understanding has shifted along with other reconceptions of what is at stake in scientific understanding.[41] Objectivity is not one generic virtue, but a contested, historically specific field, in which scientific practitioners are accountable to what has shown itself to be at stake in their reflexive engagement with the world and their own practices.

---

new way; the manifestation of natural necessity required a distinctive existential achievement.

40. In fairness, Haugeland (2000, 55) follows Heidegger in explicitly expanding familiar conceptions of truth beyond the narrowly linguistic and semantic domain to understand true comportments of all kinds.

41. Important contributions to a history of objectivity include Daston 1992, 1994; Daston and Galison 1992; Galison 1998b; Porter 1995; Wise 1995.

Another, more far-reaching consideration arises in considering the temporal orientation of scientific practices and scientific understanding. Haugeland starts out rightly by thinking of science as ongoing research rather than as the retrospective assessment of knowledge. It is the ongoing, resilient adaptation of scientists' understanding to account for newly discovered phenomena, entities, or features that characterizes their disclosure of the world. Yet Haugeland then feels compelled to express the normative accountability of such forward-looking constitutive practices as accountability to something already *there,* namely, objects and their properties. My point is not to deny the obvious and important sense in which the physical world is already there; after all, I have already emphasized the deictic character of our involvement in causal intra-action with circumstances in which we find ourselves. What is not fully determinate, however, is the practical configuration of those circumstances as a field of *intelligible* possibilities that *matter* to us.

To capture adequately the temporality of scientific understanding and its orientation toward possibilities, we need to express its accountability in appropriately futural terms. Hans-Jörg Rheinberger's discussion of the development of experimental systems in scientific research suggests how one might begin to do so. Experimental systems are close analogues to what I have been calling phenomena, although Rheinberger is talking almost exclusively about laboratory phenomena. A system in his sense is a repeatable material arrangement, for example, the introduction of radioactive-labeled amino acids into rat liver cells in tissue culture that was the proximal focus of Paul Zamecnik's research program in the 1950s. But Rheinberger emphasizes the point I made at the end of the previous chapter: the repetition of a successful experimental system is always differential, introducing variations on the established procedures and intra-actions to allow the system to behave in new ways against a more established, well-understood background. Research with experimental systems is thus directed toward the disclosure of something not yet adequately understood and articulated, yet sufficiently stabilized by the technical conditions of experimental work to present an intelligible focus of inquiry. The system, then, is oriented toward the disclosure of new possibilities. These are real possibilities rather than merely logically or nomologically possible. The difference is threefold: they are possible through an apparently feasible reconfiguration of present circumstances, they show up as significant or interesting possibilities, and their possibility is transformative rather than merely additive (they reconfigure the sense of what one was already doing and dealing with).

Rheinberger calls such possibilities "epistemic things," and he takes them, rather than already definite objects, to be what systematic scientific research is about and accountable to:

> "Epistemic things" are material entities or processes—physical structures, chemical reactions, biological functions—that constitute the objects of inquiry. As epistemic objects, they present themselves in a characteristic, irreducible vagueness. This vagueness is inevitable because, paradoxically, epistemic things embody what one does not yet know. Scientific objects have the precarious status of being absent in their experimental presence; they are not simply hidden things to be brought to light through sophisticated manipulations. (1997, 28)

Many philosophers may be tempted to construe this characteristic vagueness as "merely" epistemic: the focus of research would then be a perfectly definite object or process metaphysically, about which researchers are partially ignorant. But that would misunderstand both the modality and the normativity of epistemic things. Modally, they are not possible actualities (definite states of affairs that might or might not obtain), but actual possibilities (a present configuration of circumstances oriented toward an indefinite, partially open future). Normatively, they are possibilities in the sense that there is something at stake in their realization—not some definite, already articulated goal or value, but something that matters in ways we may not yet fully grasp.[42]

Rheinberger introduces the notion of epistemic things to express the ways in which the material systems that focus scientific research always outrun projections and conceptualizations of them. Without some technical control of their components and conceptual grasp of their configuration, experimental systems cannot even pose explorable questions; yet to

42. Scientists do sometimes have in advance a relatively clear sense of how the successful completion of their research would matter. Yet they can be mistaken about what such success amounts to and how it matters, despite having been basically correct in their practical orientation toward its achievement. The practical orientation of research toward an "epistemic thing" is thus not adequately characterized by present conceptualizations of it. Moreover, these reconceptualizations are not simply corrections of a "mistaken" preconception, but a transformation of the research situation and its stakes, which is also consequently a transformation of the researchers as scientific practitioners.

the extent that these systems are interesting or important, genuinely disclosing the world, they undo their own conceptualization. As Rheinberger notes:

> Experimental systems grow slowly into a kind of scientific hardware within which the more fragile software of epistemic things—this amalgam of halfway-concepts, no-longer-techniques, and not-yet-values-and-standards—is articulated, connected, disconnected, placed, and displaced. Certainly they delineate the realm of the possible. But as a rule they do not create rigid orientations. On the contrary, it is the hallmark of productive experimental systems that their differential reproduction leads to events that may induce major shifts of perspective within or beyond their own confines. (1997, 36)

The contrast to Haugeland's conception is sharpened by recognition that Rheinberger's concern to understand the limits of presently articulated standards and skills coincides closely with Haugeland's own fundamental philosophical concern. In characterizing the "constitutive standards" that allow objects to be authoritative for science, Haugeland recognizes that scientists' skills and conceptualizations must be resilient, open to revision and repair in the face of the recalcitrance of things. Yet he insists upon severe limits to such internal transformations, in order to allow for a clear difference between two kinds of error, incorrect performance and incorrect determinations by correct performance: "By ruling out the bulk of conceivable combinations, [constitutive standards] bind the totality of actual results within the narrow bounds of possibility, . . . and allows them to 'gang up on' isolated performances whose results are incompatible with the overwhelming majority" (Haugeland 1998, 338). His homage to Popperian falsificationism is his insistence that the constitutive standards of scientific practice be vulnerable to a prior determinacy of things, such that the enterprise itself can be in question.

Rheinberger, by contrast, notes the inherent and productive ambiguities between these two forms of accountability, to procedural norms and to the things disclosed through their performance. In discussing what Haugeland would call the normative accountability of the mundane and constitutive skills of research, Rheinberger insists upon the openness of such skills to something unprecedented and unexpected, and the consequent need *not* to allow such skills to become rigidly determinate and hence confrontational.

> The [experimental] web must not become too rigid. In deliberating upon the manner in which a system is to be handled so as to let the unknown

> intrude and invade it, Max Delbrück has spoken of a "principle of measured sloppiness." "If you are too sloppy, then you never get reproducible results, and then you can never draw any conclusions; but if you are just a little sloppy, then when you see something startling you [nail] it down." (Rheinberger 1997, 78, citing Delbrück's correspondence and oral communications as presented by Fischer and Lipson 1988)

Haugeland and Rheinberger thus agree in taking scientific practices to be fundamentally accountable to material phenomena, while disagreeing (at least implicitly) in their construal of the modality and normativity of that accountability. What I now will consider is the connection between the modality of scientific understanding as disclosing real possibilities, and the normativity of scientific practices and the natural world.

For Haugeland, the sciences become accountable to what their skills disclose by insistently holding themselves to determinate constitutive standards. Their skills (including conceptual skills) can thereby be vulnerable to the objects discovered in scientific practice.[43] In shifting the locus of such accountability from determinate objects to epistemic things (real possibilities), Rheinberger in effect takes issue with Haugeland's commitment to the "semantic" determinacy of the constitutive standards of scientific practice. He also thereby disputes Haugeland's claim that what is at stake in science is the objective correctness or incorrectness of those standards. Think back to my discussion in chapter 8 of the ambiguity of intentional interpretation as a measurement intra-action. Haugeland's existential conception of science expresses a Popperian commitment to confer a definite interpretation upon the sciences' constitutive and mundane skills (including, presumably, their conceptual-articulative skills), so that those skills can provide a definite "measure" of the world. Rheinberger claims instead that scientific research always resides within the productive ambiguity of such ongoing experimental intra-action. What is at stake there is not so much the correct discovery of objects and their properties, but the disclosure of epistemic things as binding upon us through what is at stake in their disclosure. The stakes in scientific research are not what nature is, but what it is to be nature.

43. I use the word 'discover' quite deliberately here. Haugeland (1998, 331) follows Heidegger in distinguishing ordinary or mundane discovery of specific findings within a practice, from the disclosure, or metadiscovery, that the practice is intelligible at all. In these terms, I am arguing that Haugeland still takes what is at stake in the sciences to be the correctness of their discoveries rather than how their disclosure matters more generally.

To grasp the point of this difference, consider first Rheinberger's own examples of historically shifting epistemic things. What was at stake epistemically in Peyton Rous's chicken sarcoma system, in which tumors were induced "virally" by injecting healthy chickens with the porcelain-filtered extracts from tumor cells? Initially, Rous's research seemed to be about "viruses," those apparently infectious agents that passed through the porcelain filters that were sufficient to capture unicellular bacteria and protists, and about cancer, the pathologically uncontrolled reproduction of somatic tissues. Rheinberger (1995), however, traces the shifts through which the chicken sarcoma system came to disclose instead the interplay between structure and function in the dynamics of normal cells.[44] Similarly, Paul Zamecnik's use of the uptake of radioactively labeled amino acids in rat liver tissue to study the biochemical dynamics of cancer cells dramatically transformed into an exemplification of the molecular biology of protein synthesis.[45] These dramatic shifts did not result, as Haugeland would have it, from conflicts between the researchers' mundane and constitutive skills, such that a shift in research objectives was needed to sustain their constitutive commitment to accord with the things they studied. Rather, they resulted from unexpected conjunctures and bifurcations that disclosed other possibilities, which the researchers found themselves *bound* to pursue.

It is perfectly appropriate to talk in this way about what is at stake in a practice, indeed as governing the practice, in the absence of any articulation of or agreement upon what those stakes are. This point has been made effectively by Alasdair MacIntyre in a quite different context in terms of "traditions" rather than practices:[46]

> What constitutes a tradition is a conflict of interpretations of that tradition, a conflict which itself has a history susceptible of rival interpretations. If I am a Jew, I have to recognize that the tradition of Judaism is partly constituted by a continuous argument over what it means to be a Jew. (1980, 62)

44. Rheinberger also notes the ironic recurrence through which the early electron microscope photomicrographs of sarcoma cells, which were taken (a third of a century after Rous's initial sarcoma research) in order to display submicroscopic cell structures without subjecting them to centrifugation, produced beautifully clear images of budding viruses emerging from the membranes of sarcoma cells.

45. Zamecnik's group provides the central historical case examined in Rheinberger 1997.

46. MacIntyre was also using this discussion to explicate the normative accountability of scientific research.

Being a Jew in this sense is a normative status rather than a determinate matter of fact. The normative significance of this status is sustained by its *mattering* to all parties to get it right about what is appropriately at stake here for all of them, even though they have not yet reached, and may never reach, agreement about what that is. Moreover, it matters that these stakes be binding on everyone involved.[47] The intelligibility of anyone's participation in a practice turns on there being something at stake for everyone in getting it right. That does not mean that the intelligibility of practices depends upon the possibility of ultimate agreement about and conformity to what those stakes are. Rather, it depends upon a recognition of and by those to whom the practice matters, such that they (ought to) hold themselves responsible for their different interpretations and accountable to one another.

It is crucial to understand such stakes, which I take Rheinberger's epistemic things to exemplify, as normatively binding upon practices. Albert Claude did not simply choose to drop his oncological studies (while remaining within Rockefeller Institute's Division of Cancer Research!) in order to pursue the unprecedented events disclosed by the ultracentrifugation of sarcoma cells. Instead, he pursued "a boundary program [that] allowed [him] to fluctuate between cancer research and cytomorphology [such that] the actual results *guided* the decision about what to do next" (Rheinberger 1995, 60; my emphasis). How could these results "guide" the decision of what course of research to pursue? There was no question of objective correctness here; the issue instead was what it mattered to understand, what the "epistemic thing" to be sought was. Decisions had to be made, to be sure, and there were reasons for those decisions. Rheinberger notes the conviction of Claude's director, James Murphy, that "basic research was the clue to the cancer problem" (1995, 60 n. 36), but that will not account for Claude's own full-fledged shift from oncology to cytology. Claude's move instead makes sense as a responsible responsiveness to what was disclosed within his research: "responsive" in the sense of being open

47. MacIntyre's example of the stakes in being a Jew exemplifies this point especially powerfully in the twentieth century. Even those European Jews who thought, before Hitler's ascension in Germany, that being Jewish did not fundamentally matter to them found that the issue of what it meant to be a Jew was, in those circumstances, inescapable. That does not mean that they were wrong to refuse to identify with their Jewishness; it meant that that refusal took on a new centrality and new significance in their lives, which underwent still further transformation in the aftermath of the Holocaust (which, of course, also substantially affected what was at stake in the lives of Gentiles as well).

to what emerged within the phenomena, and "responsible" in the sense that the decision mattered, and Claude was accountable for its outcome. What was disclosed was not some definite object, however, but a field of not-fully-determinate possibilities.

This element of responsibility to the possibilities of a situation can be seen more clearly by contrasts to epistemic configurations that nowadays we might find conceivable, but not (really) possible. Three quickly sketched examples drawn from recent science studies literature will highlight my point. Consider first Mario Biagioli's description (1993) of the epistemic configuration of the absolutist court culture through which Galileo sought to escape established disciplinary hierarchies that blocked mathematicians from addressing the causal questions characteristic of natural philosophy. The prince's court provided a venue in which, unlike in the universities, mathematicians were not assumed a priori to be incompetent to *address* philosophical issues, although on the other hand, no one in that venue was accorded competence to *resolve* such issues.[48] The links between objectivity, princely honor, lack of epistemic commitment, and cosmopolitan appreciation of philosophical novelty and the aesthetics of argumentation, each of which contributed to the dynamics of courtly disputes, affected what *could* be at stake there in arguing about the buoyancy of bodies in water or the nature of comets.

As a second example, consider Leibniz's proposal in "An Odd Thought Concerning a New Sort of Exhibition" (1675) for "an 'academy of sciences' [that would exhibit] all sorts of optical wonders . . . unusual and rare animals . . . extraordinary rope-dancer . . . artificial meteors . . . ballets of horses . . . museums of rarities' alongside displays of calculating machines, the air pump, telescopic observations, an anatomical theater, and the equality of the oscillations of the pendulum" (cited in Daston and Park 1998, 215). This list is astonishing to present-day sensibilities. Alongside instruments and practices recognizably central to the development of modern science are phenomena (rope dancers and horse ballets) whose relation to scientific understanding is quite difficult to grasp, and others (museums of rarities, unusual and rare animals) that Leibniz clearly accords rather different and greater scientific significance than would anyone nowadays.

A third example is the now widely accepted understanding of Newton

48. Strictly speaking, of course, the absolute prince was capable of resolving them, but the conceptual and practical links between the prince's power and the prince's epistemic indifference would have made such resolution quite extraordinary.

as having subordinated mechanics to his theological and alchemical understanding: what was at stake in the mathematics of celestial motions and the forces that drive them was not just (or even primarily) dynamical correctness, but theological insight. In Newton's vision of his science, celestial dynamics, alchemical study of the active, vegetative principles animating the world, and the historical fulfillment of biblical prophecies would be integrated into a relatively seamless understanding of God's providential involvement in his creation.[49]

What I find striking about all three examples of the normative stakes in scientific inquiry is that, while I can quite readily reconstruct the intelligibility of such norms at one level, and thus understand how someone *might* be governed by the constitutive standards such epistemic practices display, I cannot reconstruct these norms as actually compelling. These remain "merely possible" epistemic configurations that I cannot understand as making claims upon me. No decision or commitment is involved in my not being bound by these standards; I cannot intelligibly be bound by them. The normative force of princely honor, courtly *sprezzatura,* and its epistemological aesthetic of unresolved argumentative jousts, the potential scientific significance Leibniz saw in rope dancers or horse ballets, and the theological and alchemical stakes in mechanics have no hold upon me. Yet neither are these constitutive norms like the imagined cognitive norms of Frege's logical aliens. I cannot comprehend what it would be even to understand someone as answering to logical laws that permitted explicit self-contradiction. By contrast, I can quite readily make sense of these early modern epistemic configurations as possibly intelligibly binding upon someone, just not as actually binding upon me and others like me.

This difference between conceivable-but-not-compelling norms and the inconceivable norms of Frege's logical aliens has an illuminating parallel within recent discussions of alethic modalities. Sanford Shieh (2001) has shown how the difference between the modal behavior of ordinary definite descriptions and "actualized" descriptions such as "the actual number of planets" reveals that Kripke's attempt to construe names as rigid designators bound to the authority of object-identities across possible worlds depends upon a prior commitment to the intelligibility of deep necessity (necessity construed as independent of any determination of which world is

49. Dobbs and Jacob 1995, part 1, provides a succinct account of the place of Newtonian dynamics in Newton's larger theological and alchemical project.

actual). Shieh remains agnostic about the coherence of deep necessity, and about its associated two-dimensional modal semantics. In the light of my arguments earlier in this chapter, however, I take his argument to show why Kripke is mistaken in taking objects to have any normative authority apart from their occurrence within phenomena, and hence to raise fundamental problems about the very idea of rigid designation. The normative-causal structure of an actual phenomenon is the basis for identifying the boundaries of objects and the proprieties of reidentifying them as the same object or object-kind, rather than the transworld regularity of rigid designation serving as the basis for the correct identification of actual objects. My discussion here goes one step further. If deep necessity is, as I suspect, not a coherent notion, the consequent modal priority of actual circumstances turns out to be a special case of a more general normative priority. The present configuration of normative force (the stakes that can intelligibly bind current practice) governs what is even conceivable. One can ask intelligibly whether the normative authority of other ways of thinking and acting is conceivable, given what is at stake in one's own practices. One cannot ask in the same way, however, what normative configurations would be intelligible if these other norms were in fact our own. (Would the stakes in our investigations, for example, have been intelligible to Galileo or Newton?). This question is forlorn not because of a lack of imagination, a hidebound enclosure within one's own preconceptions, for one can imagine any number of more or less plausible answers to such questions. The difficulty is instead a lack of any serious constraint upon the answer. It is not clear what would make any particular answer to such questions correct or incorrect, disconnected as it would be from anything actually at stake in the question.[50]

50. The parallel argument for modal semantics would be that no determinate assignment can be given to what two-dimensional modal semanticists like Chalmers (1996) or Jackson (1998) would call the "*A*-extension" and "*A*-intension" of a term in possible world *w*, the semantic values assignable to the term if (counterfactually) world *w* were the actual world. The contrast is to the "*C*-extension" and "*C*-intension," the semantic values of the term in possible world *w* "given whatever world is in fact the actual world" (Jackson 1998, 48). Note the rhetorically clever but question-begging assignment of terms for these semantic concepts: the *A*-extension (*A* for actual) is the term's extension in some other possible world on the *counterfactual* assumption that that world is the actual world, while the *C*-extension (*C* for counterfactual) is its extension in some other possible world from the standpoint of the *actual* actual world. This assignment of terms is question-begging, because it is premised on the controversial assumption that it makes sense to consider some other possible world not merely as a possible variation on the actual world, but as the actual world of which the *actual* actual world is a merely possible variant. The difficulty

Why does this difference between two levels of intelligibility of epistemic and ontological norms matter? Haugeland rightly emphasized the importance of participatory *involvement* in allowing worldly phenomena to have authority over our performances, but he mistakenly construed such involvement as ontologically distinctive of human beings. The existential commitments through which people insistently hold themselves, others, and things accountable to norms is supposed to constitute the binding authority of objects. I am arguing that matters are reversed: it is not one's willful insistence upon upholding constitutive standards that establishes the authority of objects, but instead the normative authority of what is already at stake in one's practices that governs insistence upon standards. If we do *not* insist upon the correct application and resilient extension of scientific norms in research, we *ought* to. We ought to, because the richly and subtly articulated normativity of scientific research (which extends far beyond just "objective correctness") *matters* to us. Exactly how it matters, and what claims it makes upon us, and thus what understanding of nature is worth seeking and having are to some extent contested and in flux. We can intelligibly disagree about the relative significance and meaning of "fundamental" insight into higher-energy particle intra-actions or epigenetic regulation of gene function; interdisciplinary explorations of more complex phenomena; or directly ameliorative research into disease therapies, agricultural yields, or greenhouse gas reductions. Yet all of these possibilities make compelling claims upon us, while other merely conceivable scientific priorities make none. Undoubtedly this normative configuration of the world is continually shifting as well as contested. The many ongoing efforts to pursue particular possible research agendas, respond to what shows itself through those efforts, and articulate and resolve the many con-

---

here is an analogue to the central issue dividing Carnap and Neurath on language. For modal metaphysicians such as Chalmers or Jackson, a world (like a language for Carnap) is a formal construction rather than a causally intra-active setting. Just as no one speaks a Carnapian language, no one lives in such a world. For the *A*-extension or *A*-intension of a term to be determinately assigned, on Neurath's or my conception of a language as inextricable from causal intra-action with circumstances and fellow speakers, one would (impossibly) have to be in the actual circumstances of possible world *w*, such that there is something at stake in that world making a real difference, depending upon how one assigned semantic significance to its terms. But of course nothing makes a real difference in a merely possible world, even on the counterfactual supposition that it is actual. The semblance of a determinate interpretation to *A*-semantics can be given, however, by tacitly taking the actual circumstances and stakes of one's involvement in the world (typically expressed in the form of "intuitions" about the appropriate assignment of inferential norms from the hypothetical standpoint of world *w*) as the "metaworld" that guides the interpretation.

flicts and tensions among what is determined to matter will over time reconfigure the issues and stakes in scientific inquiry.[51] Such shifts will not simply result from choices people make, however, but also from changes in their situation that reconfigure what are intelligible and binding possible choices. We are responsible for our choices not because we constitute them, but because we are involved in them with stakes to which we are accountable, epistemically and politically.

This discussion shows more clearly what is mistaken about Cartwright's empiricist repudiation of fundamentalism as well as Haugeland's counterclaim that fundamentalist commitment to a nomological structure of impossibilities is constitutive of scientific understanding. There is something rightly compelling about each claim, for they have each seized upon real possibilities that have emerged within contemporary scientific practice. Cartwright emphasizes both the limits of precise theoretical understanding and the opportunities for making empirically reliable but theoretically circumscribed connections between scientific practices that may allow significant intervention in the world: environmental and other preventative strategies for reducing cancers, fiber optic communications, and so forth.[52] Taking up such possibilities sometimes requires forgoing the demand for a rigorous, detailed, nomologically expressed articulation of the concrete systems involved. Haugeland's insistence upon the articulation and maintenance of excluded zones also captures a legitimate scientific pull toward precise unification and consequent theoretical understanding. Yet each philosopher elevates a favored class of scientific possibilities into a general antinaturalist stance (Cartwright's empiricism, and Haugeland's existential fundamentalism). As a result, neither adequately acknowledges the force of the other's concerns. In Haugeland's case, Cartwright's arguments show that precisely articulated impossibilities are not general characteristics of scientific understanding, but are instead particular opportunities (real possibilities) for theoretical articulation. Such celebrated achievements of the supposedly impossible as the marshaling of a sufficient number of beta or neutral K-meson decays to display parity nonconservation, the identifica-

51. Politically engaged, historically and scientifically informed efforts to articulate and reconfigure what is at issue and at stake in ongoing scientific work and its wider cultural appropriations are what I have elsewhere (Rouse 1996b) called "cultural studies of science."

52. Cartwright of course recognizes and values those phenomena for which a more precise and systematic theoretical understanding can be articulated with some empirical accuracy. She only takes them to be more limited in number and extent than is usually assumed.

tion of reverse transcriptase as mapping RNA to DNA, or the discovery of genetic transposition were not occasions for "giving up" on a whole theoretical scheme (which Haugeland takes to be the only "authentic" response to persistent violations of an excluded zone). These achievements were instead a reconfiguration of the salient field of scientific possibilities, both extending and redirecting work in those fields. Cartwright is mistaken in turn, however, to regard "messy" empirical achievements as altogether outside the scope of more fundamental theories. Such theories remain authoritative for more complex phenomena even though the lack of adequate models deprives them of most of their normative force. Nothing the wind-blown $1,000 bill does can violate classical mechanics, even though the nomologically expressible constraints that mechanics imposes upon its possible motions are vague and largely irrelevant to the most interesting and likely variations in its trajectory.

These considerations are suggestive for how to understand the relations between real possibility and the more familiar modal concepts of nomological and alethic necessity. There are indeed connections between these modal concepts, but not any that would allow one concept to be straightforwardly defined in terms of another. To be sure, what is impossible alethically or nomologically is not a real possibility, but it is not always impossible in the same way. Most conceivable impossibilities are also pointless; their impossibility does not matter, because they would not be real (meaningful) possibilities even if they were alethically or nomologically possible. Some, on the other hand, are significant impossibilities, whose recognition expresses defining features of real possibilities: the impossibility of human immortality matters greatly to how we understand the various real possibilities that confront us. Still others are mistakenly apparent possibilities, such that the disclosure of their impossibility makes a difference to what one can intelligibly aspire to do.[53] The impossibility of perpetual motion machines expressed by thermodynamics, for example, led to such reconfigurations of what can be pursued intelligibly. In this latter context, the

53. I am construing real possibilities as a configuration of an agent's situation that *includes* her conceptualization of that situation, rather than as specified *by* her conceptualization. As a result, the content of such possibilities can be misdescribed by agents for whom they are nevertheless real possibilities. Thus, under circumstances in which one could intelligibly pursue the project of designing a machine whose output would include more energy usable for work than its input, that project would embody a real possibility, but not under that description. Which descriptions would correctly characterize this possibility would depend upon what was most fundamentally at stake in the effort to pursue it.

(real) possibility of expressing some necessities (impossibilities) within an articulated scheme becomes significant (the possibility of regimenting alethic-modal expressions within a formal system such as S4 or S5 would in turn have an expressive significance, characterizable in the way Brandom explicates the expressive role of other logical vocabulary and the formal relations definable in their terms). As a component of discursive practices, the formulation of nomologically and alethically modalized locutions expresses the joint incompatibility of certain commitments, but the authority that these expressions articulate only has force where the relevant collateral commitments and articulations bring particular phenomena within the scope of the expressed impossibilities.

Cartwright's discussion of nomological machines thus provides insight into the expressive significance of nomological necessity. The expression of nomological necessity articulates an open-ended practical competence in recognizing and sometimes arranging the circumstances in which a regular pattern of events would occur. The normative authority of such expressions may outrun their normative force: that is, such expressions are appropriately taken to be authoritative even under circumstances in which they can impose no definite constraints. The parallel for alethic modalities would be to recognize that logical necessity only has normative force where the relevant utterances are appropriately construed in violation of it. Its normative authority is expressed in the defeasible interpretive maxim: do not construe speakers (including oneself) as contradicting themselves. But just as there are motions with no applicable model in mechanics, there are also uses of linguistic expressions whose inferential relations to others are underdetermined by the circumstances in which they are used. In those cases, the normative authority of noncontradiction still holds, but with limited binding force. The limited applicability of logical norms on those occasions may, of course, for some purposes display a debilitating vagueness in those uses of linguistic expressions; that is part of why logically expressible clarity matters. Much more must be said to provide an adequate account of the normative expressiveness of alethic and nomological modalities, obviously, but my overall suggestion is that their expressive and articulative roles can be understood in terms of real possibilities and normative authority and force.

A different issue arises, however, when comparing the inferential structure of real possibility with that of alethic and nomological possibility. Each of the latter concepts has its own inferentially correlated concept of necessity, such that these forms of possibility and necessity are interdefinable (the modal operator $\Diamond$ can be defined as $\sim \Box \sim$, while $\Box$ in turn is equivalent to $\sim \Diamond \sim$). Real possibilities are not in this way formally interdefinable with any available conception of necessity. One can, of course, say of some proposed course of action that it is not a real possibility not to do it, but such expressions are systematically ambiguous. That might well mean simply that avoiding this course of action would be pointless or absurd.[54] But the same expression could also be used to say that this course of action was obligatory (deontically "necessary"). Again, much more needs to be said to articulate the relations between real possibility, and deontic, temporal, alethic, or nomological modalities. My point is simply to show that real possibility is a distinct modal concept, not readily reducible to or definable in terms of more familiar modalities.[55]

Normative Authority and Normative Force

Even if this account of alethic and nomological modalities has been successful, fundamental concerns may still arise concerning how to understand normativity itself. In particular, more traditional naturalists and their antinaturalist opponents may feel that the most crucial philosophical question about normativity remains unanswered in my exposition. Normativity has long been philosophically suspect, with suspicion directed both to the ground of normative authority and the source of normative force. The ground of normative authority would provide the basis for determining which adjudication of conflicting normative claims is correct. The source of normative force would explain how there has come to be normative accountability in a world in which earlier there was none, such that those norms are binding.[56] These two issues must be intimately connected, since

54. Ian Hacking has discussed this sense of possibility, whose negation expresses not a logical or nomological limitation, but a lack of the surroundings needed to confer significance or intelligibility. He adapted a wonderful phrase of Sartre's to capture the limit cases of such inconceivability, as "absolute, unthinkable, indecipherable nothingness" (Hacking 1986, 232). Hacking's examples of such limit cases are the impossibility for a medieval vassal to be a Parisian *garçon de café* or a multiple personality.

55. A. N. Prior's classic discussion (1957) of tense-logics as modal logics similarly argued that temporal modalities are distinct from other modal operators, and that they need to be articulated as an autonomous conceptual domain, even though there are significant interactions between temporal and alethic modalities. Prior's discussion may be especially relevant here because of the important temporal dimension of the notion of real possibilities. Thanks to Sanford Shieh for pointing out the parallel to Prior's discussion.

56. Naturalists, of course, require that this explication be accomplished in ways continuous with a thoroughgoing scientific understanding of the world. Rationalist or theological

the emergence of binding normative accountability from its source has to render understanding and action accountable to its grounds.

These questions arise as much for antinaturalists as for metaphysical naturalists. Indeed, here we can see especially clearly why my arguments throughout this book have been directed as much against contemporary antinaturalists, who seem to deny that the authority and force of semantic, epistemic, or moral norms can be explicated naturalistically, as against self-proclaimed metaphysical naturalists. Nowadays, most self-proclaimed antinaturalists accept the presumption that nature (as disclosed through scientific understanding) is anormative. They also join with metaphysical naturalists in accepting the legitimacy and intelligibility of the linked questions about the source and ground of normative authority and force. Their antinaturalism consists in denying that these questions can be adequately answered by their naturalist counterparts, given their common presumptions about nature as disclosed scientifically; antinaturalists then conclude that normativity must have its source in reason, natural law, a social contract, or some other divinely or humanly instituted authority.

I am arguing instead for rejecting both the presumption that nature is anormative and the demand for a final accounting of the source and ground of normative authority. The latter demand turns out to be an artifact of the former presumption: such an accounting is supposed to show how normative authority and force are explicable within an understanding of nature as anormative. Moreover, I am arguing that we should reject these assumptions for fundamentally naturalistic reasons, ultimately because we should reject arbitrary impositions on science and refuse to countenance magical or supernatural determinations. I have already argued that scientific practices disclose nature as irreducibly normative. I have also argued that this conception of scientific understanding is naturalistic: scientific practices disclose natural phenomena as patterns of causal intra-action, through their own causal intra-action with nature. I conclude my treatment by showing why a naturalist should reject the demand for an articulation of the binding source and authoritative ground of the norms to which understanding and action are accountable. The *activity* of trying to articulate sources and grounds for norms plays an important role in sustaining the normativity of thought and action, but this activity cannot

antinaturalists would invoke different conceptions of explanatory adequacy as governing their efforts to accomplish what is nevertheless a parallel task.

be brought to authoritative completion without undermining the very normativity it seeks to vindicate. This futural indeterminacy of norms is what I have been trying to express by talking about normative accountability in terms of there being something at issue and at stake in thought and action.

Wilfrid Sellars has already implicitly given compelling reasons to reject the demand for a traditional naturalistic explication of the emergence of normative force, in his acknowledgment of the unavoidably mythical character of such explications.[57] On Sellars's account, norms only acquire force through their *recognition* as binding by those who are bound by them.[58] "Prior" to such recognition, there could be nothing binding what one says or does to how things are, and consequently (I have been arguing) there could be no causal structure articulating what happens (no boundaries between objects and their circumstances, or between causes and effects). But any such recognition must understand itself as already accountable to how things are, and hence as already bound to a world "before" there is a causally articulated world to be bound to. Rebecca Kukla succinctly captures Sellars's dilemma, and his response:

> Objects may *demand of us* that we make certain claims (these claims are "wrung from us by the object"—(["Empiricism and the Philosophy of Mind"] section 16), but they can do so only in virtue of our being already committed to recognizing these demands and taking them as binding. Sellars thus finds himself in a circle, wherein epistemic authority depends upon our already being negotiators of the space of reasons who are capable of this recognition. By telling a retrospective myth, we can get this circle off the ground by constitutively misrecognizing ourselves in memory as already within the space of reasons, bound by the author-

57. Sellars (1997, 117) acknowledges this point implicitly, in construing his own arguments against empiricist accounts of the source of epistemic normativity as the "use of a myth to kill a myth." I owe my own recognition of the significance of Sellars's construal to Rebecca Kukla (2000b). I think this disclosive role that Sellars thinks must be performed mythologically (the function of myth for Sellars is not the descriptive role of giving a false history, but the practical-normative role of securing legitimate and thus binding authority) is akin to what Heidegger was trying to express in his later reflections on *Ereignis,* the "event" of opening a clearing or world within which anything can be (as accountable to some stakes in its being what it is). Needless to say, Sellars and Heidegger offer rather different philosophical responses to this shared recognition of the "ungroundedness" of normativity.

58. The relevant sense of recognition here is not epistemic but ontological. Sellars's point is akin to the problem of manifest necessity I discerned in Heidegger's and Neurath's criticisms of foundational antinaturalisms: norms must be manifest as bindingly authoritative within the concrete setting in which actual performances are subject to them.

> ity of things and committed to respecting this authority. . . . These stories by no means count as naturalizing stories in any traditional sense, since they of necessity treat normative space as already open for the purposes of explanation. (Kukla 2000b, 204–5)[59]

This constitutive "misrecognition" is a situated practical performance rather than a timeless description: it supposedly "brings about" accountability to things and what is at stake in our dealings with them. Yet it cannot be a performance *by* "us," the practically situated agents who are thereby bound to respond to what is at stake in our situation, for we are only constituted as accountable participants in the agencies of observation together with the constitution of objects and stakes as that to which we are accountable.

Sellars (and Kukla) fail to recognize, however, that there is also a parallel argument for the performative-mythological character of any construal of the ultimate ground of normative authority. To indicate the normative grounds for a particular claim (not the reasons why, in our shared setting, one of us proposes that we ought to acknowledge its claim upon us, but an authoritative, final determination of why those reasons really are the correct ones) is a prospective mythological invocation that parallels the retrospective mythology of how those grounds first emerged as binding. The parallel has to do with the "impossible" transition invoked in each case. In the case Kukla attributes to Sellars, the transition is from a supposed absence of normative binding to its presence, a transition that must already be bound by what it supposedly first accomplishes. In the case I take to be parallel, the transition is from the presence of something at issue and at stake in what is said and done to its absence, a transition that must be authoritatively governed by what it supposedly abolishes. To have something at stake in one's performances is for it to matter how things turn out, for there to be a difference it makes to what is subsequently made of these performances. This point is especially clear in Rheinberger's discussion of epistemic things, which *as such* are unattainable. The articula-

59. What Kukla calls "naturalizing stories in the traditional sense" are precisely what I have been rejecting, because such stories conceive nature (understood as causally or nomologically structured) to be the anormative source and ground of normative force and authority. Such "mythical" accounts of normativity nevertheless could still be understood as naturalistic in a different sense, the one I am advocating here. Such accounts would express how something is at issue and at stake in practices, through the ways in which those practices are caught up within the causal structure of phenomena.

tion of the epistemic thing does not reveal *it* as a definite possibility indefinitely indicated by an earlier research situation. Rather it transforms the epistemic thing into the technical conditions of a new situation, with something further at stake in its development. The same can be said of any normative assessment of a situation: to arrive at an assessment is to transform the situation, opening newly possible ways to respond to it, with something further at stake in those responses. These considerations highlight what is wrong with Brandom's, Haugeland's, and Sellars's own attempts to show that normativity is grounded in the authority of objects, to the extent that objects are understood to be definitely bounded entities. In devolving normative authority onto something already determinate, such accounts abolish the prospective character of normative authority, which requires that the definitive settlement of what is at stake in our performances is always ahead of us. Brandom recognizes this point at least implicitly, in noticing that the authoritative standpoint of objectivity can never actually be occupied by anyone within the Sellarsian space of reasons (1994, 600–601, 645–49). Objective authority is expressed prospectively by the always perspectivally configured difference between what each one (presently) takes to be the case and what (eventually) will turn out to be so, an eventuality that always remains ahead of us even as we revise our takings to accommodate it. Having implicitly recognized the futural temporality of objectivity, however, Brandom then nevertheless fails to draw the crucially correlative consequences about objects and their normative authority.

Some philosophers will hear echoes of G. E. Moore's naturalistic fallacy (1929) in this argument: there can be no definite formulation of the norms or stakes ("goods") to which we are accountable, because we can always ask whether that formulation is itself appropriate or good. But Moore's line of argument differs fundamentally from mine in at least two respects. First, Moore takes his argument to show that 'good' is a special, non-natural property, whereas I take the argument to show that 'good' is not a primitive or unanalyzable property.[60] The term 'good' and its many norma-

60. I am grateful to Sanford Shieh for very helpful clarification of the issue here. What I want to say about 'good' as a property that is not independent of the multiple ways in which particular circumstances matter closely parallels his reinterpretation of Frege's conception of truth (Shieh 2002), as a "quasi-supervenient" property that expresses the correctness of a corresponding judgment (correctness in turn has diverse grounds that vary for different kinds and instances of judgments).

tive analogues are invoked to express and acknowledge how actions and understanding are bound to circumstances, such that how those circumstances are configured, with what outcomes, matters to us. The problem is not, as Moore thought, that 'good' must be conceptually independent of any particular ways in which one's circumstances matter, but that the potentially relevant circumstances outrun any definitive specification of them. This mattering, in turn, is not a determination *by* us, but *of* us.[61] Second, as a consequence of this difference, Moore's identification of good as a predicable property also omits the temporal element that is crucial to my argument. For Moore, the good now (and always) exceeds any definite content ascribable to it. My point is that discursive articulations of a situation belong to the situation they characterize, and to the extent that such formulations matter (make a difference), they change the situation and its stakes.[62]

Other philosophers may hear in my argument echoes of Derrida's insistence (1990) that the demand for justice is inexhaustible, and cannot be determined or encoded in law.[63] But Derrida mistakenly concludes that normative authority therefore floats free from *any* determination. That is to overlook the situatedness of the normative force of what is at stake in any situation. Contra Derrida, there *is* a limited sense in which we can articulate the source and ground of normativity indexically; its source and ground is the world, *this* world, the open-ended and ongoing patterns of causal intra-action in which we always find ourselves, or perhaps better, our *belonging* to the world intra-actively. Discursive expression of normative authority cannot be finally authoritative, but that is because words

61. Remember here that we are not agents or subjects apart from these determinations. We *are* these temporally extended patterns of significance ("mattering"), and the correlative responsibility for and to what matters, in those phenomena in which we participate. For a succinct discussion of how agents are constituted within a dialogical "ethical space," see Taylor 1991. Taylor, however, would reject my extension of this conception to emphasize the irreducibly causal character of such constitution.

62. Such transformations are also retroactive, however, so that one cannot ascribe a determinate normative authority to past circumstances while denying that it is any longer applicable. A situation is temporally open-ended in both directions—it incorporates its own past and future (including how its future will retroactively incorporate it as past, and its past prospectively but indefinitely incorporated it as future). The temporal interconnectedness of causal intra-action is crucial to understanding how a normative reach always exceeds its grasp, in the sense that the authority (the stakes) to which it is accountable always outruns present understanding (indeed, outruns the present *possibilities* for understanding). Present action is always accountable to a not-yét-determinate future.

63. For an especially clear and forceful discussion of Derrida's claim, see Cornell 1992, especially chapter 6.

are never self-sufficient; they always belong to a more extensive pattern of causal intra-action, with something at stake.

A more telling precedent is Kant's insistence upon the primacy of practical reason. Kant argued that from the agent's standpoint, that of the person who must deliberate and act, the theoretical determination of one's circumstances by causal laws is unintelligibly self-defeating. I can only understand my situation as one open to deliberation and consequent action, in accord with normative authority rather than causal determination. Kant could sustain that contrast, because he took the normativity of causal ascription to be governed by explicit, determinate rules (as "necessary"). By recognizing that the boundaries that define one's causally intra-active circumstances (and the scope of any laws that would articulate their determinacy) are themselves constitutively normative, my argument collapses the distinction. The agent's standpoint, situated amid ongoing, indefinitely open patterns of causal intra-action, is the standpoint of any understanding at all. I use the term 'standpoint,' however, in light of my discussion of Brandom's conceptual perspectivalism in chapter 7: to understand from an agent's standpoint is not to be outside a situation, seeing it from a perspective, but to be within its configuration as an intelligible setting of possible intra-action. Indeed, agency is itself constituted as a locus of situated, accountable mattering. The possibilities in question are what I have been describing as real possibilities; the modality expresses the situatedness of one's normative accountability to what is at stake in their possible realization. A recognition of this practical situatedness in the midst of causal intra-action is what I take to be crucial to a defensible philosophical naturalism. Naturalists should eschew any effort to occupy a theoretical or normative standpoint that is not in medias res, affected by and responsible to the phenomena to which one belongs. These phenomena always include more than their present realization, however; they incorporate an indefinitely open field of possibilities extending ahead of them, including possible ways to go wrong.

Having rejected attempts to originate or ground normative authority outside our causally intra-active circumstances, however, what can I make of their supposed normativity? The traditional Fregean/Husserlian rejection of naturalism, after all, turned on naturalism's inability to get outside of actual circumstances so as to account for the possibility of error, and hence to express the normativity of understanding and action. What has now changed, however, is the conception of natural causality and "actual circumstances" upon which such arguments turned. Actual circumstances are only determinate as configurations of possibilities that point beyond their present realization. These are not merely logical or nomological possibilities (what is not impossible), but real possibilities that matter. I have tried to indicate their normativity (how they matter, and what it is for them to matter) by talking about what is at issue and what is at stake in our belonging to causally intra-active circumstances. The term 'at issue' indicates that agents can belong to the same situation, as a setting for real possibilities, even though the appropriate characterization of the situation and its possibilities is contested. The term 'at stake' expresses a normative accountability that outruns any definite formulation of what one is accountable to. Descriptions of those issues and stakes, or of the causally intra-active circumstances to which they belong, are themselves accountable possibilities (ways of actively configuring one's situation from within) rather than exhaustive or definitive representations of them. They express, articulate, and thereby partially reconfigure our belonging to an already meaningfully configured world. So rejecting the demand to explicate the source and ground of normativity is not to abandon all accounts of what is at issue and at stake in our performances, but to abandon a certain mistaken interpretation of what it is to give such accounts. Kukla's Sellars construes accounts of the sources of normativity as "mythological," and I have extended that construal to accounts of normative grounds. That does not mean they are false (although they can be falsely construed as determinations of fact, in the traditional sense of non-normative determinacies),[64] but only that they are *performative* expressions. They sustain accountability to what is at stake in our circumstances, by simultaneously expressing that accountability, holding us accountable (by making a claim), and further articulating what that accountability entails.[65] No

64. This claim, that explicitly normative statements are not "determinations of fact," must be interpreted very carefully. I endorse Brandom's insistence that facts are simply true claims (1994, 327–29), and that explicitly normative claims are true or false in the same sense as any other statements. So all that is being denied in this parenthetical remark is the supposedly naturalistic project of distinguishing objective facts from normative assessments. What is wrong with that project, and what I am rejecting, is its underlying conception of facts. Ironically, what is wrong with that conception of fact is that it runs afoul of its own aspiration to naturalism, by requiring that facts be expressible in a magic language that is semantically determinate independent of the world it would express.

65. I talk about "accountability" rather than "our accountability," to acknowledge Haugeland's crucial, insightful point (1998, chap. 13), that normative accountability that could sustain the boundaries of phenomena is not just accountable for one's own sayings and doings, but for the appropriately cooperative performances of all components of the

such articulation can achieve (or rightly claim) finality, because its normative significance arises from its pointing toward further possibilities (including, crucially, the possibility of *correcting* it, out of fidelity to the shared stakes invoked in its articulation).

This naturalistic strategy for understanding the normativity of practical involvement in the world directly opposes traditional philosophical strategies, which would ground normativity in some form of necessity (a priori rational necessity for early-twentieth-century antinaturalists, a posteriori nomological or objective-alethic necessity for contemporary metaphysical naturalists). My arguments turn these strategies on their heads: any account of necessities must instead be understood as an expression of normative accountability. Such modalities are expressions of normative force, of the *binding* of one's performances by the authority of what is at stake in them. My reasoning for this inversion has been twofold. First, I have argued (following Heidegger and Neurath) that the traditional strategy of beginning with necessity cannot show how such necessities are manifest as intelligibly binding upon performances within the concrete, material world.[66] In particular, I have argued that one cannot thereby make intelligible the scientific practices that supposedly disclose the causal or nomological necessities that would account for their own intelligibility. Second, I have been arguing that by starting from the normativity of causal intra-action, one *can* make sense of both normativity and other modalities. Alethic and nomological modalities are to be construed as articulated expressions of how our causally intra-active circumstances have a hold on us, by structuring our (real) possibilities. Normativity, both as explicit claims about what is good, obligatory, correct, and so forth, and as implicit in utterances and actions, has been construed as expressing and sustaining the differential reproducibility of phenomena as accountable to something at stake in their (possible) realization.

What then binds the agent (not an already definitely bounded individid-

situation. Haugeland develops the point in terms of games (one is not a player unless one assumes responsibility for the proper performance of all players and paraphernalia, not just for one's own proper performance), but the point of his argument is to extend this feature of normativity to all phenomena, notably including the natural phenomena to which scientific claims are accountable.

66. I say "intelligibly binding," because from within the space of reasons, we could understand how natural necessities are binding upon things (how things causally interact to produce definite effects), but we could not similarly understand how such necessities are binding upon the space of reasons itself. We could not, that is, make intelligible the binding authority of natural necessity.

ual, but a prosthetically embodied configuration of "agencies") to what is at stake in the phenomenon is its own possibilities of being an agent, or as Nietzsche (1967) paradoxically expressed them, the possibilities of "becoming what one is."[67] What we are is a prospective pattern of accountable mattering, of having something at stake in our continuing differential reproduction as the same person or agent. Our accountability as agents extends beyond ourselves, because the boundaries of our effective agency are more extensive. We can only become what we are in concert with other agencies and circumstances, and hence our constitutive normativity involves responsibility (through both holding ourselves responsible and being held responsible by others) to coordinated, intra-active performances, issues, and stakes.

This account of agency and normativity thus brings some closure to the line of discussion initiated in chapter 5. That chapter distinguished a normative conception of practices from more familiar conceptions of practices as regularities in the doings of some community of agents. The inadequacy of the regularist conception should by now be clear. Agency cannot be taken for granted; we are constituted as agents through the ongoing patterns of causally intra-active practice in which we are caught up, where the pattern in turn is identifiable by its accountability to what is at stake in those practices. Agency and norms cannot be fully explained by the causal interrelations of practices, because the articulated causal structure of natural phenomena and the normative stakes in practices are mutually interdependent. What is at stake in practices only acquires normative force through causally efficacious intra-action, while that causal structure is only articulated into causes and effects as interrelated components of a reproducible phenomenon, accountable to what is at stake in one's involvement. That is why the conceptual dualism of nature and normativity has been such a fundamental block to philosophical understanding. In chapter 5, I proposed that the expressive roles of the concepts of meaning, truth, knowledge, and power must be understood to be dynamically interrelated. These concepts are so crucial, so interrelated, and yet have been so philosophically troubled precisely because they hold together what has been

67. This point reiterates the substantial convergence between my account of the normativity of practical-causal intra-action, and Mark Okrent's forthcoming exposition of intentionality in terms of the teleology of agency (the telos of an agent is the reproduction of its own characteristic pattern of agency). In my terms, of course, this telos should be understood as the ongoing configuration of a phenomenon rather than as the identity of an agent.

conceived as irreconcilable. The concepts of meaning, knowledge, and truth express how what we say and do is accountable to the differentially reproducible patterns in our circumstances; the concept of power expresses how causal intra-actions reconfigure a situation as a meaningful field of possibilities for discursive practice.

My own exposition and advocacy of naturalism cannot be and is not an exception to its own account. Why should we accept, as authoritative and binding upon us, the tenets of Quinean and Nietzschean naturalism as I have construed them (no arbitrary impositions or supernatural determinations) and developed their implications? It cannot be because these are normative standards whose authority is somehow prior to or otherwise outside of ongoing inquiry and practical engagement with the world, for they are not. Rather, they mark my articulation of what is at issue and at stake in understanding the world in the midst of which we find ourselves. Scientific practices have become an integral part of that world. The sciences both express and transform what is at stake for us in much of our lives, including what is contestable within scientific practice. One of the central issues for us now is how to understand nature, ourselves as natural beings, and the scientific disclosure of nature. Thus, naturalism as I have articulated it does not appeal to an already established, determinate concept of nature or of science in order to settle supposedly philosophical questions scientifically. It rather locates us in the midst of the scientific and technological practices that continue to reshape what it is to be nature, and how we can understand ourselves and our possibilities as natural beings. These are not scientific and technological practices in a narrow sense that would distinguish them from some other kind of practice (cultural, economic, political), but in an expansive sense that recognizes the intertwining of science and technology with everything we do. Naturalism is a defeasible articulation of a field of possibilities for how to understand science, our lives and possibilities as human beings, and the world that provides the unavoidable setting for both. Whatever authority and force naturalism has will come from the ways in which it intelligibly and forcefully articulates that setting in ways that have a normative hold upon us, such that if we fail to take it into account we would partially lose our footing in this world.

REFERENCES

Ackermann, Robert. 1985. *Data, Instruments, and Theory: A Dialectical Approach to Understanding Science.* Princeton: Princeton University Press.

Addelson, Kathryn Pyne. 1993. Knowers/Doers and Their Moral Problems. In Alcoff and Potter 1993:265–94.

Addelson, Kathryn Pyne, and Elizabeth Potter. 1991. Making Knowledge. In *Engendering Knowledge: Feminists in Academe,* ed. E. Messer-Davidow and J. Hartman, 259–77. Knoxville: University of Tennessee Press.

Alcoff, Linda, and Elizabeth Potter, eds. 1993. *Feminist Epistemologies.* New York: Routledge.

Ashmore, Malcolm. 1989. *The Reflexive Thesis: Wrighting Sociology of Scientific Knowledge.* Chicago: University of Chicago Press.

Baier, Annette. 1987. The Need for More Than Justice. In *Science, Morality, and Feminist Theory,* ed. M. Hanen and K. Nielsen. Calgary: University of Calgary Press.

Barad, Karen. 1996. Meeting the Universe Halfway. In *Feminism, Science, and the Philosophy of Science,* ed. L. H. Nelson and J. Nelson, 161–94. Dordrecht: Kluwer.

———. 1998. Getting Real: Technoscientific Practices and the Materialization of Reality. *differences* 10:87–128.

———. 1999. Agential Realism: Feminist Interventions in Understanding Scientific Practices. In Biagioli 1999:1–11.

———. Forthcoming. *Meeting the Universe Halfway.*

Barnes, Barry, and David Bloor. 1982. Relativism, Rationalism, and the Sociology of Knowledge. In *Rationality and Relativism,* ed. M. Hollis and S. Lukes, 21–47. Cambridge: MIT Press.

Barnes, Barry, David Bloor, and John Henry. 1996. *Scientific Knowledge: A Sociological Analysis.* Chicago: University of Chicago Press.

Belenky, Mary, Blythe Clinchy, Nancy Goldberger, and Jill Tarule. 1986. *Women's Ways of Knowing.* New York: Basic Books.

Bennett, Jonathan. 1987. Event Causation: The Counterfactual Analysis. *Philosophical Perspectives* 1:367–86.

Biagioli, Mario. 1993. *Galileo, Courtier: The Practice of Science in the Culture of Absolutism.* Chicago: University of Chicago Press.

———, ed. 1999. *The Science Studies Reader.* New York: Routledge.

Birke, Lynda. 1986. *Women, Feminism, and Biology.* Brighton: Wheatsheaf.

Bleier, Ruth. 1984. *Science and Gender: A Critique of Biology and Its Theories on Women.* New York: Pergamon Press.

Bloor, David. 1983. *Wittgenstein: A Social Theory of Knowledge.* New York: Columbia University Press.

———. 1991. *Knowledge and Social Imagery.* 2d ed. Chicago: University of Chicago Press.

———. 1992. Left and Right Wittgensteinians. In Pickering 1992:266–82.

Boedeker, Edgar. 1998. The Concept of Showing in the Early Writings of Heidegger and Wittgenstein. Ph.D. dissertation, Northwestern University.

Bono, James. 1990. Science, Discourse, and Literature: The Role/Rule of Metaphor in Science. In *Literature and Science: Theory and Practice,* S. Peterfreund, 59–89. Boston: Northeastern University Press.

Bourdieu, Pierre. 1990. *The Logic of Practice.* Trans. R. Nice. Stanford: Stanford University Press. Originally published as *Le sens pratique* (Paris: Minuit, 1980).

Boyd, Richard. 1984. The Current Status of Scientific Realism. In *Scientific Realism,* ed. J. Leplin, 41–82. Berkeley: University of California Press.

———. Realism, Approximate Truth, and Philosophical Method. In *The Philosophy of Science,* ed. D. Papineau, 215–55. Oxford: Oxford University Press.

Brandom, Robert. 1979. Freedom and Constraint by Norms. *American Philosophical Quarterly* 16:187–96.

———. 1994. *Making It Explicit: Reasoning, Representing, and Discursive Commitment.* Cambridge: Harvard University Press.

———. 1997. Dasein, the Being that Thematizes. *EPOCHE* 5:1–38.

———. 1999. Modality and Normativity. Paper delivered at the Boston Colloquium for the Philosophy of Science, April 1999.

———, ed. 2000a. *Rorty and His Critics.* Oxford: Blackwell.

———. 2000b. Vocabularies of Pragmatism: Synthesizing Naturalism and Historicism. In Brandom 2000a:56–83.

Brandon, Robert. 1990. *Adaptation and Environment.* Princeton: Princeton University Press.

Brown, Wendy. 1995. *States of Injury: Power and Freedom in Late Modernity.* Princeton: Princeton University Press.

Buchwald, Jed. 1995. *Scientific Practice: Theories and Stories of Doing Physics.* Chicago: University of Chicago Press.

Butler, Judith. 1990. *Gender Trouble.* New York: Routledge.

———. 1993. *Bodies That Matter.* New York: Routledge.

———. 1995. Contingent Foundations. In *Feminist Contentions: A Philosophical Exchange,* ed. S. Benhabib, J. Butler, D. Cornell, and N. Fraser. New York: Routledge.

Carnap, Rudolf. 1934. *The Unity of Science.* Trans. M. Black. London: Kegan Paul, Trench, Trubner.

———. 1937. *The Logical Syntax of Language.* Trans. A. Smeaton. London: Routledge and Kegan Paul.

———. 1959a. The Elimination of Metaphysics through the Logical Analysis of Language. In *Logical Positivism,* ed. A. Ayer, 60–81. Glencoe: Free Press.

———. 1959b. Psychology in Physical Language. In *Logical Positivism,* ed. A. Ayer, 165–98. Glencoe: Free Press.

———. 1967. *The Logical Structure of the World/Pseudoproblems in Philosophy.* Trans. R. George. Berkeley: University of California Press.

———. 1987. On Protocol Sentences. Trans. R. Creath and R. Nollan. *Nous* 21: 457–70.

Cartwright, Nancy. 1983. *How the Laws of Physics Lie.* Oxford: Oxford University Press.

———. 1989. *Nature's Capacities and Their Measurement.* Oxford: Oxford University Press.

———. 1999. *The Dappled World.* Cambridge: Cambridge University Press.

Cartwright, Nancy, Jordi Cat, and Hasok Chang. 1996. Otto Neurath: Politics and the Unity of Science. In Galison and Stump 1996:347–69.

Cartwright, Nancy, Jordi Cat, Lola Fleck, and Thomas Uebel. 1996. *Otto Neurath: Philosophy between Science and Politics.* Cambridge: Cambridge University Press.

Chalmers, David. 1996. *The Conscious Mind: In Search of a Fundamental Theory.* Oxford: Oxford University Press.

Clifford, James, and George Marcus. 1986. *Writing Culture: The Poetics and Politics of Ethnography.* Berkeley: University of California Press.

Coffa, Alberto. 1991. *The Semantic Tradition from Kant to Carnap: To the Vienna Station.* Ed. L. Wessels. Cambridge: Cambridge University Press.

Code, Lorraine. 1991. *What Can She Know?* Ithaca: Cornell University Press.

Collins, Harry. 1992. *Changing Order: Replication and Induction in Scientific Practice.* 2d ed. Chicago: University of Chicago Press.

Collins, Harry, and Stephen Yearley. 1992a. Epistemological Chicken. In Pickering 1992:301–26.

———. 1992b. Journey into Space. In Pickering 1992:369–89.

Crosby, Alfred. 1986. *Ecological Imperialism: The Biological Expansion of Europe, 900–1900.* Cambridge: Cambridge University Press.

Cornell, Drucilla 1992. *Philosophy of the Limit.* New York: Routledge.

Daston, Lorraine. 1992. Objectivity and the Escape from Perspective. *Social Studies of Science* 22:597–618.

———. 1994. Baconian Facts, Academic Civility, and the Prehistory of Objectivity. In *Rethinking Objectivity,* ed. A. Megill, 37–64. Durham: Duke University Press.

Daston, Lorraine, and Peter Galison. 1992. The Image of Objectivity. *Representations* 40:81–128.

Daston, Lorraine, and Katherine Park. 1998. *Wonders and the Order of Nature, 1150–1750.* New York: Zone Books.

Davidson, Donald. 1980. *Essays on Actions and Events.* Oxford: Oxford University Press.

———. 1984. *Inquiries into Truth and Interpretation.* Oxford: Oxford University Press.

———. 1986. A Nice Derangement of Epitaphs. In *Truth and Interpretation,* ed. E. LePore, 433–46. Oxford: Blackwell.

Davidson, Donald, and Jaako Hintikka. 1969. *Words and Objections: Essays on the Philosophy of W. V. O. Quine.* Dordrecht: D. Reidel.

Dennett, Daniel. 1987. *The Intentional Stance.* Cambridge: MIT Press.

———. 1991. *Consciousness Explained.* Boston: Little, Brown.

Derrida, Jacques. 1973. *Speech and Phenomena and Other Essays on Husserl's Theory of Signs.* Trans. D. Allison. Evanston: Northwestern University Press. Originally published as *Le voix et le phenomène* (Paris: Presses Universitaires de France, 1967).

———. 1988. *Limited Inc.* Trans. Samuel Weber and Jeffrey Mehlman. Evanston: Northwestern University Press.

———. 1990. The Force of Law: The 'Mystical Foundation of Authority.' *Cardozo Law Review* 11:920–1045.

Diamond, Jared. 1997. *Guns, Germs, and Steel: The Fates of Human Societies.* New York: Norton.

Dobbs, Betty Jo Teeter, and Margaret Jacob. 1995. *Newton and the Culture of Newtonianism.* Atlantic Highlands, N.J.: Humanities Press.

Dowe, Phil. 2000. *Physical Causation.* Cambridge: Cambridge University Press.

Dreyfus, Hubert. 1972. The Perceptual Noema: Gurwitsch's Crucial Contribution. In *Life-World and Consciousness: Essays for Aron Gurwitsch,* ed. L. Embree, 135–70. Evanston: Northwestern University Press.

———. 1979. *What Computers Can't Do.* 2d ed. New York: Harper and Row.

———. 1984. Why Current Studies of Human Capacities Can Never Be Scientific. *Berkeley Cognitive Science Report* 11:1–17.

Elgin, Catherine. 1991. Understanding in Art and Science. In *Philosophy and the Arts,* ed. P. A. French, T. E. Uehling, Jr., and H. K. Wettstein, 196–208. Midwest Studies in Philosophy, vol. 16. Notre Dame: University of Notre Dame Press.

Elliott, Terri. 1994. Making Strange What Had Appeared Familiar. *Monist* 77: 424–33.

Fausto-Sterling, Anne. 1985. *Myths of Gender.* New York: Basic Books.

Fine, Arthur. 1986. *The Shaky Game.* Chicago: University of Chicago Press.

Fischer, Ernst, and Carol Lipson. 1988. *Thinking about Science: Max Delbrück and the Origins of Molecular Biology.* New York: Norton.

Fodor, Jerry. 1975. *The Language of Thought.* New York: Crowell.

———. 1981. *Representations.* Cambridge: MIT Press.

Foucault, Michel. 1970. *The Order of Things: An Archaeology of the Human Sciences.* Trans. A. Sheridan. New York: Random House. Originally published as *Les mots et les choses* (Paris: Gallimard, 1966).

———. 1977. *Discipline and Punish.* Trans. A. Sheridan. New York: Random House.

———. 1978. *History of Sexuality.* Vol. 1. Trans. R. Hurley. New York: Random House.

———. 1980. *Power/Knowledge: Selected Interviews and Other Writings, 1972–77.* Ed. C. Gordon. New York: Pantheon.

Franklin, Allan. 1986. *The Neglect of Experiment.* Cambridge: Cambridge University Press.

Fraser, Nancy, and Linda Nicholson. 1990. Social Criticism without Philosophy. In *Feminism/Postmodernism,* ed. L. Nicholson, 19–38. New York: Routledge.

Friedman, Michael. 1983. Critical Notice: Moritz Schlick, Philosophical Papers. *Philosophy of Science* 50:498–514.

———. 1987. Carnap's Aufbau Reconsidered. *Nous* 21:521–45.

———. 1988. Logical Truth and Analyticity in Logical Syntax of Language. In *Es-*

says in the History and Philosophy of Modern Mathematics, ed. W. Aspray and P. Kitcher, 82–94. Minneapolis: University of Minnesota Press.

———. 1991. The Re-evaluation of Logical Empiricism. *Journal of Philosophy* 88:505–19.

———. 1992. Epistemology in the Aufbau. *Synthese* 93:15–37.

———. 1996. Overcoming Metaphysics: Carnap and Heidegger. In Giere and Richardson 1996:45–79.

———. 2000. *A Parting of the Ways: Carnap, Cassirer, and Heidegger.* Chicago: Open Court.

Fuller, Steve. 1988. *Social Epistemology.* Bloomington: Indiana University Press.

———. 1992a. Epistemology Radically Naturalized: Recovering the Normative, the Experimental, and the Social. In *Cognitive Models of Science,* ed. R. Giere, 448–59. Minnesota Studies in the Philosophy of Science, vol. 15. Minneapolis: University of Minnesota Press.

———. 1992b. Social Epistemology and the Research Agenda of Science Studies. In Pickering 1992:390–428.

———.1993. *Philosophy of Science and Its Discontents.* 2d ed. New York: Guilford.

———. 2000. *Thomas Kuhn: A Philosophical History for Our Times.* Chicago: University of Chicago Press.

Fuss, Diana. 1989. *Essentially Speaking: Feminism, Nature, and Difference.* New York: Routledge.

Galison, Peter. 1987. *How Experiments End.* Chicago: University of Chicago Press.

———. 1997. *Image and Logic.* Chicago: University of Chicago Press.

———. 1998a. Feynman's War: Modeling Weapons, Modeling Nature. *Studies in History and Philosophy of Modern Physics* 29, ser. B: 391–434.

———. 1998b. Judgment against Objectivity. In *Picturing Science, Producing Art,* ed. P. Galison and C. Jones, 327–59. New York: Routledge.

Galison, Peter, and David Stump, eds. 1996. *The Disunity of Science: Boundaries, Contexts, Power.* Stanford: Stanford University Press.

Giere, Ronald. 1988. *Explaining Science.* Chicago: University of Chicago Press.

———. 1999. *Science without Laws.* Chicago: University of Chicago Press.

Giere, Ronald, and Alan Richardson, ed. 1996. *Origins of Logical Empiricism.* Minnesota Studies in the Philosophy of Science, vol. 16. Minneapolis: University of Minnesota Press.

Goodfield, June. 1981. *An Imagined World.* New York: Harper and Row.

Gooding, David. 1990. *Experiment and the Making of Meaning.* Boston: Kluwer.

Gooding, David, Trevor Pinch, and Simon Schaffer, eds. 1989. *The Uses of Experiment: Studies in the Natural Sciences.* Cambridge: Cambridge University Press.

Grice, Paul. 1988. *Studies in the Ways of Words.* Cambridge: Harvard University Press.

Grover, Dorothy, Joseph Camp, and Nuel Belnap. 1975. A Prosentential Theory of Truth. *Philosophical Studies* 27:73–125.

Gurwitsch, Aron. 1966. *Studies in Phenomenology and Psychology.* Evanston: Northwestern University Press.

Hacking, Ian. 1965. *The Logic of Statistical Inference.* Cambridge: Cambridge University Press.

———. 1982. Language, Truth, and Reason. In *Rationality and Relativism,* ed. M. Hollis and S. Lukes, 48–66. Cambridge: MIT Press.

———. 1983. *Representing and Intervening.* Cambridge: Cambridge University Press.

———. 1986. Making Up People. In *Reconstructing Individualism: Autonomy, Individuality, and the Self in Western Thought,* ed. T. Heller, M. Sosna, and D. Wellbery, 222–36. Stanford: Stanford University Press.

———. 1992. The Self-Vindication of the Laboratory Sciences. In Pickering 1992: 29–64.

———. 1993. Working in a New World: The Taxonomic Solution. In *World Changes: Thomas Kuhn and the Nature of Science,* ed. P. Horwich, 275–310. Cambridge: MIT Press.

Hall, Harrison. 1982. Was Husserl a Realist or an Idealist? In *Husserl, Intentionality, and Cognitive Science,* ed. H. Dreyfus and H. Hall, 169–90. Cambridge: MIT Press.

Haraway, Donna. 1989. *Primate Visions: Gender, Race, and Nature in the World of Modern Science.* New York: Routledge.

———. 1991. *Simians, Cyborgs, and Women.* New York: Routledge.

———. 1992a. Otherworldly Conversations; Terran Topics; Local Terms. *Science as Culture* 3:64–98.

———. 1992b. The Promises of Monsters: A Regenerative Politics for Inappropriate/d Others. In *Cultural Studies,* ed. L. Grossberg, C. Nelson, and P. A. Treichler, 295–337. New York: Routledge.

———. 1997. *Modest_Witness@Second_Millennium.FemaleMan©Meets_OncoMouse™.* New York: Routledge.

Harding, Sandra. 1986. *The Science Question in Feminism.* Ithaca: Cornell University Press.

———. 1991. *Whose Science? Whose Knowledge?* Ithaca: Cornell University Press.

———. 1992. After the Neutrality Ideal. *Social Research* 59:567–88.

Harman, Gilbert. 1967. Quine on Meaning and Existence. Parts 1 and 2. *Review of Metaphysics* 21:124–51, 343–67.

Hartsock, Nancy. 1984. *Money, Sex, and Power.* Boston: Northeastern University Press.

Haugeland, John. 1998. *Having Thought: Essays in the Metaphysics of Mind.* Cambridge: Harvard University Press.

———. 2000. Truth and Finitude: Heidegger's Transcendental Existentialism. In *Heidegger, Authenticity, and Modernity,* ed. J. Malpas and M. Wrathall, 43–78. Cambridge: MIT Press.

Heelan, Patrick. 1983. *Space-Perception and the Philosophy of Science.* Berkeley: University of California Press.

Hegel, Georg W. F. 1972. *Phenomenology of Spirit.* Trans. A. Miller. Oxford: Oxford University Press.

Heidegger, Martin. 1959. *An Introduction to Metaphysics.* Trans. R. Manheim. New Haven: Yale University Press. Originally published as *Einführung in die Metaphysik* (Tübingen: Max Niemeyer Verlag, 1953).

———. 1971. The Origin of the Work of Art. In *Poetry, Language, Thought,* trans. A. Hofstadter. New York: Harper and Row. Originally published as

Der Ursprung des Kunstwerkes, ed. H.-G. Gadamer (Stuttgart: Reclam, 1960).

———. 1977. *Basic Writings.* New York: Harper and Row.

———. 1985. *History of the Concept of Time.* Trans. T. Kisiel. Bloomington: University of Indiana Press. Originally published as *Prolegomena zur Geschichte des Zeitbegriffs.* (Frankfurt: Vittorio Klosterman, 1979).

———. 1990. The Self-Assertion of the German University. Trans. K. Harries. In *Martin Heidegger and National Socialism,* ed. G. Neske and E. Kettering, 5–13. New York: Paragon.

———. 1996. *Being and Time.* Trans. J. Stambaugh. Albany: State University of New York Press. Originally published as *Sein und Zeit* (Tübingen: Max Niemeyer, 1996).

———. 1998. *Pathmarks.* Ed. W. McNeill. Cambridge: Cambridge University Press. Translation of *Wegmarken* (Frankfurt am Main: Vittorio Klosterman, 1967).

Hempel, Carl. 1965. The Theoretician's Dilemma. In *Aspects of Scientific Explanation,* 173–226. New York: MacMillan.

Hesse, Mary. 1980. *Revolutions and Reconstructions in the Philosophy of Science.* Bloomington: University of Indiana Press.

Horwich, Paul. 1990. *Truth.* Oxford: Blackwell.

Horst, Steven. 1996. *Symbols, Computation, and Intentionality.* Berkeley: University of California Press.

Howard, Don. 1994. What Makes a Classical Concept Classical? Toward a Reconstruction of Niels Bohr's Philosophy of Physics. In *Niels Bohr and Contemporary Philosophy,* ed. J. Faye and H. Folse, 201–29. Dordrecht: Kluwer.

Hubbard, Ruth. 1990. *The Politics of Women's Biology.* New Brunswick: Rutgers University Press.

Hughes, Jeff. 1998. 'Modernists with a Vengeance': Changing Cultures of Theory in Nuclear Science, 1920–1930. *Studies in History and Philosophy of Modern Physics* 29, ser. B: 339–67.

Husserl, Edmund. 1969. *Formal and Transcendental Logic.* Trans. D. Cairns. The Hague: Martinus Nijhoff.

———. 1970a. *The Crisis in European Sciences and Transcendental Phenomenology.* Trans. D. Carr. Evanston: Northwestern University Press.

———. 1970b. *Logical Investigations.* Trans. J. Findlay. London: Routledge and Kegan Paul.

———. 1981. Philosophy as Rigorous Science. In *Husserl: Shorter Works,* ed. P. McCormick and F. Elliston, 166–97. Notre Dame: University of Notre Dame Press.

———. 1982. *Ideas Pertaining to a Pure Phenomenology and to a Phenomenological Philosophy.* Book 1. Trans. F. Kersten. The Hague: Martinus Nijhoff.

Jackson, Frank. 1998. *From Metaphysics to Ethics: In Defense of Conceptual Analysis.* Oxford: Oxford University Press.

Keller, Evelyn Fox. 1983. *A Feeling for the Organism.* San Francisco: W. H. Freeman.

———. 1985. *Reflections on Gender and Science.* New Haven: Yale University Press.

———. 1992. *Secrets of Life, Secrets of Death: Essays on Language, Gender, and Science.* New York: Routledge.

———. 2000. Models of and Models For: Theory and Practice in Contemporary Biology. *Philosophy of Science* 67, no. 3, supp.: S72–S86.

Kierkegaard, Søren. 1954. *Fear and Trembling and The Sickness Unto Death.* Trans. W. Lowrie. Princeton: Princeton University Press.

Kitcher, Philip. 1984. 1953 and All That: A Tale of Two Sciences. *Philosophical Review* 93:335–74.

———. 1999. Unification as a Regulative Ideal. *Perspectives on Science* 7:337–48.

Knorr-Cetina, Karin. 1981. *The Manufacture of Knowledge: An Essay on the Constructivist and Contextual Nature of Science.* Oxford: Pergamon.

Kohler, Robert. 1994. *Lords of the Fly: Drosophila Genetics and the Experimental Life.* Chicago: University of Chicago Press.

Kondo, Dorinne. 1990. *Crafting Selves: Power, Gender, and Discourses of Identity in a Japanese Workplace.* Chicago: University of Chicago Press.

Korsgaard, Christine. 1996. *The Sources of Normativity.* Cambridge: Harvard University Press.

Krieger, Martin. 1992. *Doing Physics: How Physicists Take Hold of the World.* Bloomington: Indiana University Press.

Kripke, Saul. 1980. *Naming and Necessity.* Cambridge: Harvard University Press.

Kuhn, Thomas. 1970. *The Structure of Scientific Revolutions.* 2d ed. Chicago: University of Chicago Press.

———. 1977. *The Essential Tension: Selected Studies in Scientific Tradition and Change.* Chicago: University of Chicago Press.

———. 1991. The Natural and the Human Sciences. In *The Interpretive Turn: Philosophy, Science, Culture,* ed. D. Hiley, J. Bohman, and R. Shusterman, 17–24. Ithaca: Cornell University Press.

———. 2000. *The Road since Structure: Philosophical Essays, 1970–1993, with an Autobiographical Interview.* Ed. J. Conant and J. Haugeland. Chicago: University of Chicago Press.

Kukla, Rebecca. 2000a. How to Get an Interpretivist Committed. *Protosociology* 14:180–221.

———. 2000b. Myth, Memory, and Misrecognition in Sellars' "Empiricism and the Philosophy of Mind." *Philosophical Studies* 101:161–211.

Lafont, Cristina. 2000. *Heidegger, Language, and World-Disclosure.* Cambridge: Cambridge University Press.

Lance, Mark. Forthcoming. The Logical Structure of Linguistic Commitment III: Brandomian Scorekeeping and Incompatibility.

Lance, Mark, and Philip Kremer. 1991. The Logical Structure of Linguistic Commitment I: Four Systems of Non-Relevant Commitment Entailment. *Journal of Philosophical Logic* 23:369–400.

———. 1996. The Logical Structure of Linguistic Commitment II: Systems of Relevant Commitment Entailment. *Journal of Philosophical Logic* 25:425–29.

Latour, Bruno. 1983. Give Me a Laboratory and I Will Raise the World. In *Science Observed: Perspectives on the Social Study of Science,* ed. K. Knorr-Cetina and M. Mulkay, 141–70. London: Sage.

———. 1987. *Science in Action.* Cambridge: Harvard University Press.

———. 1988. The Politics of Explanation: An Alternative. In *Knowledge and Reflexivity,* ed. Steve Woolgar, 155–76. Beverley Hills: Sage.

Latour, Bruno, and Steve Woolgar. 1986. *Laboratory Life: The Construction of Scientific Facts.* 2d ed. Princeton: Princeton University Press.

Laudan, Larry. 1981. A Confutation of Convergent Realism. *Philosophy of Science* 48:19–48.

———. 1987. Progress or Rationality? The Prospects for Normative Naturalism. *American Philosophical Quarterly* 24, ser. L: 19–31.

Levins, Richard, and Richard Lewontin. 1985. *The Dialectical Biologist.* Cambridge: Harvard University Press.

Lewis, David. 1973. *Counterfactuals.* Oxford: Blackwell.

Lewontin, Richard. 2000. *The Triple Helix: Gene, Organism, and Environment.* Cambridge: Harvard University Press.

Longino, Helen. 1990. *Science as Social Knowledge.* Princeton: Princeton University Press.

———. 1992. Essential Tensions—Phase Two: Feminist, Philosophical, and Social Studies of Science. In McMullin 1992: 198–216.

———. 1993. Subjects, Power, and Knowledge: Description and Prescription in Feminist Philosophies of Science. In Alcoff and Potter 1993: 101–20.

Lynch, Michael. 1992. From the 'Will to Theory' to the Discursive Collage. In Pickering 1992:283–300.

Mackie, J. L. 1974. *The Cement of the Universe.* Oxford: Oxford University Press.

MacIntyre, Alasdair. 1980. Epistemological Crises, Dramatic Narrative, and the Philosophy of Science. In *Paradigms and Revolutions: Applications and Appraisals of Thomas Kuhn's Philosophy of Science,* ed. G. Gutting, 54–74. Notre Dame: University of Notre Dame Press.

———. 1981. *After Virtue.* Notre Dame: University of Notre Dame Press.

MacKenzie, Donald, and Barry Barnes. 1979. Scientific Judgment: the Biometry-Mendelism Controversy. In *Natural Order: Historical Studies of Scientific Culture,* ed. B. Barnes and S. Shapin, 191–210. Beverly Hills: Sage.

Marcus, George, and Michael Fisher. 1986. *Anthropology as Cultural Critique.* Chicago: University of Chicago Press.

McDowell, John. 1994. *Mind and World.* Cambridge: Harvard University Press.

McMullin, Ernan, ed. 1992. *The Social Dimensions of Science.* Notre Dame: University of Notre Dame Press.

McNeill, William. 1976. *Plagues and Peoples.* Garden City, N.J.: Anchor.

Megill, Allan, ed. 1994. *Rethinking Objectivity.* Durham: Duke University Press.

Merleau-Ponty, Maurice. 1962. Phenomenology of Perception. Trans. Colin Smith. London: Routledge and Kegan Paul. Originally published as *Phénoménologie de la perception* (Paris: Éditions Gallimard, 1945).

Miller, Richard. 1987. *Fact and Method: Explanation, Confirmation, and Reality in the Natural and Social Sciences.* Princeton: Princeton University Press.

Minh-Ha, Trinh. 1989. *Woman, Native, Other: Writing, Postcoloniality, and Feminism.* Bloomington: University of Indiana Press.

Moore, G. E. 1929. *Principia Ethica.* Cambridge: Cambridge University Press.

Morgan, Mary, and Margaret Morrison. 1999. *Models as Mediators: Perspectives on Natural and Social Science.* Cambridge: Cambridge University Press.

Nagel, Ernest. 1961. *The Structure of Science.* New York: Harcourt Brace Jovanovich.

Nagel, Thomas. 1974. What Is It Like to Be a Bat? *Philosophical Review* 83:435–51.

Nelson, Lynn Hankinson. 1990. *Who Knows? From Quine to a Feminist Empiricism.* Philadelphia: Temple University Press.

———. 1993. Epistemological Communities. In Alcoff and Potter 1993:121–60.

Neurath, Otto. 1983. *Philosophical Papers, 1913–1946.* Ed. R. Cohen and M. Neurath. Dordrecht: D. Reidel.

Nietzsche, Friedrich 1954. *The Portable Nietzsche.* Trans. W. Kaufmann. New York: Viking.

———. 1967. *On the Genealogy of Morals and Ecce Homo.* Trans. W. Kaufmann and R. Hollingdale. New York: Random House.

Okrent, Mark. 1988. *Heidegger's Pragmatism: Understanding, Being, and the Critique of Metaphysics.* Ithaca: Cornell University Press.

———. Forthcoming. *Rational Animals.*

Pickering, Andrew. 1984. *Constructing Quarks: A Sociological History of Particle Physics.* Chicago: University of Chicago Press.

———. 1995. *The Mangle of Practice: Time, Agency, and Science.* Chicago: University of Chicago Press.

———, ed. 1992. *Science as Practice and Culture.* Chicago: University of Chicago Press.

Pinch, Trevor. 1986. *Confronting Nature.* Dordrecht: D. Reidel.

Polanyi, Michael. 1958. *Personal Knowledge: Towards a Post-Critical Philosophy.* Chicago: University of Chicago Press.

Porter, Theodore. 1995. *Trust in Numbers: The Pursuit of Objectivity in Science and Public Life.* Princeton: Princeton University Press.

Potter, Elizabeth. 2001. *Gender and Boyle's Law of Gases.* Bloomington: Indiana University Press.

Prior, A. N. 1957. *Time and Modality.* Oxford: Oxford University Press.

Proctor, Robert. 1991. *Value-Free Science? Purity and Power in Modern Knowledge.* Cambridge: Harvard University Press.

Putnam, Hilary. 1975a. The Meaning of Meaning. In *Mind, Language, and Reality,* vol. 2 of *Philosophical Papers.* Cambridge: Cambridge University Press.

———. 1975b. The Refutation of Conventionalism. In *Mind, Language, and Reality,* vol. 2 of *Philosophical Papers.* Cambridge: Cambridge University Press.

———. 1981. *Reason, Truth, and History.* Cambridge: Cambridge University Press.

———. 1983. *Realism and Reason.* Vol. 3 of *Philosophical Papers.* Cambridge: Cambridge University Press.

———. 1990. *Realism with a Human Face.* Cambridge: Harvard University Press.

Quine, W. V. O. 1960. *Word and Object.* Cambridge: MIT Press.

———. 1965. *Ontological Relativity and Other Essays.* New York: Columbia University Press.

———. 1970. Grades of Theoreticity. In *Experience and Theory,* ed. L. Foster and J. W. Swanson. Amherst: University of Massachusetts Press.

———. 1977. Facts of the Matter. In *American Philosophy from Edwards to Quine,* ed. K. Merill and R. Shahan. Norman: University of Oklahoma Press.

———. 1981. The Pragmatist's Place in Empiricism. In *Pragmatism,* ed. R. J. Mulvaney and P. J. Zeltner. Columbia: University of South Carolina Press.

Radder, Hans. 1988. *The Materialization of Science.* Assen: Van Gorcum.

———. 1996. *In and About the World: Philosophical Studies of Science and Technology.* Albany: State University of New York Press.

Ramberg, Bjorn. 2000. Post-Ontological Philosophy of Mind: Rorty versus Davidson. In Brandom 2000a:351–70.

Rawls, John. 1955. Two Concepts of Rules. *Philosophical Review* 64:3–32.

Rheinberger, Hans-Jörg. 1992. Experiment, Difference, and Writing. Parts 1 and 2. *Studies in History and Philosophy of Science* 23:305–31, 389–422.

———. 1994. Experimental Systems: Historiality, Narration, and Deconstruction. *Science in Context* 7:65–81.

———. 1995. From Microsomes to Ribosomes: 'Strategies' of 'Representation,' 1935–55. *Journal of the History of Biology* 48:49–89.

———. 1997. *Toward a History of Epistemic Things.* Stanford: Stanford University Press.

Richardson, Alan. 1990. How Not to Russell Carnap's Aufbau. In *PSA 1990: Proceedings of the 1990 Biennial Meeting of the Philosophy of Science Association,* vol. 1, ed. A. Fine, 3–14. East Lansing:Michigan State University.

———. 1992. Logical Idealism and Carnap's Construction of the World. *Synthese* 93:59–92.

Ricketts, Thomas. 1994. Carnap's Principle of Tolerance, Empiricism, and Conventionalism. In *Reading Putnam,* ed. P. Clark and B. Hale, 176–200. Oxford: Blackwell.

———. 1996. Carnap: From Logical Syntax to Semantics. In Giere and Richardson 1996:231–50.

Rorty, Richard. 1972. Indeterminacy of Translation and of Truth. *Synthese* 23: 443–62.

———. 1979. *Philosophy and the Mirror of Nature.* Princeton: Princeton University Press.

———. 1991. *Objectivity, Relativism, and Truth.* Vol. 1 of *Philosophical Papers.* Cambridge: Cambridge University Press.

———. 2000. Response to Ramberg. In Brandom 2000a:370–77.

Rosaldo, Renato. 1989. *Culture and Truth.* Boston: Beacon Press.

Rosenberg, Alexander. 1994. *Instrumental Biology or the Disunity of Science.* Chicago: University of Chicago Press.

———. 1996. A Field Guide to Contemporary Species of Naturalism. *British Journal for the Philosophy of Science* 47:1–29.

Roth, Paul. 1987. *Meaning and Method in the Social Sciences.* Ithaca: Cornell University Press.

Rouse, Joseph. 1987. *Knowledge and Power: Toward a Political Philosophy of Science.* Ithaca: Cornell University Press.

———. 1991. Policing Knowledge: Disembodied Policy for Embodied Knowledge. *Inquiry* 34:353–64.

———. 1993. Foucault and the Natural Sciences. In *Foucault and the Critique of*

Institutions, ed. J. Caputo and M. Yount, 137–62. State College: Pennsylvania State University Press.

———. 1994. Power/Knowledge. In *Cambridge Companion to Foucault,* ed. G. Gutting, 92–114. Cambridge: Cambridge University Press.

———. 1996a. Beyond Epistemic Sovereignty. In Galison and Stump 1996:398–416.

———. 1996b. *Engaging Science: How to Understand Its Practices Philosophically.* Ithaca: Cornell University Press.

———. 1998. Kuhn and Scientific Practices. *Configurations* 6:33–50.

———. 2000. Coping and its Contrasts. In *Heidegger, Coping, and Cognitive Science: Essays in Honor of Hubert L. Dreyfus,* vol. 2, ed. M. Wrathall and J. Malpas, 7–28. Cambridge: MIT Press.

———. 2002a. Kuhn's Philosophy of Scientific Practice. In *Thomas Kuhn,* ed. T. Nickles. Cambridge: Cambridge University Press.

———. 2002b. Vampires: Social Constructivism, Realism, and Other Philosophical Undead. *History and Theory* 41:60–78.

Salmon, Wesley. 1989. *Four Decades of Scientific Explanation.* Minneapolis: University of Minnesota Press.

———. 1998. *Causality and Explanation.* Cambridge: Cambridge University Press.

Sartre, Jean-Paul. 1966. *Being and Nothingness: A Phenomenological Essay on Ontology.* Trans. H. Barnes. New York: Washington Square Press. Originally published as *L'être et le néant* (Paris: Gallimard, 1943).

Scheffler, Israel. 1967. *Science and Subjectivity.* Indianapolis: Bobbs-Merrill.

Schmitter, Amy. 1996. Picturing Power: Representation and 'Las Meninas.' *Journal of Aesthetics and Art Criticism* 54:255–68.

Scott, Joan. 1988. *Gender and the Politics of History.* New York: Columbia University Press.

———. 1991. Experience. In *Feminists Theorize the Political,* ed. J. Butler and J. Scott. New York: Routledge.

Searle, John. 1969. *Speech Acts.* Cambridge: Cambridge University Press.

———. 1983. *Intentionality: An Essay in the Philosophy of Mind.* Cambridge: Cambridge University Press.

Sellars, Wilfrid. 1963. *Science, Perception, and Reality.* London: Routledge and Kegan Paul.

———. 1980. Concepts as Involving Laws, and Inconceivable without Them. In *Pure Pragmatics and Possible Worlds: The Early Essays of Wilfrid Sellars,* ed. J. Sicha. Reseda, Calif.: Ridgeview Publishing.

———. 1997. *Empiricism and the Philosophy of Mind.* Cambridge: Harvard University Press.

Shapin, Steven. 1982. History of Science and Its Sociological Reconstructions. *History of Science* 20:157–211.

Shapin, Steven, and Simon Schaffer. 1985. *Leviathan and the Air-Pump: Hobbes, Boyle, and the Experimental Life.* Princeton: Princeton University Press.

Shieh, Sanford. 2001. Meaning, Rigidity, and Modality. In *Future Pasts: The Analytic Tradition in Twentieth Century Philosophy,* ed. J. Floyd and S. Shieh. Oxford: Oxford University Press.

———. 2002. On Interpreting Frege on Truth and Logic. In *From Frege to Witt-*

genstein: Perspectives on Early Analytic Philosophy, ed. E. Reck, 96–124. Oxford: Oxford University Press.

Smith, Brian Cantwell. 1996. *On the Origin of Objects.* Cambridge: MIT Press.

Sophocles. 1969. *The Oedipus Cycle.* Trans. R. Fitzgerald. New York: Harcourt Brace Jovanovich.

Spelman, Elizabeth. 1988. *Inessential Woman: Problems of Exclusion in Feminist Thought.* Boston: Beacon Press.

Star, Leigh. 1991. Power, Technologies, and the Phenomenology of Standards: On Being Allergic to Onions. In *A Sociology of Monsters,* ed. J. Law, 27–57. London: Routledge.

Stump, David. 1992. Naturalized Philosophy of Science with a Plurality of Methods. *Philosophy of Science* 59:456–60.

Sultan, Sonia. 1995. Phenotypic Plasticity and Plant Adaptation. *Acta Botanica Neerlandica* 44:363–83.

Taylor, Charles. 1980. Understanding in Human Science. *Review of Metaphysics* 34:25–38.

———. 1985a. *Human Agency and Language.* Vol. 1 of *Philosophical Papers.* Cambridge: Cambridge University Press.

———. 1985b. *Philosophy and the Human Sciences.* Vol. 2 of *Philosophical Papers.* Cambridge: Cambridge University Press.

———. 1991. The Dialogical Self. In *The Interpretive Turn,* ed. D. Hiley, J. Bohman, and R. Shusterman, 304–14. Ithaca: Cornell University Press.

———. 1994. Charles Taylor Replies. In *Philosophy in an Age of Pluralism: The Philosophy of Charles Taylor in Question,* ed. J. Tully, 213–57. Cambridge: Cambridge University Press.

———. 1995. *Philosophical Arguments.* Cambridge: Harvard University Press.

Todes, Samuel. 1966. A Comparative Phenomenology of Perception and Imagination. Part 1. *Journal of Existentialism* 6:253–68.

———. 2001. *Body and World.* Cambridge: MIT Press.

Toulmin, Stephen. 1972. *Human Understanding.* Vol. 1. Princeton: Princeton University Press.

Traweek, Sharon. 1988. *Beamtimes and Lifetimes: The World of High-Energy Physics.* Cambridge: Harvard University Press.

———. 1992. Border Crossings: Narrative Strategies in Science Studies and among Physicists in Tsukuba Science City, Japan. In Pickering 1992:429–65.

Turner, Stephen. 1989. Depoliticizing Power. *Social Studies of Science* 19:533–60.

———. 1994. *The Social Theory of Practices: Tradition, Tacit Knowledge, and Presuppositions.* Chicago: University of Chicago Press.

Uebel, Thomas. 1991a. Neurath's Programme for Naturalistic Epistemology. *Studies in the History and Philosophy of Science* 22:623–46.

———, ed. 1991b. *Rediscovering the Forgotten Vienna Circle: Austrian Studies on Otto Neurath and the Vienna Circle.* Dordrecht: Kluwer.

———. 1992a. Neurath vs. Carnap: Naturalism vs. Rational Reconstructionism before Quine. *History of Philosophy Quarterly* 9:445–70.

———. 1992b. *Overcoming Logical Positivism from Within: The Emergence of Neurath's Naturalism in the Vienna Circle's Protocol Sentence Debate.* Amsterdam: Rodopi.

———. 1992c. Rational Reconstruction as Elucidation? Carnap in the Early Protocol Sentence Debate. *Synthese* 93:107–40.

———. 1996. The Enlightenment Ambition of Epistemic Utopianism: Otto Neurath's Theory of Science in Historical Perspective. In Giere and Richardson 1996:91–112.

Van Fraassen, Bas. 1980. *The Scientific Image.* Oxford: Oxford University Press.

Van Helden, Albert, and Thomas Hankins, eds. 1994. *Instruments: Osiris.* Vol. 9. Chicago: University of Chicago Press.

Von Wright, Georg. 1971. *Explanation and Understanding.* Ithaca: Cornell University Press.

———. 1974. *Causality and Determinism.* New York: Columbia University Press.

Wartenburg, Thomas. 1990. *The Forms of Power.* Philadelphia: Temple University Press.

Warwick, Andrew. 1992. The Uses of Theory. Part 1 of Cambridge Mathematics and Cavendish Physics: Cunningham, Campbell, and Einstein's Relativity 1905–1911. *Studies in History and Philosophy of Science* 23:625–56.

———. 1995. The Laboratory of Theory, or What's Exact about the Exact Sciences. In Wise 1995:311–51.

Wheeler, Samuel C. 2000. *Deconstruction as Analytic Philosophy.* Stanford: Stanford University Press.

Williams, Michael. 1991. *Unnatural Doubts: Epistemological Realism and the Basis of Scepticism.* Oxford: Blackwell.

Wilson, Bryan. 1970. *Rationality.* Oxford: Blackwell.

Winch, Peter. 1958. *The Idea of a Social Science.* London: Routledge and Kegan Paul.

Winsburg, Eric, Mathias Frisch, Karen Merikangas Darling, and Arthur Fine. 2000. Review of *The Dappled World,* by Nancy Cartwright. *Journal of Philosophy* 97:403–8.

Wise, M. Norton, ed. 1995. *The Values of Precision.* Princeton: Princeton University Press.

Witt, Charlotte. 1995. Anti-Essentialism in Feminist Theory. *Philosophical Topics* 23:321–44.

Wittgenstein, Ludwig. 1922. *Tractatus Logico-Philosophicus.* London: Routledge and Kegan Paul.

———. 1953. *Philosophical Investigations.* Oxford: Blackwell.

Woolgar, Steve. 1988. *Science: The Very Idea.* London: Tavistock.

Wylie, Alison. 1991. Gender Theory and the Archaeological Record: Why Is There No Archaeology of Gender? In *Engendering Archaeology: Women and Prehistory,* ed. J. M. Gero and M. W. Conkey, 31–54. Oxford: Blackwell.

———. 1992. The Interplay of Evidential Constraints and Political Interests: Recent Archaeological Research on Gender. *American Antiquity* 57:15–35.

———. 2002. *Thinking from Things: Essays in the Philosophy of Archaeology.* Berkeley: University of California Press.

INDEX